POLLUTION CONTROL IN THE ASBESTOS, CEMENT, GLASS AND ALLIED MINERAL INDUSTRIES

Pollution Control in the Asbestos, Cement, Glass and Allied Mineral Industries

Marshall Sittig

NOYES DATA CORPORATION
Park Ridge, New Jersey London, England
1975

Library of Congress Catalog Card Number: 75-10353
ISBN: 0-8155-0578-7
Printed in the United States

Published in the United States of America by
Noyes Data Corporation
Noyes Building, Park Ridge, New Jersey 07656

FOREWORD

This is the nineteenth volume of our Pollution Technology Review Series. Like several of its predecessors, it deals with industrially-caused contamination problems of the environment. Perusal of this book will convince the reader that private industry is already doing much to restrain its pollutant emissions. There is much more that can and will be done.

The present treatise attempts to provide helpful information pointing in the right direction on the arduous and costly antipollution road that industry is now forced to take under ecologic and sociologic pressures. The world over, technological and manpower resources are being directed on an increasing scale toward the control and solution of contamination and pollution problems.

In the United States, we are fortunate in receiving direct help from the numerous surveys, together with active research and development programs that are being supported by the Federal Government to help industry control its wastes and harmful emissions.

In this book are condensed vital data that are scattered and difficult to pull together. Important processes are interpreted and explained by actual case histories. This condensed information will enable you to establish a sound background for action towards combating pollution by the silicate industries.

Advanced composition and production methods developed by Noyes Data are employed to bring our new durably bound books to you in a minimum of time. Special techniques are used to close the gap between "manuscript" and "completed book." Industrial technology is progressing so rapidly that time-honored, conventional typesetting, binding and shipping methods are no longer suitable. We have bypassed the delays in the conventional book publishing cycle and provide the user with an effective and convenient means of reviewing up-to-date information in depth.

The Table of Contents is organized in such a way as to serve as a subject index and provides easy access to the information contained in this book. A bibliographic reference list is also included to provide the reader with easy access to further information on this timely topic.

CONTENTS AND SUBJECT INDEX

INTRODUCTION 1

ASBESTOS INDUSTRIES–GENERAL 2

ASBESTOS MINING AND MILLING 7
Mining Processes and Emissions 7
Emissions Control in Asbestos Mining 9
Asbestos Rock Milling and Emissions 12
Milling by Air Separation 13
Milling by Wet Method 14
Emissions Control in Asbestos Milling 14
Emissions Control Costs for Asbestos Milling 19

ASBESTOS PRODUCTS INDUSTRIES 22
Air Pollution 25
Water Pollution 31
Asbestos-Cement Pipe and Sheet 35
Asbestos Floor Tile 47
Asbestos Papers 51
Asbestos Millboard 57
Asbestos Roofing 62
Asbestos-Based Friction Materials 66
Asbestos Textiles 72
Asbestos Packing and Gaskets 78
Solid Waste Disposal 80

BRICK INDUSTRY 83
Air Pollution 85

CEMENT INDUSTRY 94
Air Pollution 98
Proprietary Control Processes 109
Water Pollution 114
In-Plant Control Measures 123
Treatment Technology 124
Cost and Reduction Benefits of Alternative Control and Treatment Technologies 131

CERAMIC CLAY PRODUCTS MANUFACTURE 136
Air Pollution 137

CONCRETE INDUSTRY 140
Air Pollution 141
Control Equipment–Wet Batching 142
Control Equipment–Dry Batching 145
Control Equipment–Central Mix Plant 145
Proprietary Control Processes 146
Water Pollution 146

FIBER GLASS INDUSTRY 159
Air Pollution 165
Proprietary Control Processes 167
Water Pollution 172
Cost Reduction Benefits of Alternate Treatment and Control Technologies 183
Proprietary Control Processes 191
Solid Waste Disposal 191

FRIT MANUFACTURING 192
Air Pollution 192

GLASS INDUSTRY 200
Air Pollution 205
Emission and Control in Batch Plants 206
Emissions and Control in Glass-Melting Furnaces 207
Emissions and Control in Glass Forming 214
Cost of Alternative Control Technologies 214
Proprietary Control Processes 219
Water Pollution 222
Flat Glass Manufacture 223
Sheet Glass Manufacture 225
Rolled Glass Manufacture 227
Plate Glass Manufacture 228
Float Glass Manufacture 235
Solid Tempered Automotive Glass Manufacture 241
Windshield Fabrication 250
Glass Container Manufacturing 259
Machine Pressed and Blown Glass Manufacturing 265
Glass Tubing Manufacturing 270
Television Picture Tube Envelope Manufacturing 274
Incandescent Lamp Envelope Manufacturing 281
Hand Pressed and Blown Glass Manufacturing 288
Solid Waste Disposal 298

MINERAL WOOL INDUSTRY 300
Air Pollution 302
Proprietary Control Processes 305

SAND AND GRAVEL INDUSTRY 306
Air Pollution 308
Water Pollution 309
Solid Waste Disposal 314

STONE QUARRYING AND PROCESSING 316
Air Pollution 317

FUTURE TRENDS 322
Asbestos Industry 322

Cement Industry 323
Fiber Glass Industry 325
Glass Industry 325
Sand and Gravel Industry 329

REFERENCES 330

INTRODUCTION

The mineral product or ceramic product industries, or silicate product industries as they might also be designated, include the asbestos, brick, cement, clay, frit, fiber glass, glass, mineral wool and stone, sand and gravel industries. They present appreciable problems of air pollution, water pollution and solid waste disposal.

The asbestos industry is faced with the twin threats of asbestosis and lung cancer as a result of airborne asbestos dust. Now, in addition, waterborne asbestos from mineral processing is being assessed as a potential hazard in drinking water.

In the cement industry, dust is a major problem; dust reclamation is an economic necessity. In spite of highly efficient collection, some dust escapes and dustfall rates of 35 tons per square mile have been recorded in areas adjacent to efficiently controlled kilns (1). A secondary air pollution problem is the effluent from the kilns which contain gaseous pollutants from the fuel and from the heating of the components. The cement and concrete industries also pose water pollution problems.

In the manufacture of glass, three chief sources of air pollution are batch drying of finely divided raw material prior to melting, gas- or oil-fired melting furnaces and glass forming. Whereas drying of raw materials presents only a dust problem, the furnaces emit particulates as well as a gas combination representing both fuel products and the melt composition. Glass forming machines generate heavy smoke form vaporization of hydrocarbon lubricants. In the fiber glass industry, phenolic emissions from binders are a major problem. Both the fiber glass and glass industries present water pollution problems as well.

In the mineral wool industry, stack emissions contain condensed fumes from the molten material, sulfur dioxide and fluorides as well as blow chamber and oven emissions consisting of fumes, oil vapors, binding agent material and wool fibers.

Finally, stone quarrying and sand and gravel production present both potential air and water pollution problems.

ASBESTOS INDUSTRIES - GENERAL

Asbestos is an air pollutant which carries with it the potential for a national or worldwide epidemic of lung cancer or mesothelioma of the pleura or peritoneum (2). Asbestos bodies have been observed in random autopsies of one-fourth to one-half of the population of Pittsburgh, Miami, and San Francisco and will probably be found in the people of every large city. Although asbestos has been shown to produce asbestosis, lung cancer, and mesothelioma in asbestos workers, the relationship between "asbestos bodies" and cancer or asbestosis has not been determined.

The latent period required to develop asbestosis, lung cancer, or mesothelioma is 20 to 40 years, and the exposure required to cause asbestosis has been estimated to be 50 to 60 mppcf-years (mppcf = million particles per cubic foot based on total dust count and 8-hr day, 40-hr week exposures). No such exposure relationship has been established between asbestos and lung cancer or mesothelioma. Asbestosis, lung cancer, and mesothelioma are all diseases which, once established, progress even after exposure to dust ceases.

Experiments with animals have shown that animals may develop asbestosis or cancer after inhaling asbestos.

Asbestosis is an asbestos-induced disease closely related to coal miners' black lung and cotton workers' brown lung; pleural and peritoneal mesotheliomas are tumor-like hardenings of the rib cage and abdominal linings, respectively. In the case of lung cancer, it has been shown that asbestos exposure and smoking are synergistic by a factor of 8. Asbestosis is not normally fatal if the exposure to asbestos is eliminated before extensive lung fibrosis develops.

Asbestos is unique among the hazardous substances, however, in that it does not have an acute toxicity. Asbestosis, in particular, is associated with long-term, primarily high-level, exposures to asbestos fibers. However, there is no known exposure threshold below which there is zero probability of one or another of these diseases.

The medical details of specific health hazards of exposure to asbestos have been detailed in a number of publications and will not be discussed at length here in the interests of holding this volume to a reasonable size. The reader is referred to publications by Sullivan and Athanassiadis (2) as well as to publications by the National Institute of Occupational Safety and Health (3), by R.E. Paddok et al (4) and by the U.S. Environmental Protection Agency (5)(6)(7).

W.E. Davis and Associates have reported (8) on the sources of asbestos emissions. Their emissions inventory was based on information obtained from production and reprocessing companies, 1968 production and use statistics from the Bureau of Mines and emission factors developed by Davis personnel. Although these emission factors are not based on quantitative data (i.e., emission tests), they are considered adequate for the purpose of identifying major sources. The estimated emissions of asbestos from sources studied by Davis Associates are as shown in Table 1. Considering this and other information, the major sources identified were asbestos mine-mill complexes.

TABLE 1: ASBESTOS EMISSIONS—1968

SOURCE CATEGORY	SOURCE GROUP	SHORT TONS
MINING AND OTHER BASIC PROCESSING	Mining and Milling	5,610
REPROCESSING		
	Friction Materials	312
	Asbestos Cement Products	205
	Textiles	18
	Paper	15
	Floor Tile	100
	Miscellaneous	28
CONSUMPTIVE USES		
	Construction	61
	Brake Linings	190
	Steel Fireproofing	15
	Insulating Cement	25
INCINERATION OR OTHER DISPOSAL		NA
NA-Data not available.	TOTAL	6,579

Source: W.E. Davis and Associates (8)

A five-year projection of estimated control costs and emission reductions was prepared for EPA in 1972 by R.E. Paddock et al (4). The estimated emissions for 1970 and projected reductions for 1977 are shown in Table 2.

TABLE 2: ESTIMATED ASBESTOS EMISSIONS FOR 1970 AND 1977

Source Category	Number of Plants (1970)	Emissions (tons/year)		
		1970 Current Control	1977 Current Control	1977 Meeting Standard*
Milling Products	9	3,860	5,440	218
Asbestos Cement	48	206	290	58
Floor Tile	18	101	142	28
Friction Material	30	314	441	88
Asbestos Paper	29	15	21	2
Asbestos Textiles	34	20	28	15

*Defined as use of a fabric filter rather than a numerical standard.

Source: R.E. Paddock, F.A. Ayer, A.B. Cole, and D.A. Le Sourd (4).

Because routine, standardized techniques for sampling and analyzing asbestos emissions have not been available, proposed standards for asbestos have often not been given in terms of numerical values. Instead, the standards have been expressed in terms of required control practices that limit emissions to an acceptable level (5). In part, control of atmospheric emissions would be achieved by:

[1] Utilizing industrial fabric filters to clean forced exhaust gases from asbestos mining, milling, and manufacturing industries and from fabricating operations that involve materials containing asbestos.

[2] Eliminating visible emissions of particulate matter from ore dumps, open storage areas, external conveyors, and tailing dumps associated with asbestos mining and milling facilities as well as from manufacturing and fabricating operations carried out with asbestos-containing materials in areas directly open to the atmosphere.

[3] Prohibiting certain applications of asbestos fireproofing and insulation by spraying processes.

Also, indirect atmospheric emissions of particulate matter would be controlled at manufacturing and fabricating sites where visible emissions normally result from operations using commercial asbestos. The maximum allowable emissions would be equivalent to those attained by either ventilating an entire work space through a fabric filter or by hooding emission sources and subsequently passing the required dust-control air through a fabric filter.

A quantitative definition of the asbestos air pollution problem is very difficult to formulate because of the lack of a dose-response relationship between levels of airborne asbestos and the resulting human diseases. Nevertheless, available evidence clearly implicates asbestos as a serious air pollution threat. This evidence includes the discovery of asbestos fibers in lungs of nonoccupationally exposed persons, the qualitative demonstration that asbestos fibers are present in ambient air, and the cited epidemiologic studies relating asbestos exposure to disease (5). However, a standard has been proposed by NIOSH.

The National Institute for Occupational Safety and Health (NIOSH) recommends that worker exposure to asbestos dust in the workplace be controlled. Control of worker exposure to the limits stated will prevent asbestosis and more adequately guard against asbestos-induced neoplasms. The standard is amenable to techniques that are valid, reproducible, and available to industry and governmental agencies. It will be subject to review and will be revised as necessary, according to NIOSH.

According to the criteria document for occupational exposure to asbestos (3), the occupational exposure to airborne asbestos dust shall be controlled so that no worker shall be exposed to more than 2.0 asbestos fibers per cubic centimeter (cc) of air based on a count of fibers greater than 5 micrometers ($>5\mu m$) in length [determined by the membrane filter method at 400 to 450X magnification (4 mm objective) phase contrast illumination], determined as a time-weighted average (TWA) exposure for an 8-hour work day, and no peak concentration of asbestos to which workers are exposed shall exceed 10.0 fibers/cc $>5\mu m$ as determined by a minimum sampling time of fifteen minutes.

This new standard becomes effective July 1, 1976 and until that date the present emergency standard is in effect. This period is believed necessary to permit installation of necessary engineering controls.

The emergency standard for exposure to asbestos dust (29 CRF 1910.93a) published in the *Federal Register,* Vol. 36, No. 234, page 23,207, Dec. 7, 1971) is as follows:

> "The 8-hour time-weighted average airborne concentration of asbestos dust to which employees are exposed shall not exceed 5 fibers per

milliliter greater than 5 microns in length, as determined by the membrane filter method at 400 to 450X magnification (4 mm objective) phase contrast illumination. Concentrations above 5 fibers per ml but, not to exceed 10 fibers per ml, may be permitted up to a total of 15 minutes in an hour for up to 5 hours in an 8-hour day."

The 1971 ACGIH tentative threshold limit value is 5 fibers/ml >5μm in length. Both are higher than the British standard of 2 fibers/cc by at least a factor of 1.5 times.

Of the methods presently being used to count dust samples in the asbestos industry, none is applicable to atmospheric asbestos air pollution. In all the asbestos monitoring methods used, microscopic counting of the fibrous particles is necessary to determine the proportion of fibrous material, and even then it is not known what fraction of the fibers is asbestos. Counting of fibers by eye under the microscope is tedious and difficult. If the number of fibers is less than 1% (less than 5 weight percent) of total dust, the other dust masks the fibers, and quantitative results cannot be obtained.

In parts of the asbestos industry where the asbestos-to-dust ratio is high (greater than 5 weight percent), it is often possible to determine the asbestos content indirectly. For example, if the proportion of asbestos in the airborne dust was known by microscopic count for a given sampling location, the concentration (at least the order of magnitude) could then be inferred from a simple measurement of the concentration of the total dust.

There are at present no proven satisfactory methods for the collection, detection, and identification of asbestos fibers in the 0.1 to 5.0 μ range in ambient air. Satisfactory sampling can probably be accomplished by use of a membrane filter-pump system. The major difficulty lies in the problem of identifying a very few asbestos fibers in the presence of relatively large numbers of a wide variety of other inorganic particulate matter found in the same air. Attempts to determine the asbestos content of urban air have revealed the need for development of new methods.

Modern analytical methods and instrumentation used in the asbestos industry are as follows (9):

Microscopic particle counting of samples on membrane filters
Thermal precipitators
Impingers
Royco particle counter
Mass concentration methods
Microsieving
Digestion
Column chromatography of organics adsorbed on the surface
X-ray diffraction
Low-temperature ashing
Atomic absorpton spectrophotometry
Electron microprobe
Neutron activation
Owens jet counter
Konimeter

Asbestos may be monitored in air by electron microscopy. The results of the Battelle Columbus Laboratories attempt to develop a method for asbestos determination in air have been summarized by R.E. Heffelfinger, C.W. Melton, D.L. Kiefer and W.M. Henry (10). The method developed included sampling, beneficiation of asbestos fiber and fibril, and determination of total asbestos by a transmission electron microscopic technique.

The problems of asbestos in water have recently been recognized. At a conference called by the National Institute of Environmental Health Sciences at Durham, North Carolina in Nov. 1973, the problem was discussed. The conference was prompted largely by the discovery earlier in 1973 that the drinking water of Duluth, Minn., and other cities on Lake Superior was heavily contaminated with asbestos, the presumed result of pollution by a mining company that has been dumping 67,000 tons of rocky waste into the lake each day for 17 years. Aqueous effluent standards for asbestos processing industries are being established by the Environmental Protection Agency.

ASBESTOS MINING AND MILLING

Chrysotile (serpentine asbestos), which accounts for over 95% of U.S. production, occurs in three types of formations: cross fiber, slip fiber, and loose. Cross fiber chrysotile is that in which the fibers span gaps (veins) in the surrounding serpentine rock formations. Slip fiber occupies similar gaps in the formations, but the fibers lie parallel to the walls of the vein. Loose fibers, which occur at only one mine site in the U.S., are premilled, occurring in loose formations near ground level mixed with various-sized aggregates.

Cross fiber chrysotile is commercially most valuable, since the highly prized spinning length fibers (¾" or longer) are most commonly found in such formations. (However, the longest known fiber bundle ever mined was from a Chinese slip fiber formation, and had a maximum fiber length of some 3½ feet). Cross fiber and slip fiber may be mined by surface (open cast or open pit) techniques or by underground methods, while loose fiber is mined by what is essentially an open cast process (4).

MINING PROCESSES AND EMISSIONS

Open cast mining involves removal of the ore by earth-moving equipment from shallow deposits, in one instance in the U.S. without the need for blasting. Generally, a shallow overburden with low concentrations of asbestos fibers must be removed. There will be emissions of asbestos fiber from the overburden dumps and exposed ores through weathering, and in concentrated amounts from drilling, blasting, overburden and ore removal, loading, and transport.

Open pit mining is similar to open cast operations except that the workings are much deeper to follow the fiber veins. Blasting and ore removal occur primarily on the sides of the pit along terraces which spiral down around the sides of the pit toward the bottom. Sources of emissions are the same as for open cast mining except that overburden removal will be proportionately smaller.

Underground mining of asbestos involves following the veins of ore with shafts, galleries, and drifts, using blasting and earth moving. However, there is no overburden removal, ore veins are not exposed to weathering, and many dusty operations take place underground. Therefore, emissions from this type of asbestos mining are much lower than for surface mining techniques. There will be significant emissions from surface ore transfer and transportation and hand cobbing of ore.

A new method of mining which has particular significance in control of emissions from open pit mining is the block caving technique, which significantly reduces the required blasting and eliminates the need for overburden removal. When the volume of rock to be mined has been determined, that volume is undercut, leaving solid support pillars to hold up the main block. As the block caves in down the chimney, ore is removed from underneath and mill and mine tailings are replaced on top to maintain the downward pressure on the block. Replacement of the tailings also reduces tailing dump emissions and space requirements. The block caving technique also reduces direct mining emissions to a level comparable to normal underground mining operations.

Mining of asbestos is limited in the U.S. to the four states of California, Vermont, Arizona, and North Carolina. A small amount of anthophyllite is mined at two locations in North Carolina; all the other mines produce varying grades and types of chrysotile. The four firms in Arizona produce a special low-magnetite cross fiber chrysotile commanding a premium for electrical applications. These firms mine asbestos at several sites using underground mining techniques.

The four mines in California use the open cast method. One of these mines is in an area where geological action has broken up the deposit, so that little or no blasting is required. This mine has some of the richest ore in the world, being in places up to 60% asbestos. (The U.S. average ore content is 4%.)

The mine in Vermont is the oldest and largest asbestos mine in the U.S. Both slip fiber and cross fiber chrysotile is mined there using the open pit process. The Vermont deposit is part of the large deposits identified as the Canadian belt in southern and southwestern Quebec. There is concern that the six Quebec mines may be sources of asbestos emissions that carry over into the United States, particularly into Vermont.

Each of the processes associated with asbestos mining that are listed above is a potential source of asbestos emissions (6). Local meteorological conditions can significantly influence the degree of emission. For example, rain, sleet, and snow are favorable influences because they result in wetting or covering exposed ore deposits in addition to scavenging the atmosphere. Conversely, strong winds that are capable of widely distributing existing emissions, in addition to entraining loosely bound asbestos fibers from material exposed to the atmosphere by mining operations, are an adverse influence.

Furthermore, the natural phenomena of earth movement, temperature cycling, wind erosion, and water erosion present opportunities for the emission of asbestos from virgin surface-ore deposits. In those surface mining operations that require blasting, the use of rotary or percussion drilling machines that incorporate air-flushing is a potential source of appreciable amounts of dust emissions. Air-flushing refers to the use of an air stream, operated by pressure, vacuum, or pressure and vacuum in combination, to cool the drill bit and lift cuttings out of the hole formed for placement of explosive charges.

Air travels down the hollow center of the drill bit as the drill cuttings move upward along the outside of the bit. Smaller dry suction drills employ an injector to exhaust air from a hood or cowl that encloses the drill bit at the hole collar. Even in a wet-drilling process, in which compressed air and water are injected in the downward-flow mode, a portion of the dust generated by drilling escapes without being converted into sludge. Further, a respirable aerosol of water droplets having entrained drilling dust can be emitted.

Detonation of explosive charges in the open-pit mining of various minerals breaks up massive deposits of asbestos-bearing rock, and the blast can produce a cloud of dust that may contain asbestos fibers. Similar emissions can occur when secondary blasting is used to reduce boulders to a size acceptable by the mill or to dislodge large rock deposits in open-cast mining.

In surface mining, the operations of removing overburden, scraping and shoveling of ore, preliminary screening of ore, conveying of ore, loading of ore into trucks, and the unloading of ore from trucks into hoppers at the mill can generate emissions of asbestos dust. Some ores have a high moisture content (as much as 20% in Fresno and San Benito counties), and, therefore, emissions from processing these ores are less than those encountered with dry ores. The emission sources associated with underground mining installations include sorting, conveying, loading, and unloading operations, which are performed outside the mines. The exhaust of ventilation air from underground mines to the atmosphere can also produce emissions.

Transporting the ore from the mine to the mill generates emissions which are generally grouped with those of mining rather than milling. Such emissions arise in large measure from open trucks, which are typically 20 to 75 tons capacity, although some 200-ton units are in use. Private mine-mill roads are frequently paved with tailings, which liberate fibers to the environment as the trucks pass.

In this context the use of 200-ton trucks has mixed significance for emission levels during transport. The larger truck capacity should reduce emissions from the transported ore, with their larger volume-to-surface ratio, and it reduces the number of trips per unit of ore. However, the road itself takes a bigger beating from each truck pass. The relative significance of the two effects has not been determined.

EMISSIONS CONTROL IN ASBESTOS MINING

Overall emissions from asbestos mining facilities are not stringently controlled at present. The absence of a higher degree of control is traceable to the fact that most operations are completely exposed to the atmosphere, with the result that emissions are diluted with ambient air over relatively large surface areas such as mining pits and roads.

The standard asbestos control technique of baghouse collection has limited application to mining processes. Portable baghouse systems can be used with good effect during drilling operations prior to blasting, since this is a localized emission source. However, this is a minor source in terms of total emissions, and some sort of wet drilling technique might be just as effective and much less costly.

Both dry centrifugal dust collectors and fabric filters have been applied to allay the dust generated during air-flushed drilling of holes for explosive charges, however. It is well known that the collection efficiency of fabric filters, expressed on a total mass basis, exceeds that of conventional cyclone collectors. In Figure 1, the application of a fabric filter of envelope type to a primary percussion drilling machine employed in asbestos mining is illustrated. Several treated synthetic filter materials, such as rayon-acetate and nylon-acetate treated with silicate, have been shown to release dust loadings readily during the cleaning cycle and to dry quickly if accidentally wetted in use.

The use of wet drilling methods to control emissions has been excluded from some asbestos mining operations because of extremely cold climates or restrictions imposed by water pollution control regulations. Other types of surface mining operations have overcome prohibitively cold weather by heating water on the drilling machine and insulating the water storage tank and all exposed piping. Even heated water, however, can freeze after discharge from the drill hole.

Since primary drill holes are often located within ten feet of the edge of a bench, which may range from 30 to 75 feet in height in asbestos quarries, the presence of ice can pose a serious occupational hazard. In warmer climates, the tendency of the drill cuttings to cement together as water seeps into asbestos seams in the fractured rock is an operational problem that limits the effectiveness of wet drilling. In the case of wet drills smaller than those used for primary drilling, the inclusion of special design features, such as front-head release ports for the venting of compressed air or an external water feed mechanism, can control the emission of unwetted dust or respirable water-dust aerosols.

The atmospheric emissions that result from primary and secondary blasting in asbestos surface mining are essentially uncontrolled at present (6). An optimum combination of amount, depth, and location of explosive charge should be sought that will produce complete combustion of the explosive compounds, along with the required loosening and breaking of a deposit, without unnecessary expulsion of material into the air. Multidelay devices for the initiation of detonation have been used successfully at limestone quarries, but incomplete combustion of multidelay charges, resulting from the highly fractured nature of the ore, has been observed at one domestic asbestos mine. Detailed technical assistance in implementing good blasting practice is available from explosives manufacturers.

FIGURE 1: FABRIC FILTER MOUNTED TO DRILLING MACHINE

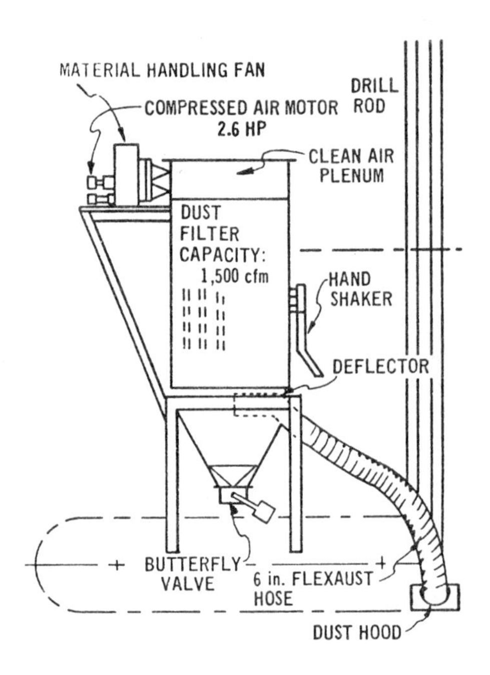

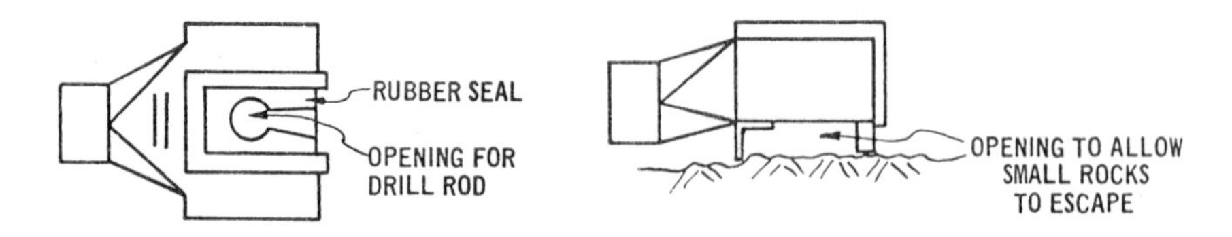

Aspiration Dust Hood

Source: J.R.M. Hutcheson (11)

The spraying of water or chemical wetting agents onto a surface prior to blasting could reduce emissions. The application of a pressurized water spray to asbestos mining would not be novel; the cleaning of deposits subsequent to the removal of overburden has been accomplished by high-pressure water sprays. The surface area of the blasted fragments, however, is so large in comparison with the surface area prior to blasting that the effect of surface wetting alone is likely to be minimal.

The use of liquid or paste stemming materials in blasting holes is a promising dust control method. In European coal mines, reductions of 20 to 80% in dust concentrations have resulted from a technique developed in France which involves placing plastic cartridges filled with water, or water in combination with a wetting agent, into holes before blasting. This technique has also been tested in copper mining operations. As an alternative to the use of liquid-containing cartridges, pastes with a cellulose or bentonite base can be employed. Container materials and wetting agents that would not interfere with the required purity of the milled asbestos must be developed, however, since plastic contamination of the ore which could not be removed by present milling methods may be a problem in applying this technique to asbestos mining.

Effective primary blasting minimizes the need for secondary blasting and is, therefore, an indirect method of controlling emissions from secondary blasting. The use of drop-ball cranes and pneumatic or hydraulic rock splitters as substitutes for secondary blasting has proved to be effective in controlling emissions from limestone quarrying, however, the extent to which the elasticity of asbestos-bearing rock might limit the effectiveness of drop-ball cranes for secondary fragmentation has not been fully evaluated.

The removal of overburden from ore deposits, shoveling of loosened ore, surface scraping of ore, preliminary screening and conveying of ore at the mine, and loading of ore into trucks produce asbestos emissions that are substantially uncontrolled at present. These operations, as well as primary and secondary blasting, should be scheduled to coincide to the maximum extent practicable with meteorological conditions favorable to the suppression of atmospheric emissions.

In particular, cognizance should be taken of seasonal variations in weather conditions. The limiting of operations to periods of favorable weather conditions may occasionally be impractical because of the large amount of equipment involved and because of safety precautions requiring that blasting be carried out on the same day that the charge is loaded. The application of water or chemical sprays can alleviate emissions from ore loading in some cases. Limiting factors are the possible freezing of the water or the introduction of chemicals that would interfere with the end use of the asbestos.

Control of dust generated during ore loading does not appear to be feasible and is not known to be practiced in any asbestos mining operations (4). Dusts generated during transport may be controlled by [1] wetting the ore surface, [2] covering the truck body with canvas or a more rigid sealing cover, [3] wetting road surfaces where they are covered with tailings. Wetting the ore surface in the trucks bed is not known to be used. In hot, dry weather it would have to be done frequently to be effective during long trips to the mill. Canvas covers are currently being used at several sites. Wetting road surfaces is used in some places. Constant wetting of the surface with water is effective but causes slippery road surfaces and requires much labor and water.

Wetting cannot be used in freezing weather. Good results have been obtained (in increasing order of effectiveness) with road oil, a 10 to 25% water solution of liquid sulfonate, and emulsified asphalt. Where the ore body is utilized as a mine road chemical dust suppressants may cause unacceptable ore contamination. It is expected that the overall control level from careful application of the best techniques would be about 80% of the potential road dust emissions.

Asbestos emissions that result from the dumping of ore from trucks at the mill site can be abated by the use of water sprays or by the application of capture hoods or enclosures combined with gas-cleaning devices. Some domestic mills currently use partial enclosure and water spraying techniques. Attempts have been made to stabilize mine overburden dumps where the waste rock, sand, and clays of hard-rock asbestos deposits are chemically neutral. These efforts have been successful to the extent that grasses and trees have been established over the surface of some waste dumps. Most areas exposed by open-pit mining, other than steep slopes, can probably be revegetated.

ASBESTOS ROCK MILLING AND EMISSIONS

Once asbestos ore has been mined, the asbestos fibers must be separated from it. When the asbestos has been separated, it is graded on the basis of content of various lengths of fibers. In general, for a particular source of chrysotile, the larger percentage of longer length fibers in the final mix bring the higher prices. The most expensive grade of fiber is called crude, which is not milled, but hand-separated (cobbed) from the surrounding rock into bundles of fiber with aggregate fiber length of ¾" or more. Crude (long) fiber is valued for weaving of asbestos textiles, for which shorter grades are not suitable.

In order to maintain maximum fiber length and promote maximum recovery, it is desirable to hold mechanical working of fibers to a minimum. Although asbestos fibers have very high tensile strength per unit weight, the individual fibers are so fine that they are rather easily broken. Compounding the recovery problem is that asbestos fibers have the same density and chemical composition as the surrounding rock. The solution applied in all but one of the asbestos mills in the U.S. is to use mechanical means to free the fibers from the rock, but accomplish the actual removal from the ore by an air aspiration system. In order to reduce fiber losses and industrial hygiene problems, it has become common to convey the fibers by air also.

The result is a requirement for seven to ten tons of process air for every ton of fiber produced, or a volume ratio of about 1,600 to one. The large volumes of air required, plus the floatability of asbestos fibers, leads to significant potential emissions.

The one exception to the air aspiration milling system in the U.S. is found in the mill which processes the loose fiber ore found in California. At this plant a proprietary wet separation process which presumably involves some sort of flotation technique is used to separate the fibers from the ore tailing. As a result, both industrial hygiene problems and air pollution potential from the fiber extraction process are significantly reduced.

The milling of asbestos ore by a dry process requires an extensive amount of handling and subdividing of the material in both a damp and a dry state. Consequently, there are numerous potential sources of asbestos emissions at a milling facility. The dumping of mine ore from trucks onto a wet-ore stockpile or into receiving hoppers is a potential emission source at the mill site. Further, asbestos-containing dust at the surface of an ore pile is susceptible to varying degrees of atmospheric entrainment, depending upon the moisture content of the ore and the strength of local winds.

The separating, cleaning, and grading of asbestos fibers requires large volumes of air, which are ventilated through fabric filters before being exhausted to the atmosphere or recirculated to mill buildings. Because makeup air is drawn in to replace the exhausted air process areas of a mill are frequently under negative pressure. When the volume of air exhausted to the atmosphere is sufficient for the entire mill to be under negative pressure, emissions to the atmosphere are reduced.

As asbestos ore, asbestos fibers, and asbestos-containing tailings are transported among the numerous processing devices of the mill by belt conveyors, the jostling motion, combined with the large surface area of material exposed to the environment, can produce significant asbestos emissions either directly into the exterior atmosphere or into the surrounding work space. Examples of such emission sources are transportation of material from a wet-ore stockpile to a dryer, from a dryer to a grading screen, from one vibrating air aspiration screen to another, and from the undersized side of a vibrating air aspiration screen to a tailings conveyor. Potential for severe emissions exists whenever asbestos-containing materials are handled at transfer points of conveyor systems.

The severe fracturing of rock by primary and secondary crushers frees additional asbestos fibers from the ores; the accompanying mixing action of the crushers facilitates the emission of asbestos-containing dusts to the interior spaces of the equipment. Because feed and discharge ports must be provided for crushers, an opportunity exists for the emission of asbestos to the exterior environment.

A primary source of emissions from asbestos mills is the effluent from ore dryers. The mechanical agitating action of the dryer and the necessity for contacting the ore with large volumes of air contribute to the entrainment of asbestos-containing dust in the heated gas stream. In addition to contaminants from the ore, the dryer exhaust contains a significant amount of moisture and the products of combustion from the air-heating device. The effluent temperature varies widely and can range from 140° to 500°F.

The vibratory or oscillating motion of grading screens and the resulting sifting action of the screens as the asbestos-containing material is separated into a range of sizes expose large surface areas of material to the surrounding air; the surface of a typical screen measures 5 feet by 11 feet. Accordingly, this process results in appreciable quantities of airborne dust. If there are no provisions for capturing and containing the dust, it is emitted directly into the mill work space.

Even though the packaging of asbestos fibers by machine minimizes handling and exposure of the material to the atmosphere, emissions can occur at the interface between the material and the package during the filling and sealing of containers. The packaging of fibers into coarsely woven bags or otherwise non-dust-tight containers can yield emissions during further handling operations.

Large quantities of dry, finely divided rock that contain asbestos dust must be removed from most asbestos mills as waste material. The transfer of this rock by a moving-belt device or by vehicle to an exposed tailings dump can generate emissions to the atmosphere. Emissions can also result from the placement of tailings onto an existing dump, from the leveling of the dump to permit further deposition of wastes, and from direct entrainment of surface dust by ambient air currents.

Milling by Air Separation

The incoming ore is first unloaded from the arriving truck where a cloud of dust and fiber is generally produced. The incoming coarse ore is then typically crushed by a jaw crusher to a size that depends upon the mill. Oversize rock is separated by rotating cylindrical trammel screens and crushed in a secondary crusher, usually a cone type. The ore streams in most plants are then conveyed to a dryer (a rotary kiln in larger installations) where moisture in the ore (up to 30% by weight) is removed.

The dried ore is then stored, with large amounts being held to allow for variations in fiber demand and mine production over time. At least one company departs from this procedure by storing the ore wet, thereby smoothing out drier operations and reducing emissions from ore storage and handling. This company further reduces emissions and homogenizes its product by stripping ore for further milling from the underside of the storage pile via a moving conveyor systems housed in a tunnel chamber underneath the storage area.

Dried ore is conveyed to an additional crushing step and then through a series of milling, shaking, and aspirating steps. The milling, done by either hammer mills (fiberizers) or crushers, serves to separate the fibers from the rock and from each other. The shaking is accomplished on progressively finer screens, where small rocks and fiber bundles pass through for further treatment, larger rocks are retained for conveying to tailing dumps or further crushing, and the freed fibers are removed by air flow through powerful suction hoods.

Separated fibers are caught in dry cyclones and conveyed to grading screens. After grading, the fibers are sent to storage bins by grade (length). The final operation is removal of fibers from storage bins, blending in a mixer to produce the desired final grade, and bagging for shipment.

Residual rock, which contains a small amount of unremoved fiber and dust, is usually transported by some means to a tailing dump. If the block caving technique is used at the associated mine, the tailings may be returned to the mine to assist in the caving.

Every one of the processes described above leads to some loss of fibers to the air either inside or outside the plant. Emissions to in-plant air are usually kept to a minimum consistent with maintaining the industrial hygiene Threshold Limit Value. This is most effectively accomplished by operating as many pieces of equipment as possible in a closed condition and under a slightly negative pressure. This is done, however, at the expense of increasing the plant ventilation load and greater potential atmospheric emissions.

The major emissions potential at asbestos mills arises from the cyclone collector exhaust. The series of process cyclones normally used are of a relatively large-diameter design to limit damage to the fibers as much as possible. The resultant overall capture efficiency for the series is in the range of 90%. The asbestos fibers which escape are predominantly the smaller, respirable and wind transportable ones. The overall emissions by the asbestos milling industry, which has a partial usage of fabric filters, is 100 pounds per ton of asbestos produced, or a rate of loss of about 5%.

The source of emissions in the mill which is more expensive than the others to control is the ore dryer. The exhaust stream of the dryer is typically at high humidity and about 250°F. The associated baghouse must therefore be insulated to avoid condensation and resultant blinding, and the bags need to be of temperature-resistant materials, such as Orlon or Dacron. This means, in practice, a separate baghouse or baghouse section, and higher investment and maintenance costs than for other emission sources in the mill.

Milling by Wet Method

The details of operation of the one wet mill asbestos plant in the U.S. are not known, since it is a proprietary process. However, two important points relevant to emissions are known. First, because of the loose nature of the ore at the source of this mill's asbestos, little or no ore crushing is required. Second, the product is available as either loose fiber or as compressed pellets or balls of ¼" or ½" diameter.

However, because of the nature of the ore more emissions are to be expected during transport and unloading, because so much more of the asbestos is free fiber at this stage. Emission potential during milling is virtually zero since the fibers are wet. Drying of the fiber, however, has great emission potential since the fiber is believed to be loose at this time.

If the fibers are pelletized before drying, however, the emission potential would be comparable to that from conventional ore drying processes. Emission potentials from bagging and shipping of loose fiber would be the same as for conventional plants, but much less in the case of pelletized fiber. Emissions from tailing dumps are expected to be less, especially if the tailings are discarded wet, because of the low fraction of ore that becomes tailings. Water polution potential is of course quite high.

EMISSIONS CONTROL IN ASBESTOS MILLING

To control asbestos emissions from the surface dusting of ore stockpiles, water can be sprayed onto the material from adjacent towers (6). This technique has also been successfully applied in the control of particulate emissions from exposed limestone stockpiles. In a typical limestone application, water is sprayed at a rate of 500 gallons per minute from towers 40 feet high; the spray covers a circle 200 feet in radius. For asbestos applications, it may be necessary to use the lowest feasible flow rates in order to avoid the discharge of asbestos-containing water from the facility and to comply with applicable water pollution control regulations.

It is technically feasible to house exterior belt and bucket conveyor systems in completely enclosed galleries to prevent asbestos emissions from material in transit and from the emptied return side of the systems. Furthermore, the attainment of safe occupational asbestos exposure levels may require the enclosure of in-plant conveyor systems. The asbestos milling industry is currently applying these control techniques to a limited extent. Points at which asbestos ore, asbestos fiber, and asbestos-containing waste materials are transferred between process equipment and conveyor systems, as well as conveyor system transfer points, can be hooded and ventilated to gas-cleaning devices to control emissions. A schematic diagram of this technique, as applied to the transport of asbestos ore from a crusher to a storage bin, is shown in Figure 2.

FIGURE 2: CONTROL OF EMISSIONS FROM TRANSPORT OF ORE

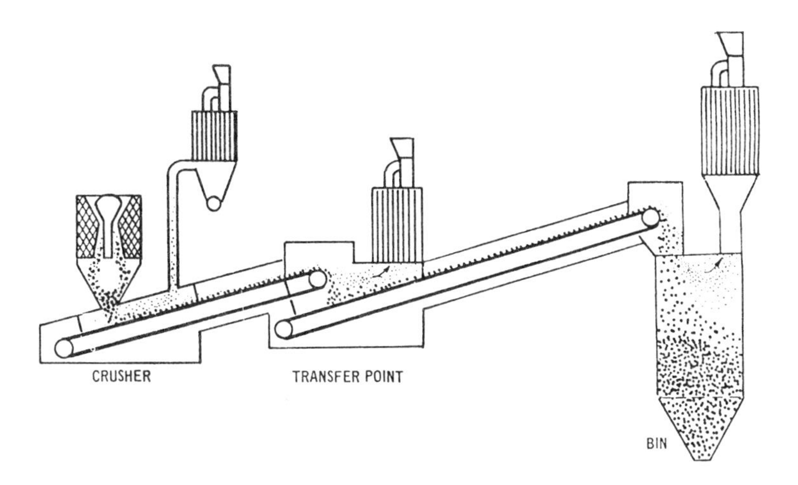

Source: J.R.M. Hutcheson (11)

The feed and discharge ports of ore crushers can be fitted with dust capture hoods to control asbestos emissions; the hoods should be ventilated to a gas cleaning device such as a fabric filter. Figure 3 illustrates a device of this type, having an air flow capacity of 3,000 cubic feet per minute, attached to the inlet of a 48-inch by 60-inch jaw crusher. A hinged suspension permits convenient displacement of the hood to provide access in cleaning ore blockages from the crusher.

Historically, cyclone collectors have been applied more widely than any other type of gas-cleaning device to control asbestos emissions from ore dryers, largely because of the relatively low initial cost, simplicity of construction, and low maintenance cost of these devices. Also, the dust collection efficiency of cyclones is relatively insensitive to variations in process gas temperature and to the condensation of moisture within the collector; however, the fact that the efficiency of these dry centrifugal collectors is considerably less than that attainable with some other widely employed gas-cleaning devices has prompted attempts to gain increased collection efficiency. For example, one milling facility has employed 200 small-diameter cyclones, each with a capacity of 100 ft^3/min as a substitute for a single cyclone of 20,000 ft^3/min air-handling capacity. Partial plugging of the small collector elements occurred, possible as a result of internal water condensation, with the result that collection efficiency was greatly decreased, rather than increased.

As a compromise between the commonly applied 10-foot-diameter cyclones and the potentially more efficient small-diameter devices, twin cyclones 4 feet in diameter were chosen. In recognition of the relatively low efficiency of cyclones for the collection of finer particulates, the Canadian asbestos industry is seeking control techniques that exceed the performance of dry centrifugal collectors.

FIGURE 3: DUST CAPTURE HOOD FITTED TO ORE CRUSHER

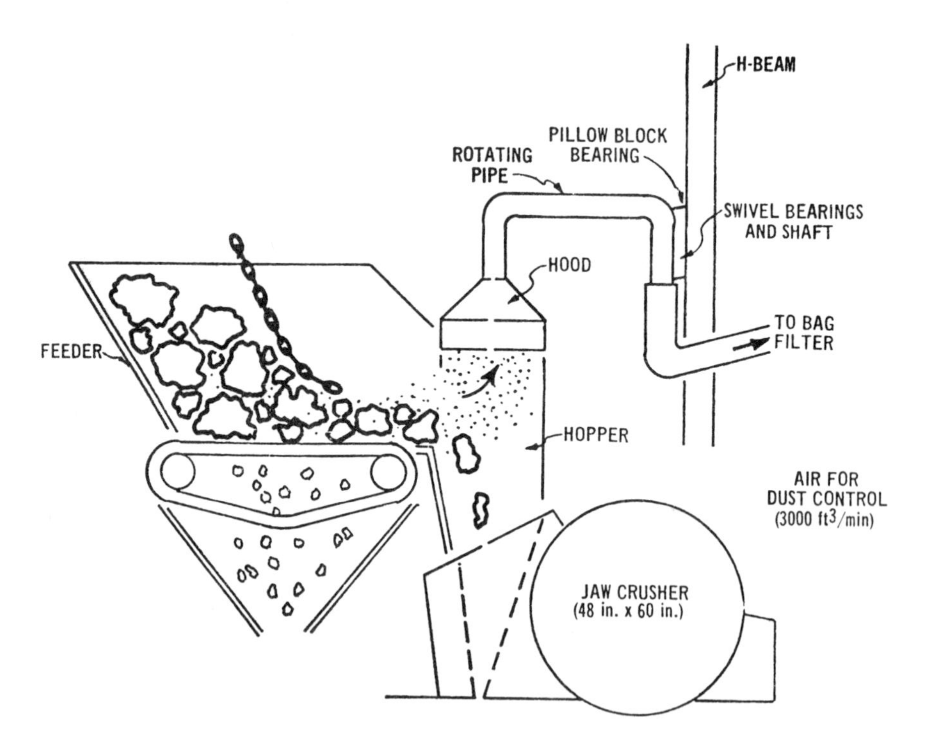

Source: J.R.M. Hutcheson (11)

Wet collectors are presently employed by the Canadian asbestos industry on at least two ore-drying installations; the process gas flow rates are 100,000 and 65,000 cubic feet per minute. In these two collectors, the particulate-laden gas stream passes through a water spray and then enters into a centrifugal fan that dynamically separates dust and particulate water from the stream as air is drawn through the blades; the asbestos-containing particulates are removed as a slurry. Corrosion resulting from sulfur oxides present in the dryer effluent, and the limited collection efficiencies of 85 to 95% are significant disadvantages of these wet collection devices.

In spite of low pressure loss and the theoretically high collection efficiencies, electrostatic precipitators are not widely applied to the control of asbestos emissions from ore dryers. This is a consequence of the necessity for maintaining close control of gas velocity, gas temperature, and particulate moisture content in order to realize design collection efficiency. An electrostatic precipitator of 170,000 cubic-feet-per-minute capacity is now in operation at a Canadian asbestos mill.

Fabric filters have been successfully applied to the control of emissions from asbestos ore dryers, and it is reported that asbestos emission levels of approximately 2 x 10^6 particles per cubic foot (ppcf) have been realized. Two new units were placed into operation in Canada in 1971; Figure 4 shows an asbestos ore dryer of the fluidized-bed type and accompanying bag filter installation that was installed at a Canadian mill in 1972.

FIGURE 4: CONFIGURATION OF FABRIC DUST COLLECTOR FOR ORE DRYER

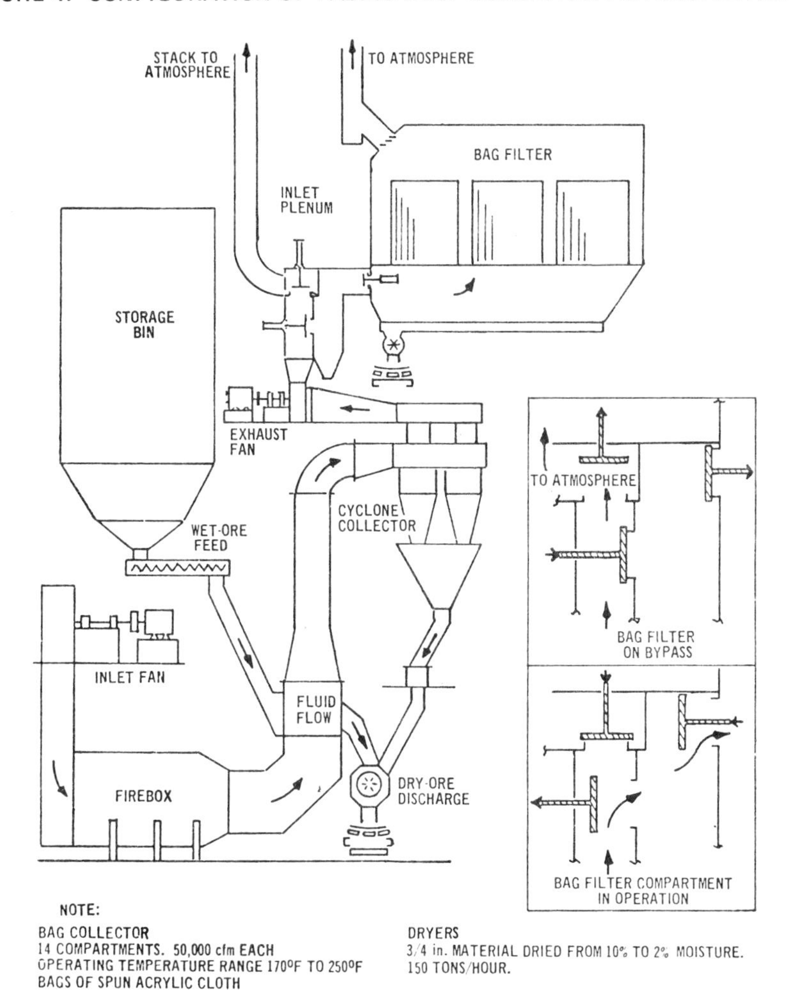

Source: J.R.M. Hutcheson (11)

The filtering chambers of these baghouses are thermally insulated to prevent excessive cooling of the effluent gas streams and the possible condensation of water; the occurrence of condensation could irreversibly cement adhering dust cakes. Orlon, Dacron, Nomex, Teflon, Terylene, or Fiberglas, which can withstand the high temperatures of the gas streams, are required as filter materials.

Additional protection against excessively high temperatures or condensation of moisture during short time periods can be provided by the use of bypass arrangements. For effective control of asbestos emissions, engineering design and operational procedures should minimize the duration and number of periods in which bypass devices are utilized.

Figure 5 provides relative comparisons of asbestos-containing dust emissions from ore dryers subsequent to cleaning of the effluent stream by one of five control devices. These calculated estimates of emissions are based upon operating experience of the Quebec asbestos milling industry. The emission rates are based upon an assumed value of 1 pound per hour for fabric filter collectors.

FIGURE 5: DUST EMISSIONS FROM ORE DRYERS

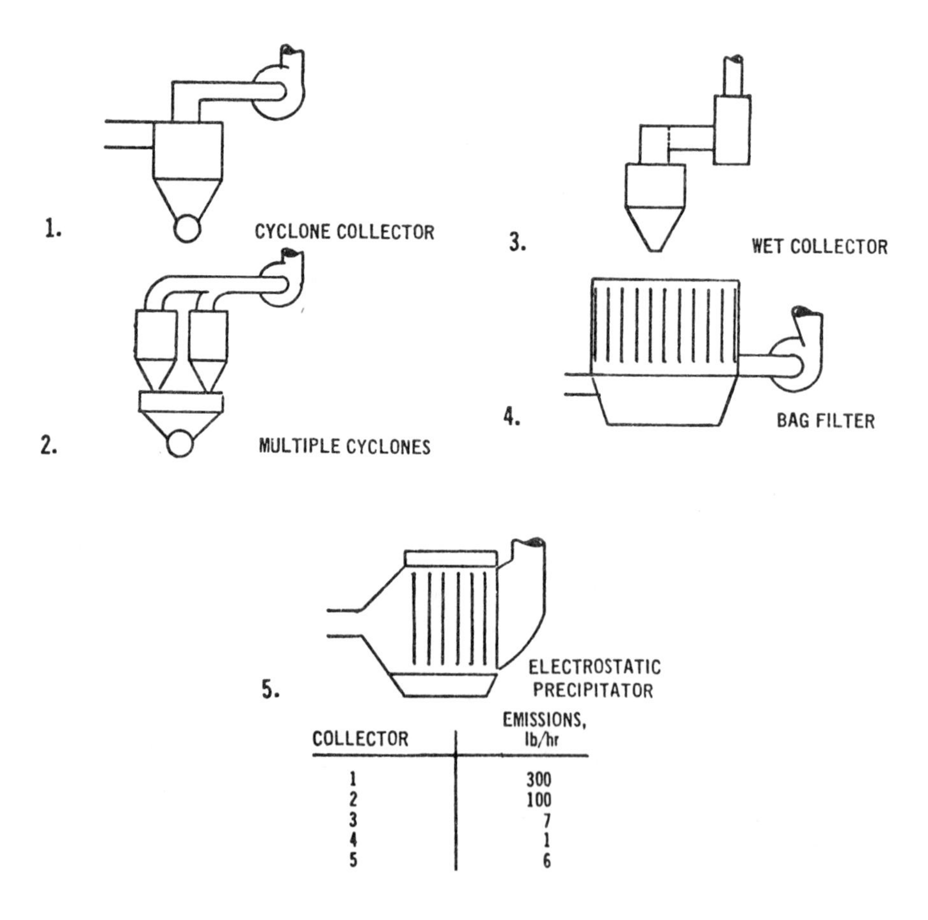

COLLECTOR	EMISSIONS, lb/hr
1	300
2	100
3	7
4	1
5	6

Source: J.R.M. Hutcheson (11)

Asbestos emissions from the bed of a vibrating grading screen can be controlled by covering the screen, with a dust capture hood, as completely as practicable without interfering with the required screen motion. The hood exhaust streams are passed through a fabric filter to remove the entrained dust after asbestos fiber has been deposited in cyclone-type collectors.

Quantitative tests of a rotary, air-swept screen have shown that refinements in dust shielding and ventilation of the screen can reduce material emissions from 36.9 pounds per day to less than 0.5 pounds per day; local dust counts were diminished from 12×10^6 ppcf to less than 2×10^6 ppcf.

Asbestos emissions that accompany the bagging of fibers can be controlled by installing high-volume, low-velocity ventilation hoods over packing operations. Further, low-volume, high-velocity systems can, during packaging, collect dust in the immediate vicinity of bag-filling valves and on bag support platforms.

As one method of controlling emissions when dry, asbestos-containing mill tailings are placed on a relatively flat disposal pile, a mobile dumper is used at the end of a belt conveyor that transports the wastes. As disposal proceeds, the location of the dumper is periodically changed in order to maintain the tailings pile as nearly level as possible and thereby minimize emissions caused by shifting the tailings with earth-moving equipment. An inverted funnel mounted to the dumper discharges the wastes in close proximity to the surface of the dump in order to assist in reducing emissions at the point of deposition; however, the elimination of visible emissions at the point of deposition may also require that a water or chemical spray be used.

In other milling complexes, mixtures of water and wetting agents have been applied to tailings during their discharge onto waste piles, and this has proved to be moderately successful. Visible emissions generated by the dumping of tailings have been totally eliminated at one domestic asbestos mill by the mixing of tailings with water prior to deposition. This control technique is promising for mills that have access to sufficient water and that can overcome the problem of freezing conditions.

In some cases, asbestos mill tailings form large mounds across which long belt conveyor systems with several transfer points are deployed. The transfer points can be enclosed and ventilated to gas-cleaning devices to provide emission control. Potential emissions from segments of the conveyor system between transfer points can be controlled by enclosing the equipment.

Emissions from the surfaces of tailings dumps can be controlled by providing a protective covering or seal. Because of the large surface areas involved, most of the control methods are expensive. Wherever the eventual surface of the dump is reasonably level, soil can be spread as a sealing medium. The establishment of vegetation on dumps is hindered by the high alkalinity (pH = 9) of the tailings.

In preliminary tests, grass has been grown on tailings by first mixing them with the acidic tailings of a copper mine across a soil depth of about 2 inches. Chemical agents that can be sprayed onto waste dumps to form a protective surface crust that is permeable to water are commercially available. The penetration of moisture through the crust controls the potential erosion and disintegration of the cover by heavy rainfall.

In some instances, tailings piles from the milling of long-fiber asbestos ores are somewhat self-stabilizing because of the relatively low percentage of very fine dust, the tendency of meteorological conditions to form a layer of larger particles that protect the interior of the pile, and the consolidation of the pile by freezing during long periods of the year.

EMISSIONS CONTROL COSTS FOR ASBESTOS MILLING

The estimated 1970 emissions and emissions after full controls are instituted are shown in

Table 3 along with the associated control costs. There are two categories shown for control costs—small plants and large plants. This is done because the average control costs per ton of asbestos produced were fairly uniform within these categories, and because it was necessary to avoid identifying individual plants.

However, such a classification does hide the variation in costs among plants. Additional control cost per ton varies among plants from zero to $5.96 per ton. The additional annual investment varies from zero to $183,000. The control costs in Table 3 are based on a confidential plant examination describing the control equipment needed, and the data in Table 4. Minimum costs are assumed. Units are added rather than totally replaced, no land is acquired, and ducting is minimal. Table 5 estimates the cost of controlling emissions from asbestos milling for 1970 and 1977.

TABLE 3: EMISSIONS AND CONTROL COSTS ASSOCIATED WITH ASBESTOS MILLING

Plant Size	Number of Plants	Total Emissions 1970 Actual	If Fully Controlled	Control Costs* Investment	Annual	Per Ton of Capacity
Large	5	3,750	150	333,480	124,189	0.44
Small	4	110	4	8,340	3,579	1.38
Totals	9	3,860	154	341,820	128,189	0.78
Average		430	17	37,980	14,243	0.78

*acfm = actual cubic feet per minute

TABLE 4: UNIT COSTS FOR ASBESTOS MILLING EMISSION CONTROL

Control Unit	Investment (dollars/acfm)*	Operating and Maintenance (dollars/acfm-year)
Cyclone collector	$0.55	$0.13
Baghouse		
Low temperature (cotton sateen bags, no insulation)	3.00	0.90
High temperature (Orlon bags, insulated)	6.00	1.30

*acfm = actual cubic feet per minute

TABLE 5: COST ESTIMATES FOR CONTROLLING EMISSIONS FROM THE ASBESTOS MILLING INDUSTRY

Year	Estimated Capacity (Tons)	Required Investment	Annualized Cost
1970	163,600	$341,820	$128,189
1977	176,000	367,840	137,280

Source: R.E. Paddock, F.A. Ayer, A.B. Cole, and D.A. LeSourd (4)

Gas flow data required to produce total costs for Table 3 from the unit costs in Table 4 were either obtained directly from an OAP/EPA survey of plant sites or were based on capacity data obtained in the survey combined with the seven tons of air per ton of asbestos criterion for air flows. The seven ton value is for milling only; air flows for drying of ore is highly dependent on ore asbestos concentration and water content. In any case, actual air flow rates for ore dryers needing control were directly available.

The basic processing of domestic asbestos ores is carried out in nine mills in the U.S. with production capacities ranging from 200 to 65,000 tons per year. These mills produce approximately one-sixth of the total consumption of asbestos in the United States. The estimated additional annualized costs required of these existing sources for compliance with the proposed standards range from zero to $5.96 per ton of asbestos fiber produced per year; this represents a range of zero to 6.7% of the average selling price per ton of domestically produced asbestos in 1969. The average investment, for the entire industry, is estimated to be $0.78 (0.9%) per ton of asbestos fiber produced per year. Actual investments range from $2,780 for an essentially uncontrolled mill of 200-ton/year capacity to $183,000 for a partially controlled mill of 40,000-ton/year capacity (5).

Standardized conveyor housings that cover the carrying runs of conveyor belts and thereby shield exterior belts and the material being transported from atmospheric precipitation are commercially available. A measure of emission control is also provided by protecting the material from the winds. These housings are typically in the form of curved sections of corrugated sheet metal, one side of which is hinged to the conveyor system. This type of construction permits each section of the belt housing to be lifted so that access is provided to potential blockages of the conveyor system. The additional equipment cost of such housings, above that for completely exposed conveyors, is approximately $10 to $15 per lineal foot of conveyor system, depending upon the width of the conveyor belt.

Complete enclosure of external conveyor systems can furnish more positive control of atmospheric emissions than is possible with conveyor housings. This can be accomplished by providing roof and sidewall coverings for standardized commercial conveyor systems of gallery construction. In this type of system, a truss is employed to support the conveyor system and adjacent maintenance walkways across long spans.

The additional equipment cost of an enclosed gallery section is approximately $125 per lineal foot of conveyor in excess of the cost of a corresponding fully exposed system. A standard open belt conveyor, with walkway along one side, is priced at approximately $200 per lineal foot.

Chemical coatings formulated with a nontoxic organic base, and commercially available, can be sprayed onto exterior material storage or waste piles, such as asbestos mill tailings dumps, to prevent the entrainment of material by ambient winds. Temporary coatings that provide protection for 1 month require from 17 to 170 gallons per acre of surface area at a material cost of $25 to $50 per acre.

A typical semipermanent (1 year or longer) coating requires 44 gallons per acre at a material cost of $480 to $1,140 per acre of area, depending upon the quantity of material purchased and the particular coating formulation that is compatible with the material to be encrusted. The cost of application, estimated to be $70 per acre per application, is an important factor to be considered in the determination of which type of coating to utilize (6).

ASBESTOS PRODUCTS INDUSTRIES

Although known as a curiosity since biblical times, asbestos was not used in manufacturing until the latter half of the 19th century. By the early years of the 20th century, much of the basic technology had been developed, and the industry has grown in this country since about that time. Canada is the world's largest producer of asbestos, with the USSR and a few African countries as major suppliers. Mines in four states, Arizona, California, North Carolina, and Vermont, provide a relatively small proportion of the world's supply.

Asbestos is normally combined with other materials in manufactured products, and consequently, it loses its identity. It is a natural mineral fiber which is very strong and flexible and resistant to breakdown under adverse conditions, especially high temperatures. One or more of these properties are exploited in the various manufactured products that contain asbestos.

Asbestos is actually a group name that refers to several serpentine minerals having different chemical compositions, but similar characteristics. The most widely used variety is chrysotile. Asbestos fibers are graded on the basis of length, with the longest grade priced ten to twenty times higher than the short grades. The shorter grades are normally used.

On a worldwide basis, asbestos-cement products materials and pipe currently consume about 70% of the asbestos mined (13). In the United States in 1971, the consumption pattern was reported to be:

Asbestos-cement pipe & sheet	25%
Floor tile	18%
Paper	14%
Friction products	10%
Textiles	3%
Packing and gaskets	3%
Sprayed insulation	2%
Miscellaneous uses	25%

These figures do not accurately reflect the production levels of these products because the asbestos content varies from about 10 to almost 100 percent among the different manufactured products.

With the exception of roofing and floor tile manufacture, there is a basic similarity in the methods of producing the various asbestos products. The asbestos fibers and other raw materials are first slurried with water and then formed into single or multilayered sheets as most of the water is removed. The manufacturing process always incorporates the use of save-alls (settling tanks of various shapes) through which process wastewaters are usually routed. Water and solids are recovered and reused from the save-all, and excess overflow and underflow constitute the process waste streams. In all of these product categories, water serves both as an ingredient and a means of conveying the raw materials to and through the forming steps.

At most plants, only one asbestos product is manufactured. There are three reported locations that manufacture more than one category of asbestos products in the same plant in a manner that results in a combined wastewater flow. Since the wastewaters from all the asbestos products categories, except roofing and floor tile, have many common characteristics, they are generally treatable by the same types of control technology. Consequently, the combined wastewaters from the manufacture of multiple asbestos products do not present significant additional problems in control.

Of more significance from a water pollution control point of view is the manufacture of nonasbestos products with confluent waste streams at some of the locations. The most common combinations are the manufacture of plastic pipe at asbestos-cement pipe plants and the manufacture of organic (cellulose fiber) paper at asbestos paper plants. Plastic pipe manufacture is not likely to result in the discharge of significant pollution, other than waste heat. Organic paper manufacturing wastewaters, however, are significantly stronger and of different character than those from asbestos paper production. The raw materials are often paperstock (salvaged paper) as well as virgin pulp and the wastes are highly colored, turbid, and high in oxygen demand.

Forty-five manufacturing plants representing 30 different companies or corporations were contacted directly in an EPA study of the asbestos textile, friction material and sealing device industry (12). Information was collected from six additional plants through a questionnaire. This coverage is believed to include better than 80% of all the plants that are properly within the two SIC classes, and it represents an accurate picture of this segment of the asbestos manufacturing industry. A total of ten plants were found that discharge process-related wastewaters. These plants are described individually in Table 6.

In reviewing Table 6, it should be noted that the discharges from the two plants using the dispersion process for making asbestos yarn are included, even though these operations are experimental and not yet classified as subcategories of this industry. Of the remaining eleven waste streams, seven result from air pollution control equipment and only four from manufacturing and associated operations.

The asbestos manufacturing industry grew rapidly in the first two-thirds of the 20th century. Many observers expect that growth will be less rapid in the future. Environmental and health considerations, plus competition from fiber glass, silicone products, aluminum sheet, and other materials, are among the factors contributing to the slowdown in growth. Many of the plants visited in an EPA study were not operating at full capacity. New uses and markets for asbestos may be more difficult to develop in the future. Despite the decline in the rate of growth, asbestos has unique characteristics, and its use in manufacturing can be expected to continue to a significant degree in the foreseeable future.

The most significant effect of the recently increased concern about the industrial hygiene aspects of asbestos is the trend toward substitution of other materials, especially among users of textile products.

New uses and markets for asbestos will be more difficult to develop in the future unless means are found to reduce the potential hazards. Despite the declines in some areas, however, the unique characteristics of asbestos plus new developments within the industry make the outlook for future growth favorable in the textile, friction materials, and sealant manufacturing segments of the industry.

TABLE 6: GENERAL DESCRIPTION OF KNOWN WASTEWATER SOURCES OF SOME ASBESTOS PRODUCTS MANUFACTURING PLANTS

Plant	Product	Waste Water Source	Treatment Provided	Effluent Discharged To
A	Textiles	Coating	None	Municipal Sewer
B	Textiles	Coating	Two-Stage Lagoon	Municipal Sewer
B	Textiles	Fume Scrubber	Two-Stage Lagoon	Municipal Sewer
C	Textiles	Dispersion Process	Filtration	Municipal Sewer
D	Textiles	Dispersion Process	None/Lagoon	Municipal Sewer/ Surface Water
D	Sheet Gasketing	Solvent Recovery	Lagoon	Surface Water
E	Friction Materials	Dust Scrubber	Lagoon	No Discharge
F	Friction Materials	Dust Scrubber	Sedimentation	Surface Water
G	Friction Materials	Dust Scrubber	Two-Stage Lagoon	Surface Water
H	Friction Materials	Dust Scrubber	Lagoon	Surface Water
I	Friction Materials	Solvent Recovery	None	Municipal Sewer
I	Friction Materials	Dust Scrubber	Lagoon	No Discharge
J	Friction Materials	Dust Scrubber	Chemical Precipitation with Other Wastes	Surface Water

Source: EPA (12)

AIR POLLUTION

There has long been concern about the industrial hygiene aspects of the dust and fiber emitted to the air in mining, processing, transportation, and manufacturing operations. This concern has recently been expanded to include the general public. Asbestos is among the first materials to be declared a hazardous air pollutant under the Clean Air Act amendments of 1970. Stringent regulations have also been promulgated to control exposure of workers in the industry.

Because asbestos is exceptionally resistant to thermal degradation and chemical attack, settled particles are persistent in the environment and subject to reentrainment into the atmosphere. It can readily be mechanically subdivided into fibers of submicron diameter, which can remain airborne for long periods of time. These factors, coupled with the presence of large numbers of emission sources, as noted above, would indicate the presence of a background level of asbestos in the atmosphere. Semiquantitative data confirm this conjecture and show that urban background concentrations are significantly larger than nonurban ones (5).

The only significant potential air pollution problem associated with the application of wastewater treatment and control technologies at a typical asbestos manufacturing plant is the release of asbestos fibers and other particulates from improperly managed solid residues. Exposed accumulations of dried solids may serve as sources of air emissions upon weathering.

The biodegradable organic matter content of asbestos solids is low or nonexistent. The solids do not undergo appreciable microbial breakdown and there are no odor problems associated with asbestos wastes. There are no unusual or uncontrollable sources of noise associated with application of the treatment and control technologies.

In most of the plants in the asbestos textile, friction material and sealing device industry, particulate emissions are controlled by baghouses or other dry devices. In a few wet plants, wet dust collectors are used and a wastewater results.

Where small quantities of solvents are used, they may be wasted rather than be recovered. Among the techniques used for controlling the emissions of vaporized materials is absorption in water, which may result in a wastewater effluent. The type of air pollution control equipment used in this industry provides a useful basis for categorization.

Three of the four water pollution control subcategories in the textile, friction material and gasket segments of this industry (solvent recovery, vapor absorption and dust collection) relate totally or partially to control of pollutant emissions to the atmosphere. The use of the substituted dry control devices would effect equal, or better, control of the pollutants of interest. The only significant potential air pollution problem associated with the application of the control technologies at a typical plant is the release of materials from improperly managed solid residues. For example, exposed accumulations of dust from friction materials plants may serve as sources of air emissions.

There are no significant odor problems associated with implementation of the wastewater control and treatment technologies. Neither are there any unusual or uncontrollable sources of noise associated with the control measures.

The manufacture of asphalt roofing felts and shingles involves saturating fiber media with asphalt by means of dipping and/or spraying. Although it is not always done at the same site, preparation of the asphalt saturant is an integral part of the operation. This preparation, called blowing, consists of oxidizing the asphalt by bubbling air through the liquid asphalt for 8 to 16 hours. The saturant is then transported to the saturation tank or spray area. The saturation of the felts is accomplished by dipping, high-pressure sprays, or both. The final felts are made in various weights: 15, 30 and 55 pounds per 100 square feet (0.72, 1.5, and 2.7 kg/m^2). Regardless of the weight of the final product, the makeup is approximately 40% dry felt and 60% asphalt saturant. The major sources of particulate emissions from asphalt roofing plants are the asphalt blowing operations and the felt saturation. Another minor source of particulates is the covering of the roofing material with roofing granules. Gaseous emissions from the saturation process have not been measured but are thought to be slight because of the initial driving off of contaminants during the blowing process.

A common method of control at asphalt saturating plants is the complete enclosure of the spray area and saturator with good ventilation through one or more collection devices, which include combinations of wet scrubbers and two-stage low voltage electrical precipitators, or cyclones and fabric filters. Emission factors for asphalt roofing are presented in Table 7.

TABLE 7: EMISSION FACTORS FOR ASPHALT ROOFING MANUFACTURING WITHOUT CONTROLS*

Operation	Particulates**		Carbon monoxide		Hydrocarbons (CH_4)	
	lb/ton	kg/MT	lb/ton	kg/MT	lb/ton	kg/MT
Asphalt blowing	2.5	1.25	0.9	0.45	1.5	0.75
Felt saturation						
Dipping only	1	0.5	–	–	–	–
Spraying only	3	1.5	–	–	–	–
Dipping and spraying	2	1	–	–	–	–

*Approximately 0.65 unit of asphalt input is required to produce 1 unit of saturated felt. Emission factors expressed as units per unit weight of saturated felt produced.

**Low-voltage precipitators can reduce emissions by about 60 percent; when they are used in combination with a scrubber, overall efficiency is about 85 percent.

Source: EPA (15)

As shown in Figure 6, a typical line for the manufacture of asphalt felt includes the following sequence of machines for processing the felt roll: unwind stand, dry floating looper, saturator spray section, saturator dip section, drying-in drums, wet looper, cooling rolls, finish floating looper, and roll winder. Auxiliary equipment includes a heater and storage tank for the asphalt and hot asphalt pumps. Saturators are designed to saturate felt by spraying or by dipping or a combination of spraying followed by dipping (14).

Equipment lines for manufacturing shingles, mineral-surfaced roll roofing, and built-up smooth rolls employ in sequence: an unwind stand, dry looper, asphalt saturator, drying-in drums, and wet looper, as described for the manufacture of asphalt-saturated felt, plus additional machines for applying filled asphalt coating, sand, mica, and rock granules.

A typical line, illustrated in Figure 7, includes after the wet looper: a coater, granule applicator, backing applicator, press section, water-cooled rolls, finish floating looper, and either a roll winder or a shingle cutter and stacker, depending upon the product being made.

Heating asphalt saturant to 400° or 450°F under agitation results in the vaporization of low-boiling point hydrocarbon oils in the form of dense white emissions varying in opacity from 50 to 100%. Sources of these emissions are asphalt-saturant storage tanks, saturator, drying-in drums, and wet looper.

Emissions of excessive opacity also may occur from the cooling drums used in the manufacture of asphalt-saturated felt. The heated asphalt used in coating produces dense white emissions emanating from the storage tanks, filled-coating mixer, and coater.

Although the opacities of the emissions from this equipment are less than those of emissions from equipment containing saturant, they may exceed 40% opacity (14).

In the saturator, removal of moisture from the paper felt as water vapor or steam in effect distills the lower boiling hydrocarbons and produces a great number of small particles. Many of these particles have diameters that match the wavelengths of visible light (0.4 to 0.77 micron) and are, therefore, most effective in blocking light to produce a high percentage of opacity.

The concentration of oil particulates exhausted from different saturator machines varies from 0.416 to 0.768 grain per standard cubic foot (gr/scf). Emission rates as well as opacities from saturators and coaters are affected by: line speed, product manufactured, felt moisture content, air temperature and humidity, asphalt composition and temperature, number of sprays and gates, and gas exhaust rate. Generally, emissions increase directly with increasing line speed, felt moisture content, air temperature and humidity, number of spray headers and gates in operation, and the asphalt temperature. Line speeds for some installations exceed 500 fpm. As expected, emission rates and opacities are highest during the manufacture of roofing materials requiring the highest asphalt saturating rates.

FIGURE 6: SCHEMATIC DRAWING OF LINE FOR MANUFACTURING ASPHALT-SATURATED FELT

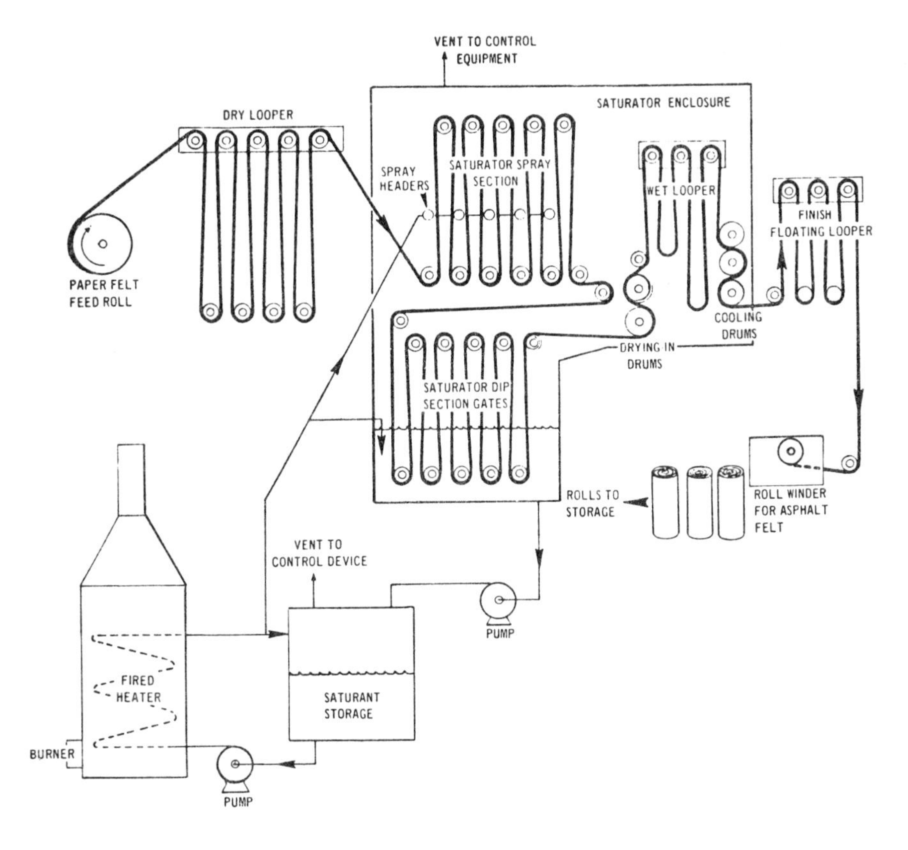

Source: J.A. Danielson (14)

FIGURE 7: MANUFACTURE OF ASPHALT SHINGLES, MINERAL-SURFACED ROLLS AND SMOOTH ROLLS

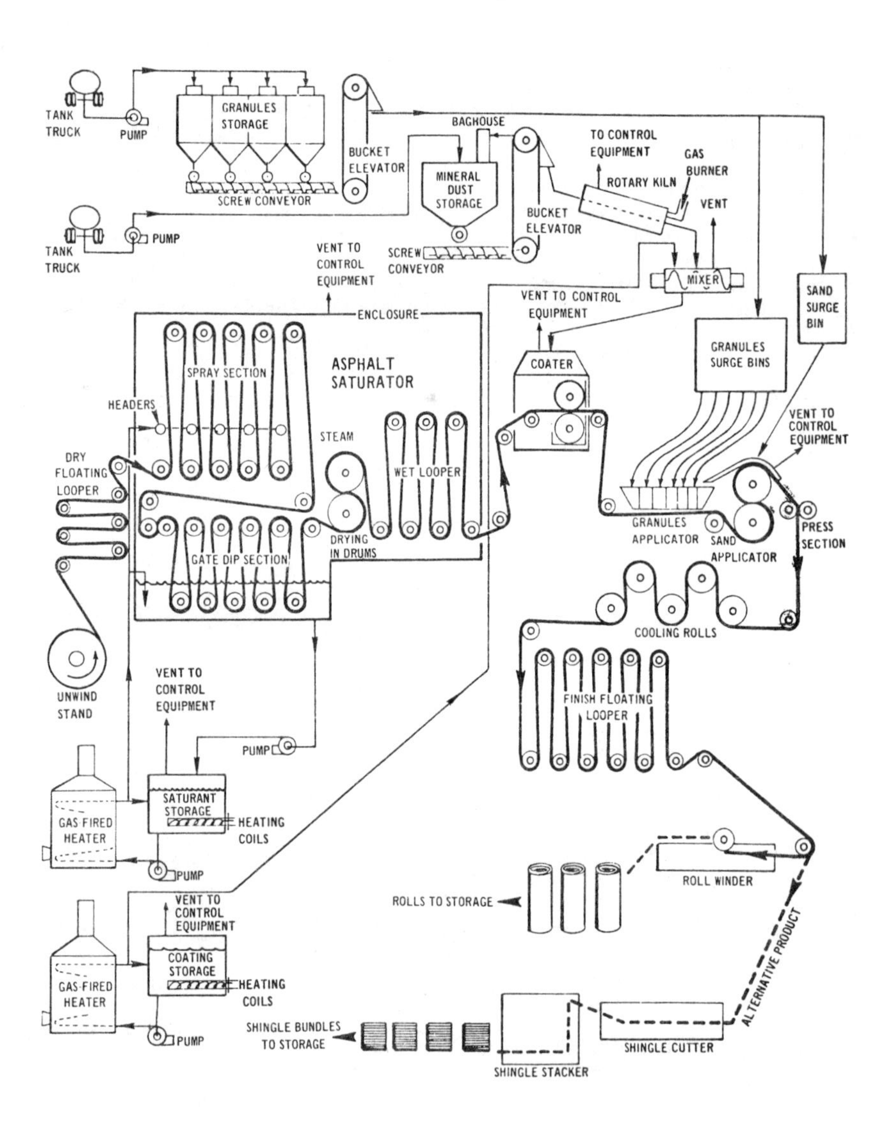

Source: J.A. Danielson (14)

Emission rates and opacities are also characteristic of the particular crude used for making asphalt saturant or asphalt coating. Generally, asphalts with the highest content of volatile oils produce maximum emissions. A number of observations of the effluent from an electrical precipitator venting a saturator revealed a reduction of opacities by about 15 percentage points when an asphalt containing lesser amounts of volatile components was used.

While these observations were being made, process variables such as roofing product, saturant temperature, line speed, and air temperature and humidity were held as constant as possible. In addition, source test results showed that particulate concentration in the exhaust gas from the precipitator also were reduced from 0.07 to 0.03 gr/scf. Reductions in particulate emissions are not always accompanied by corresponding reductions in opacity. Conditions exist wherein the number of submicron-size particles, which cause dense opacities, but which constitute only a small fraction of the total weight of all the particles emitted, are not reduced in proportion to the larger, heavier particles.

Typical specifications for roofing asphalts include: gravity, API; flash point, Cleveland Open Cup; viscosity, 210°F Saybolt Furol Seconds; solubility in CCl_4; amount of sediment; and water content. Unfortunately, these specifications do not measure the gaseous emission potential of an asphalt. Weight loss on heating by the ASTM D-6-67 cup method and the ASTM 1754-67T plate method also should be used with caution in determining volatile potential since some asphalt samples can air-polymerize and gain weight at the same time they lose weight by volatilization.

Sometimes saturators are operated at such high speeds as to cause hot asphalt to be flung from the top rollers of an asphalt dip section and thereby atomize oil particles. This condition can be corrected by installing curved adjustable shields about one inch from the rollers to quickly capture the hot asphalt spray and reduce the volatilization of oil particles.

Dust is generated along with oil mists and aerosols in the manufacture of shingles, mineral-surfaced rolls, and built-up (smooth) rolls. The sources for these dust emissions, which result in excessive opacities and particulate weight losses, are conveying and storage equipment for mineral dust, sand, mica, talc, asbestos, and other materials. Excessive dust is emitted from the mineral dust dryer, the mixer for hot mineral dust and coating asphalt, and the applicators for sand and mica.

Generally, only slight quantities of dust are emitted from the conveying and storage equipment and applicators for granules. This is because typical granules are of relatively large size; less than 2% by weight, pass a Number 35 U.S. mesh screen. The granules also are treated with oil to prevent dusting and promote bonding of the granules to the filled asphalt coating.

On the other hand, sand, mica, and mineral dust have significant amounts of particles less than 10 microns, which create dust problems during handling. For example, large quantities of dust are emitted when storage bins or surge bins are being filled with these materials by pneumatic conveyor.

Perhaps the largest single source of dust is from the direct-fired rotary kiln for drying mineral dust. A typical mineral dust may contain about 4% by weight of particles under 10 microns. Considerably less dust is generated if a heated flight screw conveyor is used instead of the rotary kiln.

Oil mists and aerosols as well as dust are emitted in excess of those quantities allowed by Los Angeles County air pollution regulations from the mixer for producing filled asphalt coating.

Nearly all asphalt saturators, drying-in drums, and wet loopers in Los Angeles County, for example (14), have been equipped with canopy-type hoods. With canopy hoods, the volume of air to be exhausted through the hood is in the range of 10,000 to 20,000 cfm.

By designing a relatively large room enclosure around an asphalt felt saturator, drying-in drums, wet looper, and cooling drums, the volume of contaminated air to be controlled can be drastically reduced. Such a room enclosure should include enough space for the operators to remove broken felt from the equipment. The only openings into this enclosure would be two inch high slots for the inlet and outlet of the felt web and automatic doors for access to the machinery by operating personnel. These slot openings should be as close to the floor as possible. Indraft velocities through the slots for the web and one open door should be designed for at least 20 ppm so that only about 3,000 to 5,000 cfm of contaminated air ventilates the saturator and its associated equipment.

A separate small enclosure venting to control equipment should be built tight to the coater to allow the line operator to control the product by hand adjustments to the coater. The operator can stand outside of this enclosure, where he does not have to breathe contaminated air. All openings into this enclosure should be kept to a minimum size. Design ventilation rates for the coater should not exceed 300 cfm, with indraft velocities ranging from 150 to 200 fpm.

Oil mists and aerosols emitted from the storage tanks for saturant and coating asphalt are usually vented to the same air pollution control equipment that vents the saturator drying-in drums and wet looper. Design ventilation rates for each storage tank and from each mixer average about 100 cfm.

Baghouses, low pressure scrubbers, and two-stage electrical precipitators have been used to control emissions from saturators, coaters, and asphalt storage tanks. Most of the earlier control devices vented relatively large volumes (10,000 to 20,000 cfm) of air contaminated with oil mists and aerosols. Although these devices have proved capable of controlling the particulate emissions measured on the basis of weight per effluent volume, scrubbers as the only control devices have in most instances failed to collect the large number of submicron-size particles that create the opacity problem. Even baghouses and electrical precipitators employing scrubbers as precleaners have had difficulties in controlling opacities.

Extensive modifications of the earlier devices have been necessary to keep these air pollution control systems in compliance. Opacities of emissions from baghouses and electrical precipitators following modifications have usually ranged from 15 to 35% white.

An approach to this control problem involves building a tight enclosure around the saturator drying-in drums and wet looper, and a separate enclosure around the coater to reduce the total volume of contaminated air to less than 5,000 cfm. At this flow rate, afterburners of the size that normally operate without visible emissions and with control efficiencies over 95% based upon measurement of total carbon can be installed. Afterburners also are superior to other control devices in abating odors, which in some instances are strong enough to create a nuisance.

Baghouses have been used almost exclusively to control the dust emissions. Line blenders or mixers for filled asphalt coating omit dust with only trace amounts of oil particulates and essentially no moisture. These emissions have been successfully controlled by baghouses. Large batch blenders, however, emit moisture from undried mineral dust along with dust particles and relatively large quantities of oil particulates. These emissions should be controlled by venturi scrubbers.

A recent development in the control of effluent from roofing plant saturators consists of passing contaminated effluent through a slowly moving unrolled mat of glass fiber. At velocities of 500 to 700 fpm through the mat, over 90% of the liquid particulates is removed, and there is a correspondingly high reduction in odor level. Pressure drop through the mat can be increased from 16 to 25 inches of water column for efficient collection of submicron-size particulates. At higher pressure drops, the stack discharge is reported to be virtually free of visible pollutants. The spent mat is rerolled and discarded. Operating costs are comparable or lower than those of other methods of filtration.

WATER POLLUTION

The increased concern with the health effects of asbestos fibers in the air has produced changes that affect, to some degree, the water pollution control aspects of the industry. The principal change has been conversion of dry processes into wet processes and the use of water sprays to allay dust from mining operations and slag piles. This shifting is expected to continue in the future.

While there has been considerable interest and much research on the health effects of asbestos in air, there has been little study of the effects of fibers in water. The first major investigations of this possible problem are now being initiated. The impetus for these studies was supplied by the finding of asbestos-like material in the drinking water of Duluth, Minnesota.

Water is commonly used in asbestos manufacturing as an ingredient, a carrying medium, for cooling, and for various auxiliary purposes such as in pump seals, wet saws, pressure testing of pipe, and others. Water is used only for cooling in the manufacture of asbestos roofing and floor tile products. In the discussion below, these two categories are not included unless specifically mentioned. In most asbestos manufacturing plants the wastewaters from all sources are combined and discharged in a single sewer.

Asbestos manufacturing in almost all cases, involves forming the product from a dilute water slurry of the mixed raw ingredients. The product is brought to the desired size, thickness, or shape by accumulating the solid materials and removing most of the carriage water. The water is removed at several places in the machine and it, together with any excess slurry, is piped to the save-all system.

The mixing operations are carried out on a batch, or semicontinuous basis. Water and materials are returned from the save-alls as needed during mixing. Excess water and, in some cases, materials are discharged from the save-all system. Fresh water and additional raw materials are added during mixing. The fresh water is often used first as vacuum pump seal water before going into the mixing operations.

The major source of process wastewater in asbestos manufacturing is the machine that converts the slurry into the formed wet product. It is not practical to isolate individual sources of wastewater within the machine system. The water is commonly transported from the machine to the save-all system and back to the machine in a closed system. To measure the quantity of water flowing in the system involves a monitoring program beyond the scope of this study. Only one manufacturing plant provided data on in-plant water flows that were more than rough estimates. This information is presented in Figure 11 (p 40). The relative amount of internal recycling in all asbestos manufacturing plants is significant and of roughly the same relative proportion as detailed for this pipe plant.

An important factor influencing both the volume and strength of the raw wastewaters is the save-all capacity in the plant. Save-alls are basically settling tanks in which solid-liquid separation is accomplished by gravity. Their purpose is first to recover raw materials (solids) and, second, water. The efficiency of separation is primarily dependent upon the hydraulic loading on the save-all. Plants with greater save-all capacity have greater flexibility in operation, more water storage volume, and a cleaner raw wastewater leaving the manufacturing process. In many asbestos manufacturing plants, the solids in the save-alls are dumped when the product is to be changed or when it is necessary to remove the accumulated waste solids at the bottom. It may also be necessary to dump the save-alls when the manufacturing process is shut down.

While seemingly similar when described by the common collective parameters (suspended solids, oxygen demand, etc.), the wastewaters from the different product categories exhibit some important differences. The differences relate both to the in-plant and end-of-pipe control measures and to the speed with which the category can be brought to the point where pollutants are not discharged. In general the raw waste load and volumes differed for each product subcategory. No great differences existed between the asbestos-cement pipe and asbestos-cement sheet subcategories, nor between the two asbestos paper subcategories. However, the evidence shows that asbestos-cement sheet plants will be able to achieve zero discharge sometime before asbestos-cement pipe plants. The same is true for asbestos paper (starch binder) versus asbestos paper (elastomeric binder).

When classified in terms of the major wastewater pollutants, those segments of the asbestos manufacturing industry involving textile, friction material and gasket production fall into two groups: [1] textile coating, solvent recovery, and vapor absorption; and [2] wet dust collection. The wastewaters from the first group contain significant levels of organic materials in solution. The raw wastes from textile coating may also contain suspended materials that will settle in quiescent conditions. The wastes from wet dust collectors are entirely suspended solids with minimal dissolved organic content. Some of the in-plant control measures apply to both groups, but the end-of-pipe treatment technologies are basically different.

Sedimentation, with various auxiliary operations, yields an effluent of low pollution potential when properly applied to asbestos manufacturing wastewaters. The settled solids are inert, dense, and appropriate for landfill disposal. While present practices within the industry are not achieving the best possible results in all cases, they can be upgraded without major technical problems.

Treatment beyond sedimentation and pH control is not appropriate for wastes from the major product categories in the asbestos manufacturing industry. The only pollutant constituent remaining at significant levels, other than temperature, is dissolved solids. While these levels may be at undesirably high levels for certain industrial water uses, they do not present serious hazards to human health or to aquatic life.

To remove the dissolved solids burden in these wastewaters would require advanced treatment operations techniques, e.g., reverse osmosis, electrodialysis, or distillation. The initial and annual costs associated with these advanced treatment operations are so high that alternative solutions, namely, complete recycle of wastewaters, will be implemented by the industry instead of further treatment.

Within the textile, friction material and gasket manufacturing segments of this industry, the only end-of-pipe treatment technology in use is sedimentation, normally in lagoons. While this operation may be adequate for wastewaters from wet dust collectors, it is inappropriate as the sole method of treatment for the first group of subcategories. It should be pointed out that some friction materials manufacturing plants provide treatment beyond sedimentation. These are primarily for wastes from nonasbestos manufacturing, e.g., metal finishing operations, and wastes from the wet dust collectors are treated in the same facility.

The control technologies recommended here are addressed at the principal pollutant parameters, namely COD, suspended solids, and pH. There are insufficient data available to ascertain the need for additional control measures for such dissolved pollutants as heavy metals, phenols, and plant nutrients. In most of the known cases, the costs of end-of-pipe treatment technologies more advanced than those recommended here are so high that alternative solutions will be used, e.g., substitution of baghouses for wet dust scrubbers in friction materials plants. At some of the plants, such a substitution program, on a phased schedule, has already been initiated.

Based on the available information, the in-plant control measures and end-of-pipe treatment technology outlined below can be implemented as necessary within the appropriate subcategories of the industry. Factors relating to plant and equipment age, manufacturing process and capacity, and land availability do not generally play significant roles in determining whether a given plant can make the changes. Because so few plants are actually affected today, the recommended technology has been defined with all of the known plants in mind. Implementation of a particular control or treatment measure will involve approximately the same degree of engineering and process design skill and will have the same effects on plant operations, product quality, and process flexibility at all locations. Each plant is unique, however, and the possibility of peculiar requirements should not be ignored.

Many asbestos manufacturing plants incorporate some in-plant practices that reduce the release of pollutant constituents. These practices have resulted in economic benefits, e.g., reduced water supply or waste disposal costs, or both. Few plants include all of the control measures that are possible, however.

Raw materials are normally stored indoors and kept dry. There are no widespread water pollution problems related to improper raw materials storage practices.

In all cases, sanitary sewage should be disposed of separately from process wastewaters. Public health considerations as well as economic factors dictate that sanitary wastes not be combined with asbestos process wastes.

Other nonprocess wastewaters are often combined with manufacturing wastes in asbestos plants. A careful evaluation should be made in each plant to determine if some or all of these wastes could be segregated and recirculated. Such reduction in waste volumes might result in smaller, more economical waste treatment facilities.

Except for roofing and floor tile plants, housekeeping practices do not greatly influence the wastewater characteristics. The use of wet clean-up techniques are common to control fiber and dust air emissions. In view of the alternative, continuation of the proper use of such wet methods should not impair the efficiency of end-of-pipe treatment facilities.

Fresh water should be used first for pump seals, steam generation, showers, and similar uses that cannot tolerate high contaminant levels. The discharges from these uses should then go into the manufacturing process as make-up water and elsewhere where water quality is less critical. Water conservation equipment and practices should be installed to prevent overflows, spills, and leaks. Plumbing arrangements that discourage the unnecessary use of fresh water should be incorporated.

Plans should be made for complete recirculation of all wastewaters. This will permit the installation of new equipment and the making of the plant alterations as part of an integrated, long-range program. In some cases, it may be more economical for a given plant to move directly toward complete recirculation rather than install extensive treatment facilities. In line with water use practices, evaluation of the benefits of increased save-all capacity should be made at some plants. This would provide more in-plant water storage, permit greater operating flexibility, and reduce the level of pollutant constituents in the raw wastewaters discharged from the plant.

In some friction materials plants, water is recirculated in the wet dust collectors. This is the only in-plant control measure that is generally used in this industry. Other in-plant measures, as described below, have been implemented at individual plants to eliminate the generation or discharge of process-related wastewaters. Raw materials are normally stored indoors and in containers. There is no widespread water pollution problem related to improper or inadequate raw material storage practices.

In all cases, sanitary sewage should be discharged separately from process-related wastewaters. Public health considerations as well as economic factors dictate that sanitary wastes not be combined with process-related wastes for on-site treatment.

In all four subcategories, the wastewaters originate at one point in the process or the auxiliary operation. The wastes, therefore, can be isolated for separate control. In many plants, the wastes are diluted with cleaner waters, such as spent cooling water and steam condensate. By mixing these streams, the entire discharge becomes, by definition, a process-related waste subject to control. These clean water discharges should be segregated and managed separately.

The only subcategory where housekeeping practices influence the quality of the wastewater is textile coating. Since the waste results primarily from clean-up of equipment and dumps, changes here can result in significant improvements in the quality of the wastewaters.

Attention should be directed toward water conservation in all subcategories. In the clean-up operations in asbestos textile coating, there is a tendency to use more water than is required. The water used in the vapor absorption and dust collection equipment should be reduced to the minimal level dictated by air quality requirements. Spent cooling water, where available, can be used for these operations. As described below for three of the four individual subcategories, water usage can be eliminated through substitution of alternative procedures or equipment.

Most asbestos manufacturing plants currently provide some form of treatment of the raw wastewaters before discharge to receiving waters. In virtually all cases, this treatment is sedimentation. At several plants, the treatment facilities are small and of simple design. Fortunately the waste solids are dense and almost any period of detention will accomplish major removal of the pollutant load.

Sedimentation is the oldest of all treatment unit operations in sanitary engineering practice. It is well understood and its costs, ease of operation, efficiency, and reliability make it ideally suited for industrial application. Sedimentation is an appropriate form of treatment for asbestos manufacturing plant wastewaters regardless of the plant size and capacity, manufacturing process, or plant and equipment age. Design is based on the hydraulic discharge and plants with smaller effluent volumes can use smaller units. The treatment system can be sized to accommodate surges and peak flows efficiently. Because waste asbestos solids are inert biologically, overdesign does not result in solids management problems.

If necessary, complete settling facilities large enough to treat the waste flows from any asbestos manufacturing plant can be placed on an area of 0.1 hectare (0.25 acre) or less. If more land is available, larger units that provide solids storage may be constructed. Such units would result in lower operating costs. This design is especially appropriate for wastewaters from asbestos-cement products manufacture because the solids are inert. Solids with significant BOD levels may require more prompt reuse or dewatering and disposal.

The land requirements for asbestos solids disposal are not excessively high. Some plants have disposed of solids within relatively limited boundaries for decades. While this practice results in problems it does serve to indicate that land disposal, if properly carried out, is an appropriate means of disposing of waste solids from asbestos manufacturing.

The recommended end-of-pipe technology for the industry is sedimentation, with ancillary operations as necessary. The subsequent control technology recommended is complete recirculation of all process wastewaters from all categories of asbestos manufacturing covered by this document. In most cases, complete recycle will require that the save-all system be expanded or supplemented to provide higher quality water for some in-plant uses. The wastewater treatment facility could very readily serve this function.

Consequently, the recommended end-of-pipe control technology would represent part of an overall long-term control program to achieve zero discharge of pollutant constituents at most locations.

Included with end-of-pipe treatment technologies are those in-plant modifications that are more than control measures, e.g., substitution of dry air pollution control equipment for wet scrubbers. This is regarded as a logical arrangement because the changes are separate from the manufacturing processes, major equipment installation is required, and both relate to protection of environmental quality through treatment.

The recommended control and treatment technologies are believed to be applicable to the appropriate subcategories, as outlined below, and are based on the limited data available. It is conceivable that unknown factors would render a particular technology inoperative at a given plant. The steps described here cannot, therefore, be applied without careful analysis of each plant's wastes and particular requirements.

The control and treatment technologies recommended here can be applied regardless of plant size and capacity, the manufacturing process, or plant and equipment age. The design can be altered to fit the plant's needs, and the wastes from both large and small plants can be managed efficiently using these technologies.

All of the recommended control and treatment technologies require relatively little land area; less than 0.1 hectare (0.25 acre) in all cases. If more land is available at a given plant, larger facilities may be employed to reduce operating costs.

The additional land required for disposal of containerized liquid wastes resulting from the technologies described here are not large. The wastewater volumes are relatively small when compared to many industries, and the volumes of waste generated for land disposal are also relatively small.

In some categories, the Level I technologies (1977) are based on treatment to reduce the pollutants to acceptable levels prior to discharge, while the Level II technologies (1983) involve substitution of equipment so that no wastewater is generated. The two levels are incompatible in that the money spent in implementing the Level I controls is lost when the Level II controls are installed. Whether to stop at Level I or move directly to Level II is a management decision for each plant. Since half of the plants known to be generating process wastewaters now discharge to municipal sewerage systems, the decision takes on added dimensions.

Asbestos-Cement Pipe and Sheet

Asbestos fibers in asbestos-cement products serve the same role as steel rods in reinforced concrete, i.e., they add strength. Portland cement and silica are also major ingredients of these products.

Asbestos-cement pipe is manufactured for use in high pressure and low pressure applications in diameters from 7.6 to 91.5 cm (3 to 36 inches) and in lengths up to 4 meters (13 feet). It is used to carry wastewaters, water supplies, and other fluids and in venting and duct systems. Asbestos-cement flat and corrugated sheets are used for exterior sheathing, siding and roofing, interior partitions, packing in cooling towers, laboratory bench tops, and many other specialty applications.

The largest single use category of asbestos fibers in the United States is the manufacture of asbestos-cement products. The pipe segment is the largest component of this product category.

Asbestos-cement products contain from 10 to 70% asbestos by weight, usually of the chrysotile variety. Crocidolite and other types are used to a limited extent depending upon the properties required in the product. Portland cement content varies from 25 to 70%. Consistent cement quality is very important since variations in the chemical content or fineness of the grind can affect production techniques and final product strength. The remaining raw material, from 5 to 35%, is finely ground silica. Some asbestos-cement pipe plants have facilities for grinding silica as an integral part of their operations. Finely ground solids from damaged pipe or sheet trimmings are used by some plants as filler material. A maximum of 6% filler can be used in some products before strength is affected.

The interwoven structure formed by the asbestos fibers in asbestos-cement products functions as a reinforcing medium by imparting increased tensile strength to the cement. As a result, there is a 70 to 80% decrease in the weight of the product required to attain a given structural strength. It is important that the asbestos be embedded in the product in a completely fiberized or willowed form, and the necessary fiber conditioning is frequently carried out prior to mixing the fiber with the cement and silica. In some cases, however, this fiber opening is accomplished while the wet mixture is agitated by a pulp beater, or hollander.

Asbestos-cement sheet products are manufactured by the dry process, the wet process, or the wet mechanical process. Figures 8 through 10 illustrate the sequence of steps in each of these manufacturing processes with the sources of wastes indicated. Products having irregular shapes are formed by molding processes which account for only a very limited production today. Extrusion processes are not widely used in the United States.

In the dry process (Figure 8), which is suited to the manufacture of shingles and other sheet products, a uniform thickness of the mixture of dry materials is distributed onto a conveyor belt, sprayed with water, and then compressed against rolls to the desired thickness and density. Rotary cutters divide the moving sheet into shingles or sheets which are subsequently removed from the conveyor for curing. The major source of process wastewater in this process is the water used to spray clean the empty belt as it returns.

FIGURE 8: ASBESTOS-CEMENT SHEET MANUFACTURING OPERATIONS, DRY PROCESS

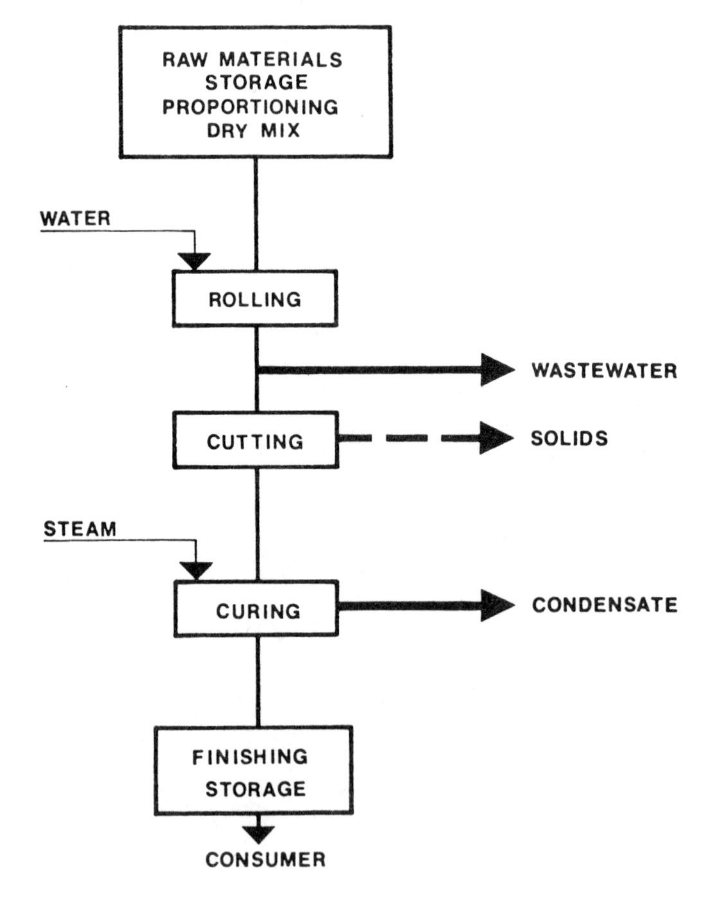

Source: EPA (13)

The wet process (Figure 9) produces dense sheets, flat or corrugated, by introducing a slurry into a mold chamber and then compressing the mixture to force out the excess water. A setting and hardening period of 24 to 48 hours precedes the curing operation. The large, thick monolithic sheets used for laboratory bench tops are manufactured by this process. The grinding operations used to finish the sheet surfaces produce a large quantity of dust which may be discharged with the process wastewaters. This affords a means of reducing and controlling air emissions.

FIGURE 9: ASBESTOS-CEMENT SHEET MANUFACTURING OPERATIONS, WET PROCESS

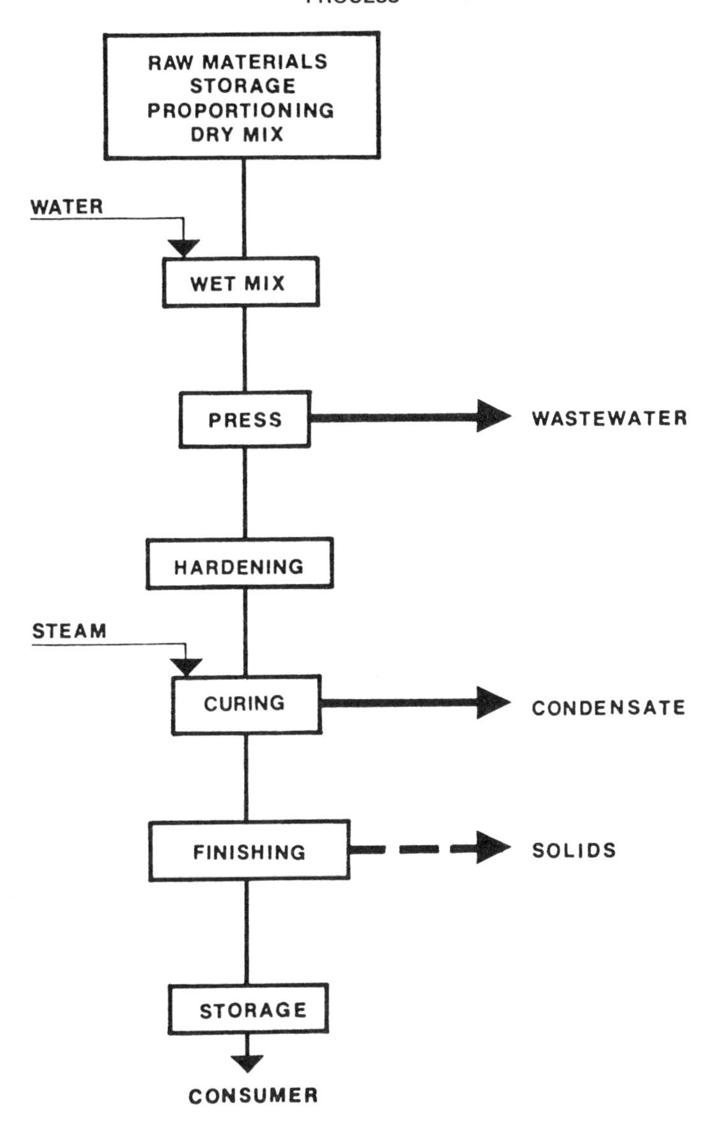

Source: EPA (13)

The wet mechanical process, which is shown for sheet manufacture but which is also used for the manufacture of asbestos-cement pipe (Figure 10), is similar in principle to some papermaking processes. The willowed asbestos fiber is conveyed to a dry mixer where it is blended with the cement, silica, and filler solids. After thorough blending of the raw materials, the mixture is transferred to a wet mixer or beater. Underflow solids and water from the save-all are added to form a slurry containing about 97% water. After thorough mixing, the slurry is pumped to the cylinder vats for deposition onto one or more horizontal screen cylinders. The circumferential surface of each cylinder is a fine wire mesh screen that allows water to be removed from the underside of the slurry layer picked up by the cylinder. The resulting layer of asbestos-cement material is usually from 0.02 to 0.10 inch in thickness. The layer from each cylinder is transferred to an endless felt conveyor to build up a single mat for further processing.

FIGURE 10: ASBESTOS-CEMENT SHEET MANUFACTURING OPERATIONS, WET MECHANICAL PROCESS

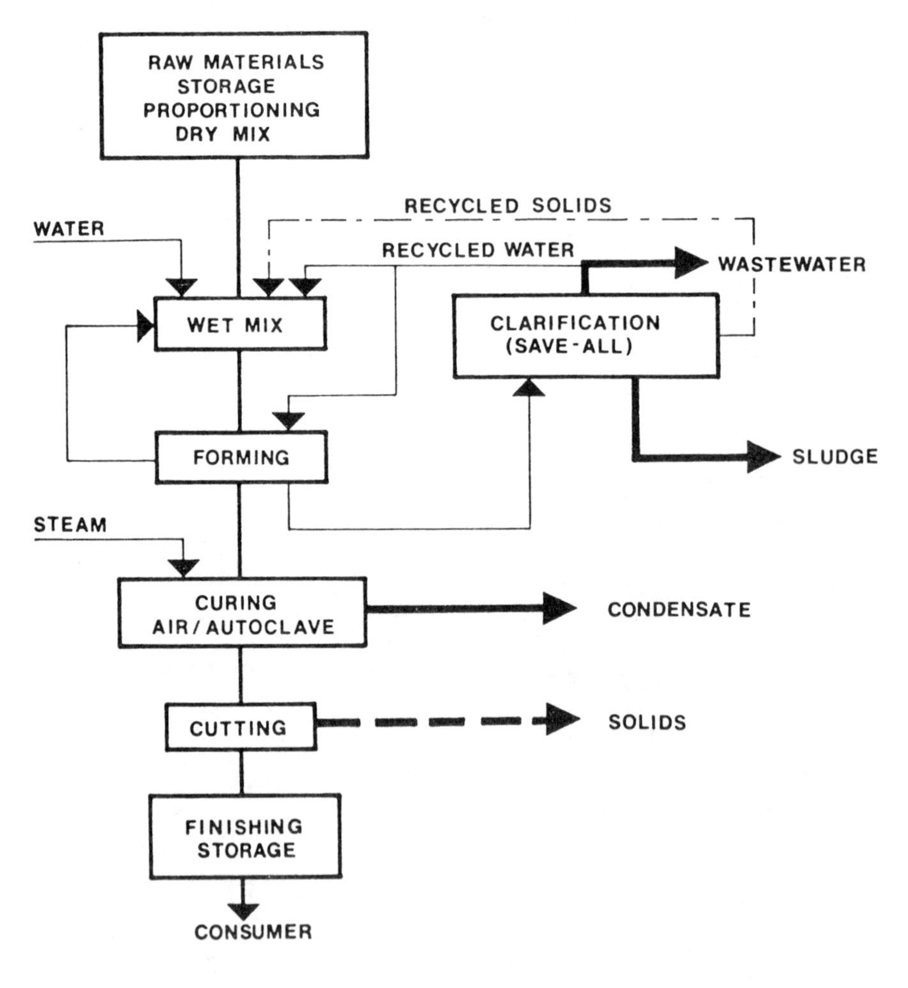

Source: EPA (13)

A vacuum box removes additional water from the mat prior to its transfer to mandrel or accumulator roll. This winds the mat into sheet or pipe stock of the desired thickness. Pressure rollers bond the mat to the stock already deposited on the mandrel or roll and remove excess water. Pipe sections are removed from the mandrel, air cured, steam cured in an autoclave, and then machined on each end.

In the manufacture of sheet products by the wet mechanical process, the layer of asbestos-cement on the accumulator roll is periodically cut across the roll and peeled away to form a sheet. The sheet is either passed through a pair of press rollers to shape the surface and cut the sheet into shingles, formed into corrugated sheet, or placed onto a flat surface for curing. The asbestos-containing water removed from the slurry or mat is recycled to the process. Very little asbestos is lost from the manufacturing process.

The cylinder screen and felt conveyor must be kept clean to insure proper operation. Cylinder showers spray water on the wire surface after the mat has been removed by the felt. Any cement or fiber particles are washed out of the holes in the screen to prevent blinding.

The cylinders, mandrels, and accumulator rolls are occasionally washed in acetic or hydrochloric acid to remove cement deposits. This cleaning may be carried out while the machine is in operation or the component, especially cylinder screens, may be removed to a separate acid washing facility.

The felt washing showers are a row of high-pressure nozzles that, with the aid of a whipper, wash fiber out of the felt after the mat of fiber has been picked up by the mandrel or accumulator roll. Fiber build-up in the felt can prevent vacuum boxes from removing excess water from the mat.

Asbestos-cement product plants recycle the majority of their water as a means of recovering all useable solids. All water serving as the carrying agent, 80 to 90% of the water in the process, passes through a save-all after leaving the machine vat. Solids that settle out and concentrate near the bottom of the save-all are pumped to the wet mixer to become part of a new slurry. Much of the clarified overflow from the save-all can be used for showers, dilution, and various other uses depending upon the efficiency of the save-all.

The save-all overflow may be discharged from the plant or may be treated and returned to the plant for whatever uses its quality justifies. This may include water for wet saws, vacuum pump seals, cooling, hydrotesting, or makeup water for plant startup. If any of these uses cannot be served by treated water, fresh water must be used since the quality and temperature of save-all overflow water is rarely acceptable without additional clarification.

At most asbestos-cement product plants, part of the products that are damaged or unacceptable for other reasons, are crushed, ground, and used as filler in new products. The remainder is crushed and added to a refuse pile or landfill.

Asbestos-cement sheet plants trim the edges of the wet sheets as they come off the accumulator roll. The trimmings are immediately returned to the wet mixer. At this stage, the cement has not begun to react and the trimmings can be an active part of the new slurry. Asbestos-cement pipe plants typically operate 24 hours a day and five or six days a week. Sheet plants may operate two shifts a day rather than three depending upon market demand.

The water balance at one asbestos-cement pipe plant was provided by the plant personnel. The values were verified in this study as far as possible. The balance is outlined diagrammatically in Figure 11. The fresh water going into the pipe manufacturing machine is only about one-quarter of the total used. The rest is water recycled from the save-all. The percentage figures in Figure 11 are in terms of the total water entering the manufacturing system, i.e., the fresh water and that returned from the save-all.

Fresh water is used for wet saws, hydrotesting, cooling, sealing vacuum pumps, and making

FIGURE 11: WATER BALANCE DIAGRAM FOR A TYPICAL ASBESTOS-CEMENT PIPE PLANT

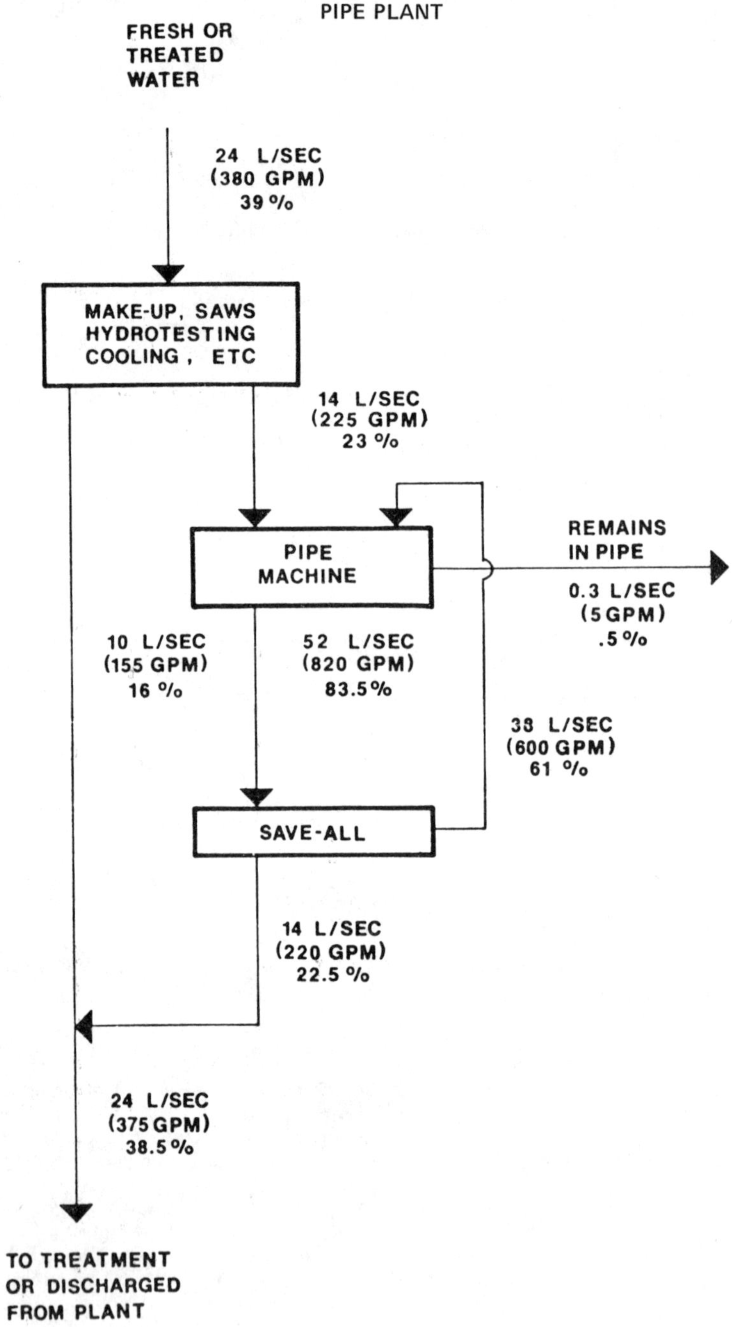

Source: EPA (13)

steam for the autoclave as well as makeup in the mixing unit. Water is used with the saws to control dust and fiber emissions to the air. This is in contrast to the normally dry lath operations that finish the pipe ends.

Hydrotesting is a routine procedure in which the strength of the pipe is tested while full of water under pressure. At some plants, the hydrotest water is reused.

A pipe plant must remove solids from the bottom of the save-alls to prevent their hardening into concretions. At some plants, this dumping and clean up is carried out when the manufacturing operations are shut down for the weekend. At other plants, dumping occurs more frequently.

The reported wastewater discharge from 10 of the 14 asbestos-cement pipe plants ranges from 76 to 2,080 cubic meters per day (0.02 to 0.55 mgd). The plants with minimal effluent volumes discharge about 5.0 to 6.3 cubic meters per metric ton (1,200 to 1,500 gallons per ton) of product. The accuracy of these values is not known. At a few locations, there is reduced discharge because of evaporative losses from lagoons. Discharge records for a period of a year or more were available at two pipe plants. At one plant the minimum flow was 65% of the average and the maximum was 145%. The flow figures included cooling water from the manufacture of plastic pipe, however. The maximum discharge at the other plant, which produced only asbestos-cement pipe, was 670% of the average. The standard deviation in 403 values at this plant was of the same magnitude as the average flow.

The characteristics of raw wastewaters from asbestos-cement pipe manufacturing were developed from sampling data from three plants and reported values from one plant that provides minimal treatment. Two of the plants recirculated water from the external treatment system back into the plant. These plants tended to use much more water and the dissolved (filterable) solids levels were much higher in the wastewaters from these plants.

The manufacture of asbestos-cement pipe in a typical plant increases the levels of the major constituents in the water by the following approximate amounts:

	mg/l	kg/kkg	lb/ton
Total solids	1,500	9	18
Suspended solids	500	3.1	6.3
BOD (5 day)	2	0.01	0.02
Alkalinity	700	4.4	8.8

The dissolved salts are reported to be primarily calcium and potassium sulfates with lesser amounts of sodium chloride. The magnesium levels are not known to be high. The alkalinity is primarily caused by hydroxide with a small carbonate contribution. The pH ranges as high as 12.9, but is generally close to 12.0, or slightly lower.

The temperature fluctuations at a given plant are smaller than the differences between plants. The maximum raw waste temperature measured in this study was 40°C. This plant recirculated some water from its treatment facility. The average temperature at two other pipe plants were 10° to 15°C hotter than the intake water.

The oil and grease content of raw waste samples taken at pipe plants was below detectable levels. Reported data indicate that at some plants there are measurable oil and grease levels in the final plant effluent. This is believed to be from the equipment rather than the process.

The organic content of pipe plant wastewaters is normally low. Some plants use organic acids (acetic) to clean the mandrels and to remove scale in the plant. This could contribute BOD to the waste stream. The waste acid is neutralized when mixed with the highly alkaline process waste stream. The high pH precludes the presence of any biological forms.

The measured and reported average levels of the plant nutrients nitrogen and phosphorus in pipe plant effluents were below 2.5 mg/l and 0.05 mg/l, respectively. There are unconfirmed peak values at individual plants of Kjeldahl nitrogen values as high as 12 mg/l and total phosphorus levels of 0.4 mg/l.

The information on other constituents was derived from reported data from a few individual plants. Most plants did not have data on every constituent. Among the constituents reportedly measured in the effluents from some asbestos-cement pipe plants are chromium, cyanide, mercury, phenols, and zinc. Based on the limited data available, the levels were not judged to be significant.

The raw wastewaters from pipe manufacture are very turbid and of a gray-white color. When the solids are removed, the water has no color. The variations in raw waste loadings from a typical plant are not known. No plant measures or records the characteristics of the raw wastewaters. The wastewater treatment systems are designed on hydraulic principles and their operational efficiency is largely independent of the strength of the influent wastewater.

The changes in waste characteristics associated with start-up of a pipe plant are minor and less than the normal fluctuations associated with operation. When a pipe plant is shut down and the save-alls dumped, there is released a heavy charge of suspended solids in a short period of time. Other parameters remain the same or decrease slightly because of dilution by the flush water. Grab samples of raw pipe wastewaters collected during clean-up at one plant gave results in the following ranges: 1,400 to 3,100 mg/l, total solids; 300 to 2,900 mg/l, suspended solids; and 540 to 2,000 mg/l, alkalinity. Fluctuations in raw wastewater quality should not cause serious problems in the physical treatment facilities appropriate for pipe plant wastes.

No information is known to be available about the internal water balance in an asbestos-cement sheet plant. It is expected that the percent recycle from the save-alls is roughly the same as for asbestos-cement pipe (Figure 11).

The reported wastewater discharge from 4 of the 13 known sheet plants ranges from 280 to 2,040 cubic meters per day (0.07 to 0.54 mgd). The raw waste flows from the three sheet plants sampled during this study were 570, 650, and 920 cubic meters per day (0.15, 0.17, and 0.24 mgd). The largest of the three values was from a plant that discharges no effluent and, consequently, may use relatively more water. The minimal effluent volume from a plant was 7.5 cubic meters per metric ton (1,800 gallon per ton) of production.

There are no known monitoring records of discharge from asbestos-cement sheet plants and no estimate of the minimum, maximum, and variability of the flow from a plant can be made.

The characteristics of raw wastewaters from asbestos-cement sheet manufacturing were developed from sampling data from two plants. No other data were available except that reported by one plant using the wet press forming technique to make high-density sheet. Since this product may include pigments and other additives and since it is produced at only two known locations, neither of which have adequate data, it is not properly included in this category.

The manufacture of asbestos-cement sheet products in a typical plant increases the level of constituents in the water by the following approximate amounts:

	mg/l	kg/kkg	lb/ton
Total solids	2,000	15	30
Suspended solids	850	6.5	13
BOD (5 day)	2	0.015	0.03
Alkalinity	1,000	7.5	15

Little information is available on the dissolved salts in sheet wastewaters, but they should be similar to those from asbestos-cement pipe manufacture. The alkalinity is caused primarily by hydroxide with a pH averaging 11.7 and ranging from 11.4 to 12.4 in all reporting plants.

Meaningful temperature data was available from only one sheet plant. With a flow of 920 cubic meters per day (0.24 mgd), the temperature was increased 13°C in the sheet manufacturing process. The reported peak summer temperatures of wastewaters discharged from asbestos-cement sheet plants was 50°C.

The presence of oil and grease in wastewaters from sheet plants has not been reported. No measurable oil and grease was found in the samples analyzed in this study.

The discussion regarding organic content, plant nutrients, other chemicals, turbidity and color, and fluctuations of the characteristics of asbestos-cement pipe wastewaters applies to those from asbestos-cement sheet.

As regards in-plant control measures, some pipe plants completely recirculate the water used in the hydrotest operation. Some plants reuse part of the autoclave condensate directly. Consideration should be given to piping wastewaters from wet saws to the save-all system. At least one pipe plant recycles a major fraction of the effluent from its waste treatment facility back into the manufacturing process.

No plant making only asbestos-cement pipe has accomplished complete recirculation. A reported experimental attempt to do so by one company was not successful. The raw wastewater flow from asbestos-cement pipe manufacture is typically in the range of 4.1 to 5.2 cubic meters per metric ton (1,200 to 1,500 gallons per ton) of product.

Many of the in-plant control measures described above for pipe plant could be incorporated in sheet plants. The raw wastewater flow from sheet manufacture is typically in the range of 5.2 to 6.2 cubic meters per kilokilogram (1,500 to 1,800 gal/ton).

One asbestos-cement sheet plant achieves complete recirculation most of the time. The manufacturing process is so balanced that the fresh water intake equals the amount of water in the wet product. Fresh water enters the system only for boiler make-up and as part of the vacuum pump seal water. This plant is connected to a municipal sewer and excess flows caused by upsets and process shut-downs are discharged intermittently. With sufficient holding capacity to accommodate these surges, discharge to the sewer could be eliminated.

The benefits of complete recycle at this plant include reduced water cost and sewer service charges, minimal asbestos loss and, reportedly, a somewhat stronger product. The major problem encountered in complete water recycle at this plant is scaling. Spray nozzles require occasional unplugging, the water lines are scoured regularly with a pneumatically driven cleaner, and fine sand is introduced into the pumps to eliminate deposits.

While one sheet plant has accomplished almost complete recirculation, this is not regarded as fully demonstrated technology. This plant makes only a few asbestos-cement sheet products. The intermittent discharge to the sewer does provide some blowdown relief to the system. Whether such complete recirculation could be applied to plants making sheet products with more stringent quality specifications is not known. The progress at this plant does indicate that complete recirculation is a realistic goal for the future.

The applicable end-of-pipe technology for wastewaters from the manufacture of asbestos-cement pipe and sheet products is sedimentation and neutralization. Designs based on total detention periods of 6 to 8 hours or loading levels of 24 cubic meters per day per square meter (600 gallons per day per square foot) of surface area yield effluent suspended solids levels of 30 mg/l or lower.

Neutralization to a pH level of 9.0 or below has been achieved at two locations in the industry by adding sulfuric acid or on-site generated carbon dioxide. At both of these locations, sedimentation precedes and follows neutralization.

The solids removed by the settling units are best dewatered by gravity thickening. They are dense and biochemically inert and are suitable for disposal by proper landfill disposal techniques.

To achieve complete recirculation of process wastewaters, surge capacity will have to be added to the water system. A sedimentation unit cannot function in this capacity. A water storage tank or reservoir would be required in the system. With complete recycle, the neutralization operation will not be required. Its function is to protect the receiving water. High pH levels are not a problem in the manufacture of asbestos-cement products. As noted in a previous section, additional scale control measures are necessary when complete recycle is implemented.

As noted above, complete recirculation of asbestos-cement sheet process water has been demonstrated partially. Problems with product strength have been reported in one effort to completely recycle wastewater from asbestos-cement pipe manufacture. Additional research is needed to achieve this level of control.

A variety of alternatives have been explored by EPA (13) for wastewater control in asbestos-cement pipe manufacture as follows.

Alternative A — No Waste Treatment or Control

Effluent waste load is estimated to be 3.1 kg/kkg (6.3 lb/ton) of suspended solids, 4.4 kg/kkg (8.8 lb/ton) of caustic (hydroxide) alkalinity, and 6.3 kg/kkg (12.6 lb/ton) of dissolved solids for the selected typical plant at this minimal control level. The pH of the untreated waste is 12.0. In-plant use of save-alls is assumed, as this is universally practiced in the industry.

Costs — None

Reduction Benefits — None

Alternative B — Sedimentation of Process Wastes

This alternative includes settling of all process wastewaters. Some form of sedimentation is applied at almost all plants in the industry. Costs include land disposal of dewatered sludge. Effluent suspended solids load estimated to be 0.19 kg/kkg (0.38 lb/ton). Alkalinity, pH, and dissolved solids remain high.

Costs — Investment costs are approximately $124,000.

Reduction Benefits — Effluent suspended solids reduction of approximately 94%.

Alternative C — Sedimentation and Neutralization of Process Wastes

This alternative includes settling of all process wastewaters before and after neutralization to pH 9.0 or below. This alternative is practiced presently by about 30% of the pipe plants. Effluent suspended solids load of less than 0.19 kg/kkg (0.38 lb/ton), caustic alkalinity removed, and dissolved solids reduced somewhat.

Costs — Incremental costs are approximately $77,000 over Alternative B; total costs are $201,000.

Reduction Benefits — Reduction of effluent suspended solids of at least 95%, caustic alkalinity of almost 100%, and an indeterminate reduction in dissolved solids.

Alternative D — Complete Recycle of Process Water

This alternative includes complete recycle of all process power wastewater back into the manufacturing processes and other in-plant uses. Fresh water taken into plant equals quantity leaving in wet product and other evaporative losses. Complete control of pollutant constituents without discharge is effected. No plant making only pipe presently recycles all of the process wastes.

Costs — Incremental costs are approximately $104,000 over Alternative C; total costs are $305,000.

Reduction Benefits — Reduction of all pollutant constituents, including suspended and dissolved solids and alkalinity, of 100%.

The annual costs and resulting effluent quality for each of the four treatment alternatives for asbestos-cement pipe are summarized in Table 8.

TABLE 8: WATER EFFLUENT TREATMENT COSTS FOR MANUFACTURING ASBESTOS-CEMENT PIPE

Treatment or Control Technologies	Alternatives			
	A	B	C	D
Investment*	-	$124	$201	$305
Annual Costs:*				
Capital Costs	-	9.9	16.1	24.4
Depreciation	-	6.2	10.1	15.3
Operating and Maintenance Costs (excluding energy and power costs)	-	63.8	87.8	98.3
Energy and Power Costs	-	2.8	7.0	11.9
Total Annual Cost*	-	82.7	121	149.9

*Costs in thousands of dollars

Effluent Quality:

Effluent Constituents Parameters (Units)	Raw Waste Load	Resulting Effluent Levels			
Suspended Solids - kg/MT	3.1	do	0.19	0.19-	0
Caustic Alkalinity - kg/MT	4.4	do	4.4	0	0
pH	12	do	12	9.0	0
Dissolved Solids - kg/MT	6.3	do	6.3	6.3-	0
Suspended Solids - mg/l	500	do	30	30-	0
Caustic Alkalinity - mg/l	700	do	700	0	0
Dissolved Solids - mg/l	1000	do	1000	1000-	-

Source: EPA (13)

Similarly, a variety of alternatives have been explored by EPA (13) for wastewater control in asbestos-cement sheet manufacture as follows.

Alternative A — No Waste Treatment or Control

Effluent waste load is estimated to be 6.5 kg/kkg (13 lb/ton) of suspended solids, 7.5 kg/kkg (15 lb/ton) of caustic (hydroxide) alkalinity, and 8.5 kg/kkg (17 lb/ton) of dissolved solids for the selected typical plant at this minimal control level. The pH of the untreated waste is 11.7 or higher. In-plant use of save-alls is assumed, as this is universally practiced in the industry.

Costs — None

Reduction Benefits — None

Alternative B — Sedimentation of Process Wastes

This alternative includes settling of all process wastewaters. Some form of sedimentation is applied at most plants in the industry. Costs include land disposal of the dewatered sludge. Effluent suspended solids load estimated to be 0.23 kg/kkg (0.45 lb/ton). Alkalinity, pH, and dissolved solids remain high.

Costs — Investment costs are approximately $56,000.

Reduction Benefits — Effluent suspended solids reduction of approximately 96%.

Alternative C — Sedimentation and Neutralization of Process Water

This alternative includes settling of all process wastewaters before and after neutralization to pH 9.0 or below. This alternative is used by 10% or less of the sheet plants. Effluent suspended solids load of less than 0.23 kg/kkg (0.45 lb/ton), caustic alkalinity removed, and dissolved solids reduced somewhat.

Costs — Incremental costs are approximately $36,000 over Alternative B; total costs are $92,000.

Reduction Benefits — Reduction of effluent suspended solids of at least 96%, caustic alkalinity of almost 100%, and an indeterminate reduction in dissolved solids.

Alternative D — Complete Recycle of Process Water

This alternative includes complete recycle of all process wastewaters back into the manufacturing processes or other in-plant uses. Fresh water taken into plant equals quantity leaving in wet product and other evaporative losses. Complete control of pollutant constituents without discharge is effected. One sheet plant is known to accomplish complete recycle during routine operation.

Costs — Incremental costs are approximately $59,000 over Alternative C; total costs are $151,000.

Reduction Benefits — Reduction of all pollutant constituents, including suspended and dissolved solids and alkalinity, of 100%.

The annual costs and resulting effluent quality for each of the four technology or control alternatives for asbestos-cement sheet products are presented in Table 9.

TABLE 9: WATER EFFLUENT TREATMENT COSTS FOR MANUFACTURING ASBESTOS-CEMENT SHEET

Treatment or Control Technologies	Alternatives			
	A	B	C	D
Investment*	-	$56	$92	$151

(continued)

TABLE 9: (continued)

Treatment or Control Technologies		Alternatives A	B	C	D
Annual Costs:*					
Capital Costs		-	4.5	7.3	12.1
Depreciation		-	2.8	4.6	7.5
Operating and Maintenance Costs (excluding energy and power costs)		-	41.4	53.3	92.4
Energy and Power Costs		-	2.8	4.2	7.0
Total Annual Cost*		-	51.5	69.4	119.0
Effluent Quality:					
Effluent Constituents Parameters (Units)	Raw Waste Load	Resulting Effluent Levels			
Suspended Solids - kg/MT	6.5	do	0.23	0.23-	0
Caustic Alkalinity - kg/MT	7.5	do	7.5	0	0
pH	11.7	do	11.7	9.0	0
Dissolved Solids - kg/MT	8.5	do	8.5	8.5-	0
Suspended Solids - mg/l	850	do	30	30 -	0
Caustic Alkalinity - mg/l	1000	do	1000	0	0
Dissolved Solids - mg/l	1150	do	1150	1150-	0

*Costs in thousands of dollars

Source: EPA (13)

Asbestos Floor Tile

The shortest grades of asbestos fibers are used in vinyl and asphalt floor tile manufacture. The fibers are used to provide dimensional stability. Today, vinyl asbestos floor tile accounts for most of the asbestos used in this category, with asphalt tile serving some special applications and where darker shades are permissible.

Most floor tile manufactured today uses a vinyl resin, although some asphalt tile is still being produced. The manufacturing processes are very similar and the water pollution control aspects are almost identical for the two forms of tile.

Ingredient formulas vary with the manufacturer and the type of tile being produced. The asbestos content ranges from 8 to 30% by weight and usually comprises very short fibers. Asbestos is included for its structural properties and it serves to maintain the dimensional stability of the tile. PVC resin serves as the binder and makes up 15 to 25% of the tile. Chemical stabilizers usually represent about 1%. Limestone and other fillers represent 55 to 70% of the weight. Pigment content usually averages about 5%, but may vary widely depending upon the materials required to produce the desired color.

The tile manufacturing process, shown in Figure 12, involves several steps: ingredient weighing, mixing, heating, decoration, calendering, cooling, waxing, stamping, inspecting, and packaging. The ingredients are weighed and mixed dry. Liquid constituents, if required, are then added and thoroughly blended into the batch. After mixing, the batch is heated to about 150°C and fed into a mill where it is joined with the remainder of a previous batch for continuous processing through the rest of the manufacturing operation.

The mill consists of a series of hot rollers that squeeze the mass of raw tile material down to the desired thickness. During the milling operation, surface decoration in the form of small colored chips of tile (mottle) are sprinkled onto the surface of the raw tile sheet and pressed in to become a part of the sheet. Some tile has a surface decoration embossed and linked into the tile surface during the rolling operation. This may be done before or after cooling.

After milling, the tile passes through calenders until it reaches the required thickness and is ready for cooling. Tile cooling is accomplished in many ways and a given tile plant may use one of several methods. Water contact cooling in which the tile passes through a water bath or is sprayed with water is used by some plants. Others use noncontact cooling in which the rollers are filled with water.

FIGURE 12: ASBESTOS FLOOR TILE MANUFACTURING OPERATIONS

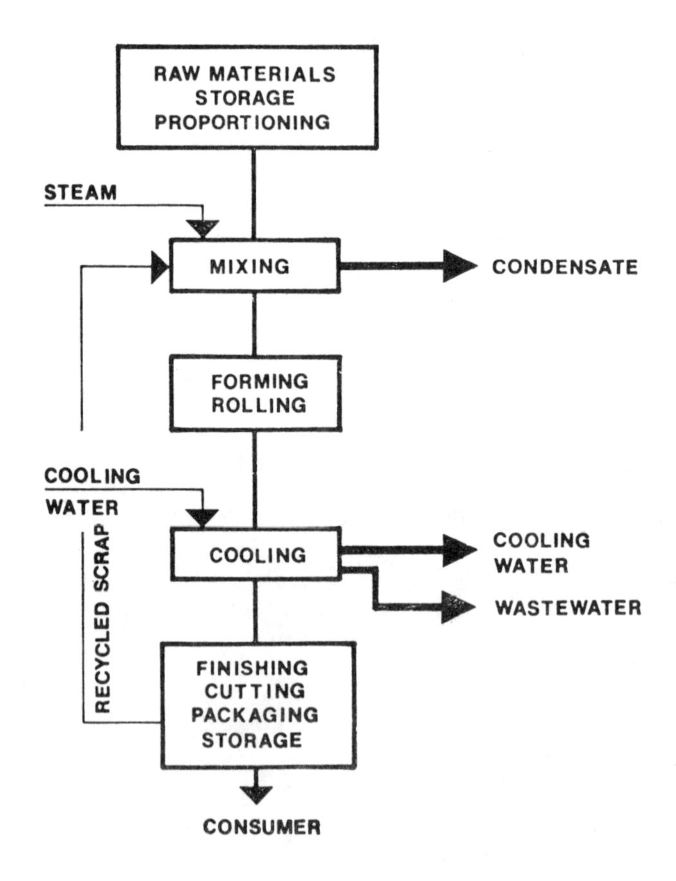

Source: EPA (13)

In some plants, the sheet of tile passes through a refrigeration unit where cold air is blown onto the tile surface. After cooling, the tile is waxed, stamped into squares, inspected, and packaged. Trimmings and rejected tile squares are chopped up and reused.

Water serves only as a heat transfer fluid. It is used in the form of steam to heat the batches and the hot rollers. Fresh water is required for boiler make-up, but only in quantities large enough to replace leakage and boiler blowdown water. Noncontact cooling water remains clean and can be reused continually if cooling towers or water chillers are available to remove the heat picked up from the hot tile.

Make-up water is required only to replace water that leaks from the system. Direct contact cooling water from the cooling baths or sprays does not become contaminated from direct contact with the tile but may pick up dust or other materials. This water may be reused if facilities are available to clean the water and remove the heat. Fresh water is required to replace leakage and water that evaporates. Leakage from all sources collects dirt, oil, grease, wax, ink, glue, and other contaminants. This represents a serious potential for pollution if discharged to a receiving water. Floor tile plants typically operate 24 hours a day on a five or six day per week schedule.

From a water use and wastewater characterization point of view, vinyl and asphalt tile manufacturing both produce the same result. Like roofing, water is used only for cooling purposes. Both contact and noncontact cooling are usually employed. Water does not come into contact with the tile until it has been heated and rolled into its final form. In this stage it is completely inert to water.

Cooling water usage information was available from six floor tile plants with an average daily production of about 400,000 pieces. The reported discharges ranged from about 80 to 1,700 liters (21 to 450 gallons) per 1,000 pieces with an average of 1,130 liters (300 gallons).

The wide range reflects differences in intake water temperatures, whether or not the water is recirculated, and whether both contact and noncontact waters are included in the figures. Because the water is used for cooling, fluctuations within a given plant should not be large and should primarily be the result of changes in production levels or seasonal temperature changes, or both.

Despite that floor tile itself is inert in water, the contact cooling water becomes contaminated with a diverse variety of materials including wax, inks, oil, glue, and miscellaneous dirt and debris. The material has a high organic content although the limited data available indicate that it is not readily biodegradable.

The added waste constituents in a typical floor tile plant are as follows:

	mg/l	kg/1,000 pc*	lb/1,000 pc*
Suspended solids	150	0.18	0.40
BOD (5 day)	15	0.02	0.04
COD	300	0.36	0.80

*Pieces of tile, 12" x 12" x 3/32"

The reported pH of tile plant wastewaters ranges from 6.9 to 8.3, averaging 7.3.

The reported temperature data are inconsistent among the few plants reporting. Some plants with large per unit flow volumes show a larger temperature increase than plants with much smaller flows per 1,000 pieces. Oil and grease is reportedly present in tile plant effluents, with an average concentration of 5.5 mg/l after treatment.

The COD is believed to be largely associated with the suspended solids with much of it being wax. The limited data on plant nutrients indicate that the increased total nitrogen and phosphorus levels should be less than 5.0 and 1.5 mg/l, respectively.

Trace amounts of phenols and chromium were each reported by one plant. The levels were judged not to be significant. Data on the color and turbidity of wastewaters from floor tile manufacture are not available. The wastes do have measurable levels of both parameters, however. There are no known data by which to assess the variations in constituent concentrations in wastewaters from floor tile plants.

There are several in-plant measures that should be used in floor tile plants to control the release of pollutant constituents. Raw materials should be stored, measured, and mixed in an area completely isolated from the cooling water systems. Only after the ingredients are made into tile are they insoluble in water. Toxic materials should be eliminated from the tile ingredients.

If possible, contact water cooling operations should be eliminated. If this is not feasible, the contact cooling water should be protected from contamination. Bearing leaks should be controlled and escaping water protected from contact with wax, oils, glue, and other dirt.

If the contact cooling water and the noncontact cooling water that escapes were prevented from becoming contaminated, it would be much easier to treat. This contamination is unnecessary and the resulting process wastewater is costly to treat.

The applicable end-of-pipe technology for floor tile manufacturing wastewaters is sedimentation with coagulation and skimming to remove suspended solids. It is believed that the high COD levels associated with some tile plant wastes are caused by insoluble materials. Properly designed and operated facilities should reduce suspended solids levels to 30 mg/l and COD to 75 mg/l or less.

The wastes from different tile plants are somewhat different and the precise technology required to achieve these levels cannot be predicted. At present, treatment beyond plain sedimentation and skimming is not practiced by the industry. Sorption on activated carbon following filtration should remove soluble organic materials to an acceptable level.

Complete elimination of the discharge of pollutants will necessitate either cooling and reuse, or the use of noncontact cooling water systems. No information is available by which to determine the nature of the treatment best suited for the former method.

A variety of alternatives have been explored by EPA (13) for wastewater control in asbestos floor tile manufacture as follows:

Alternative A — No Waste Treatment or Control

Effluent waste load is estimated to be 0.18 kg (0.38 lb) of suspended solids, 0.017 kg (0.04 lb) of BOD, and 0.34 kg (0.75 lb) of COD per 1,000 pieces of tile manufactured at the selected typical plant at this minimal control level.

Costs — None

Reduction Benefits — None

Alternative B — Coagulation and Sedimentation of Process Wastes (Contaminated Cooling Water)

This alternative includes polyelectrolyte coagulation and sedimentation with skimming as necessary to remove suspended matter. The percentage of tile plants applying this alternative is not known, but is expected to be less than 25%. The effluent load is estimated to be 0.04 kg (0.08 lb) of suspended solids and 0.09 kg (0.19) of COD per 1,000 pieces of tile manufactured. The BOD load may be reduced somewhat.

Costs — Investment costs are approximately $52,000.

Reduction Benefits — Estimated reduction of effluent suspended solids of 80% and COD of 75%.

Alternative C — Complete Recycle of Process Water (Contaminated Cooling Water)

This alternative includes additional treatment by filtration, cooling, and reuse of process wastewaters (contaminated cooling water). No process wastes are discharged and complete control of pollutant constituents is effected.

Costs — Incremental costs are approximately $58,000 over Alternative B; total costs are $110,000.

Reduction Benefits — Reduction of suspended solids, BOD, and COD and all other pollutant constituents of 100%.

The annual costs and resulting effluent quality for each of the three treatment or control technology alternatives for asbestos floor tile are summarized in Table 10.

TABLE 10: WATER EFFLUENT TREATMENT COSTS FOR MANUFACTURING ASBESTOS FLOOR TILE

Treatment or Control Technologies	Alternatives		
	A	B	C
Investment*	-	$52	$110
Annual Costs:*			
Capital Costs	-	4.2	8.8
Depreciation	-	2.6	5.5
Operating and Maintenance Costs (excluding energy and power costs)	-	11.0	10.8
Energy and Power Costs	-	1.8	3.0
Total Annual Cost*	-	19.6	28.1

*Costs in thousands of dollars

Effluent Quality:

Effluent Constituents Parameters (Units)	Raw Waste Load	Resulting Effluent Levels		
Suspended Solids - kg/1000 pc	0.18	do	0.04	0
BOD (5-day) - kg/1000 pc	0.017	do	0.017-	0
COD - kg/1000 pc	0.34	do	0.09	0
Suspended Solids - mg/l	150	do	30	0
BOD (5-day) - mg/l	15	do	15-	0
COD - mg/l	280	do	75	0

Source: EPA (13)

Asbestos Papers

Asbestos papers have a high fiber content and are manufactured with a variety of binders and other additives for many applications. These include pipe coverings, gaskets, thermal linings in heaters and ovens, and wicks. Heavier papers are commonly used for roofing materials and shingles. Millboard is a heavier, stiffer form of paper that includes clays, cement, or other additives. It is used for stove lining, filament supports in toasters, and several other high temperature applications.

Asbestos paper has a great variety of uses and ingredient formulas vary widely depending upon the intended use of the paper. The purchaser frequently specifies the exact formula to insure that the paper has the desired qualities.

Asbestos paper usually contains from 70 to 90% asbestos fiber by weight, usually the short grades. A mixture of the various varieties of asbestos fiber is used with chrysotile as the principal type. The binder content of asbestos paper accounts for 3 to 15% of its weight. The content and type varies with the desired properties and intended applications of the paper. Typical binders are starch, glue, cement, gypsum, and several natural and synthetic elastomers.

Asbestos paper used for roofing paper, pipe wrapping, and insulation usually contains between 5 and 10% kraft fiber. Mineral wool, fiber glass, and a wide variety of other constituents are included to provide special properties and may represent as much as 15% of the weight.

Asbestos paper is manufactured on machines of the Fourdrinier and cylinder types that are similar to those which produce cellulose (organic) paper. The cylinder machine is more widely employed in the industry today. The overall manufacturing process is shown in Figure 13 with waste sources indicated.

The mixing operation combines the asbestos fibers with the binders and any other minor ingredients. A pulp beater or hollander mixes the fibers and binder with water into a stock which typically contains about 3% fiber. Upon leaving the stock chest, the stock is diluted to as little as one-half percent fiber in the discharge chest. The amount of dilution depends upon the quality of the paper to be produced.

The discharge chest of a Fourdrinier paper machine deposits a thin and uniform layer of stock onto an endless moving wire screen through which a major portion of the water is drawn by suction boxes or rolls adjacent to the sheet of paper. The sheet is then transferred onto an endless moving felt and pressed between pairs of rolls to bring the paper to approximately 60% dryness. Subsequently, the continuous sheet of paper passes over heated rolls, while supported on a second felt, to effect further drying. This is followed by calendering, to produce a smooth surface, and winding of the paper onto a spindle.

The operation of a cylinder paper-making machine includes a mixing operation for stock as indicated for the Fourdrinier machine. Cylinder-type paper machines usually have four to eight cylinders instead of two as in most asbestos-cement pipe machines.

The stock is pumped to the cylinder vats of the machine. Each vat contains a large screen-surfaced cylinder extending the full length of the vat. The stock slurry flows through the screen depositing a thin layer of fiber on the surface of the rotating cylinder before flowing out through the ends of the cylinder. The layer of fiber is then transferred to a carrier felt moving across the top of the rotating cylinders. The layers picked up from the cylinders are pressed together becoming a single homogeneous sheet as the felt passes over each successive cylinder.

Vacuum boxes draw water out and pressure rolls squeeze water out of the sheet and felt until the sheet is dry enough to be removed from the felt. After leaving the felt, the sheet is dried on steam rolls and in ovens. The paper is then calendered to produce a smooth surface and wound onto a spindle. The width of the paper sheet is regulated by the deckles, a row of nozzles located at each end of the cylinder screens. The deckles spray water on the screen at the edge of the sheet and wash off excess fiber.

FIGURE 13: ASBESTOS PAPER MANUFACTURING OPERATIONS

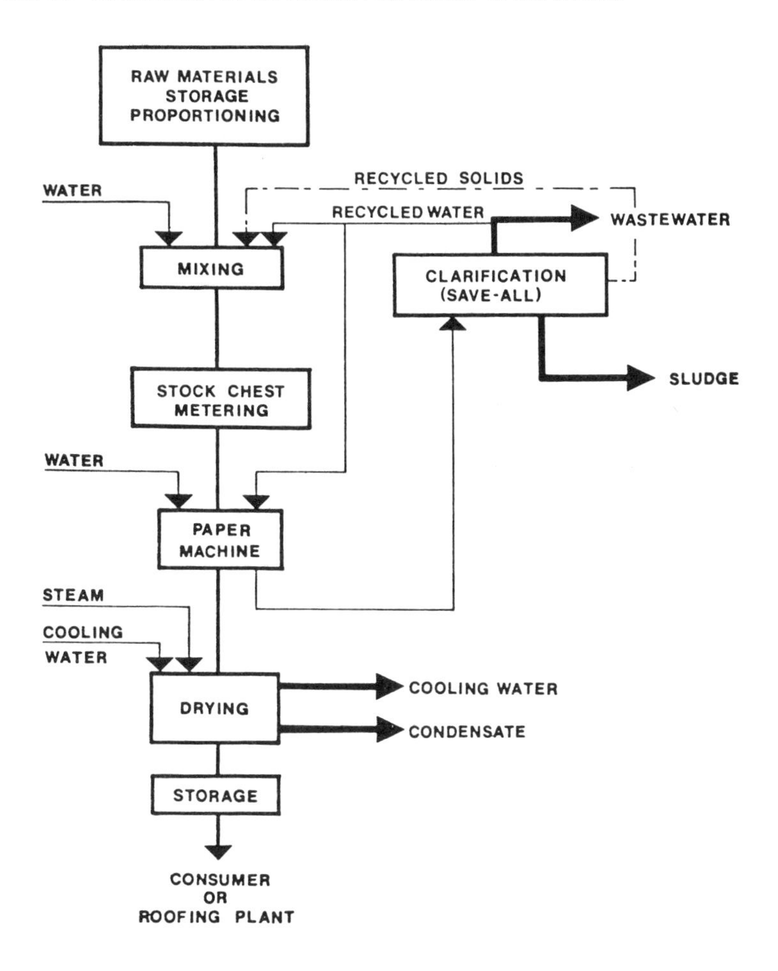

Source: EPA (13)

The cylinder showers are a row of nozzles that spray water on the surface of the cylinder screens after the paper stock mat has been removed by the felt. They wash any remaining fiber and binder out of the holes in the screens to prevent a build-up of fiber from blinding the screen and stopping the flow of water required to deposit a layer of fiber on the surface of the cylinder.

The felt washing operations are carried out using high pressure nozzles as in asbestos-cement pipe manufacture. The asbestos-containing water, or white water, which is removed from the stock prior to passage across the heated drying rolls is recycled to the process.

Water serves three basic purposes in the asbestos paper manufacturing process: ingredient carrier, binder wetting agent, and heat transfer fluid. Other uses include water for showers, deckles, pump seals, plant make-up, boiler make-up, and cooling.

Fresh water enters the system as boiler make-up, process make-up, pump seal water, and shower water. Boiler make-up water provides steam for heating the paper stock and drying the finished paper. The steam used to heat the stock slurry becomes a part of the slurry and must be replaced. Condensate from the drying rolls is recovered and returned to the boiler. Fresh water must be used to cool the dried paper unless a cooling tower is available. Save-all overflow and other plant water is usually too hot for such purposes. Large quantities of fresh water are required during plant start-up to fill the system. This occurs infrequently, however.

Small quantities of water are required continuously to replace that which evaporates during drying and that which becomes a permanent part of the paper. The characteristics of some paper products are such that fresh water must be used for part, or all, of the beater make-up water.

Cylinder and felt washing showers usually require fresh water because save-all overflow water is rarely clean enough to be used in the high pressure shower nozzles without causing plugging. Fresh water is used for the pump shaft seal water because the presence of dirt in the seal water will cause plugging and can cause scoring of the shaft. Although the cooling water and part of the pump seal water may be discharged from the plant after a single use, most of the fresh water introduced into the plant enters the ingredient carrying system, and therefore, the paper machine save-all loop.

The majority of the water in a paper plant serves as an ingredient carrier and continually circulates in a loop through the paper machine and the save-all. All water flowing out of the cylinder screen and that drawn by vacuum out of the wet paper sheet is pumped to the save-all. The solids settle to the bottom of the save-all and are pumped to the stock chest of the beater. Occasionally, the solids from the save-all must be discharged from the plant due to a product change, rapid setup of the binder, or a plant shutdown. Save-all overflow water is used for beater makeup, dilution, deckle water, and occasionally shower water.

Excess overflow water must be discharged from the plant or sent to a wastewater treatment facility for additional treatment before it can be reused. Trimmings from the edge of the paper, defective paper, and other waste paper can usually be returned to the beater and repulped for recycling. Asbestos paper manufacturing plants typically operate 24 hours a day and 7 days a week.

The reported total wastewater discharges from 5 of the 12 asbestos paper manufacturing plants range from 490 to 4,900 cubic meters per day (0.13 to 1.3 mgd). The accuracy of these values is not known. The volumes of raw wastewater discharged to the treatment facility at two plants visited in connection with this study were 1,700 and 2,700 cubic meters per day (0.45 and 0.72 mgd). Many plants recirculate water and solids from the wastewater treatment facility to the papermaking process and the effluent volume is considerably less than the raw wastewater discharge. An effluent flow of 13.8 meters per metric ton (3,300 gallons per ton) was reported at the exemplary plants.

Information about variability of flow is available from one plant only. This is the monitoring record of the treated effluent over a recent eight month period. The average flow was 490 m^3/day (0.13 mgd) with minimum and maximum values of 430 and 755 m^3/day (0.14 and 0.20 mgd), respectively. The standard deviation of the 113 readings taken during the period was 53 m^3/day (0.014 mdg). The exact quantities of water recycled from the save-all system and from the waste treatment facility at this plant are not known.

The raw wastewater characteristics from asbestos paper manufacturing were developed from sampling data at two plants. Both plants provide high levels of wastewater treatment with low volumes of effluent discharge. Consequently, the use of water within these two plants may be higher than in plants that do not recycle treated wastewater.

The manufacture of asbestos paper in a typical plant increases the levels of the constituents in the water by the following approximate amounts:

	mg/l	kg/kkg	lb/ton
Total solids	1,900	26	52
Suspended solids	680	9.5	19
BOD (5 day)	110	1.5	3
COD	160	2.2	4.4

The pH of raw wastewaters from asbestos paper manufacturing is 8.0 or lower.

The highest reported summer temperature value for treated effluent is 32°C. It is believed that heated water is used in mixing the raw materials at most plants, although at least one uses cold water. Recycled water tends to have a higher temperature.

Oil and grease was detected in only one of the samples collected at the two paper manufacturing plants. The level was low, 1.2 mg/l, and was believed to be from plant equipment. This type of material is not part of the product ingredients.

The oxygen demand is believed to be largely due to the organic binders, i.e., starch or synthetic elastomers. These latter include several materials of different chemical compositions. The total nitrogen levels reported in effluents from a few paper plants averaged 16 mg/l, with the Kjeldahl fraction about 11 mg/l. Phosphorus levels ranged from 0.25 to 1.0 mg/l.

Trace amounts of copper, mercury, and zinc were reported to be in the wastes from individual asbestos paper plants. The levels were judged not to be significant. The clarified wastewaters are known to have some color. The levels at two plants were 10 and 15 units.

There was greater variability among the data from the two paper plants than observed in most other asbestos manufacturing operations. There are no data on the variations in quality of raw asbestos paper wastewaters other than the sampling results and these were from too limited a period of time to be of value. Results from the monitoring program at one paper plant refer to treated effluent. They provide some indication of the variability of the wastewater characteristics, as follows:

	Minimum	Average	Maximum	Standard Deviation
Total solids	500 mg/l	685 mg/l	870 mg/l	260 mg/l
Suspended solids	32 mg/l	64 mg/l	95 mg/l	44 mg/l
BOD (5 day)	22 mg/l	57 mg/l	91 mg/l	48 mg/l

Unlike asbestos-cement products plants, asbestos paper plants do not use Portland cement and the solids in the save-alls do not tend to form concretions. Shut-down is less regular and the plants tend to operate around the clock. Shut-downs are sometimes necessary when changing products. Since the elastomeric binders are not always compatible, the save-all solids may be dumped at these times. There were no routine shut-down or start-up operations while the paper plants were being sampled in this study and there is no information on the characteristics of the raw wastewaters during these periods.

The in-plant control measures outlined above for asbestos-cement pipe can be applied in part in asbestos papermaking plants. One paper plant has been able to close up its process water system when making paper with a starch binder. Such operation is not possible when elastomeric binders are used and excess water is then discharged to the municipal sewer.

An asbestos paper plant that practices partial recycle of water from its waste treatment unit typically discharges within 30% of 11 m^3/kkg (3,300 gal/ton).

Partial recycle of water and underflow solids from the wastewater treatment facility is not uncommon in the asbestos paper industry. Complete recirculation and zero discharge has not been demonstrated on a continuing basis at any plant making only paper. It is likely that paper could be manufactured using a closed system if only starch binders were used. Total and continuous recycle of water and solids when using elastomeric binders cannot be accomplished today. Since some paper plants use both types of binders, a guideline based on the type of binder used would be impractical.

The applicable end-of-pipe technology for wastewaters from the manufacture of asbestos paper is sedimentation preceded, as necessary, by grit removal and coagulation with polyelectrolytes. This treatment has been demonstrated at three or more locations. Units designed for a loading of 24 cubic meters per day per square meter (600 gallons per day per square foot) have achieved suspended solids and BOD reductions to 25 mg/l or less.

Most of the settled solids as well as part of the clarified water should be recycled from the settling unit to the manufacturing process at paper plants. The waste solids, which are normally kept to a minimum, may be stored for later use or dewatered for land disposal with the grit. Waste solids result, in part, from the incompatibility of certain synthetic binders.

To achieve complete recycle of all process wastewaters at asbestos paper plants, surge capacity will be required. A water storage tank will be required because the sedimentation unit cannot provide this function.

As noted above, complete recirculation of asbestos paper process water has been demonstrated partially when starch is used as the binder. Additional research is needed to achieve this level of control when using elastomeric binders.

A variety of alternatives have been explored by EPA (13) for wastewater control in asbestos paper (using both elastomeric binders) manufacture as follows:

Alternative A — No Waste Treatment or Control

Effluent waste load is estimated to be 9.5 kg/kkg (19 lb/ton) of suspended solids, 1.5 kg/kkg (3 lb/ton) of BOD, and 16.5 kg/kkg (33 lb/ton) of dissolved solids for the selected typical plant at this minimal control level. In-plant use of save-alls is assumed, as this is universally practiced in the industry.

Costs — None

Reduction Benefits — None

Alternative B — Sedimentation of Process Wastes

This alternative includes settling of all process wastewaters. Some form of sedimentation is applied at approximately 70% of plants in the industry. Costs include land disposal of dewatered sludge. Effluent load estimated to be 0.35 kg/kkg (0.7 lb/ton) of suspended solids and of BOD and 16.5 kg/kkg (33 lb/ton) of dissolved solids.

Costs — Investment costs are approximately $237,000.

Reduction Benefits — Estimated reduction of effluent solids of 96% and BOD of 75%. Dissolved solids remain unchanged.

Alternative C — Complete Recycle of Process Water

This alternative includes complete recycle of all process wastewaters back into the manufacturing processes and other in-plant uses. Fresh water taken into plant equals quantity leaving in wet product and other evaporative losses. Complete control of pollutant constituents without discharge is effected. One paper plant is known to achieve complete recycle when using starch binder under routine conditions.

Costs — Incremental costs are approximately $57,000 over Alternative B; total costs are $294,000.

Reduction Benefits — Reduction of all pollutant constituents, including suspended and dissolved solids and BOD, of 100%.

The estimated annual costs and effluent quality for each of the alternatives for asbestos paper manufacturing wastewaters are given in Table 11.

TABLE 11: WATER EFFLUENT TREATMENT COSTS FOR MANUFACTURING ASBESTOS PAPER

Treatment or Control Technologies	Alternatives A	Alternatives B	Alternatives C
Investment*	-	$237	$294
Annual Costs:*			
Capital Costs	-	19	24
Depreciation	-	12	15
Operating and Maintenance Costs (excluding energy and power costs)	-	16	44
Energy and Power Costs	-	16	16
Total Annual Cost*	-	63	99

*Costs in thousands of dollars

Effluent Quality:

Effluent Constituents Parameters (Units)	Raw Waste Load	Resulting Effluent Levels		
Suspended Solids - kg/MT	9.5	do	0.35	0
BOD (5-day) - kg/MT	1.5	do	0.35	0
Dissolved Solids - kg/MT	16.5	do	16.5	0
Suspended Solids - mg/l	700	do	25	0
BOD (5-day) - mg/l	110	do	25	0
Dissolved Solids - mg/l	1200	do	1200	0

Source: EPA (13)

Asbestos Millboard

Asbestos millboard is considered by some to be a very heavy paper and is in fact very much like thick cardboard in texture and structural qualities. It can easily be cut or drilled and can be nailed or screwed to a supporting structure.

Millboard formulas vary widely depending upon the intended use of the product. Purchasers frequently specify the ingredients and composition of the millboard to insure that the product meets their particular requirements. Asbestos content ranges between 60 and 95% with the higher content for products that will be in close or direct contact with high temperature materials. Portland cement and starch are the most common binders used and represent 5 to 40% of the product. Clay, lime, mineral wool, and several other materials are frequently used as fill material or to provide special qualities. Water is also an important ingredient in millboard.

The manufacturing steps in asbestos millboard production with waste sources indicated are shown in Figure 14. Millboard is produced on small cylinder-type machines similar to those used for making asbestos-cement pipe. The machines are equipped with one or two cylinder screens, conveying felt, pressure rolls, and a cylinder mold. After the ingredients are mixed in a beater, the slurry is transferred to a stirring vat or stock chest from which it is diluted and pumped to the cylinder vats of the millboard machines. Each cylinder vat contains a large screen surfaced cylinder extending the full length of the vat.

The slurry flows through the screen depositing a mat of fiber on the surface of the rotating cylinder before flowing out through the ends of the cylinder. The mat of fiber is then transferred to a carrier felt moving across the top of the rotating cylinder. On two-cylinder machines, the mats from the first and second cylinders are pressed together becoming a single homogeneous sheet as the felt picks up the mat from the second cylinder. Pressure rolls above the felt squeeze water from the mat as it is picked up from the cylinder. Some millboard machines have vacuum boxes adjacent to the felt that draw water out of the mat of fibers. Additional pressure rolls remove more water from the mat as it is wound onto the cylinder molds.

The cylinder mold is a drum about four feet wide and usually about four feet in diameter. As the carrier felt passes the cylinder mold, the mat is transferred to the cylinder because the adhesion to the wet cylinder surface is greater than the adhesion to the felt. The cylinder mold rotates, collecting successive layers of fiber until the desired thickness is obtained.

FIGURE 14: ASBESTOS MILLBOARD MANUFACTURING OPERATIONS

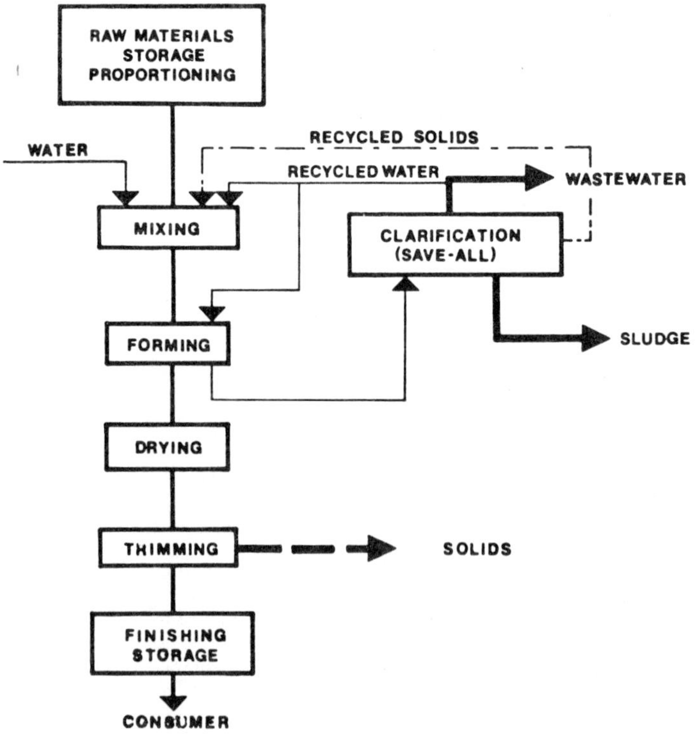

Source: EPA (13)

The cylinder is then momentarily stopped and the mat of fiber cut along a notch on the surface of the cylinder parallel to the cylinder axis. The sheet of millboard is removed as the cylinder starts rotating to build up another sheet. The wet millboard, containing about 50% water, is air dried or moved into an autoclave or oven for rapid curing. Finished millboard usually contains 5 to 6% water.

The operation of the deckles, cylinder showers, and felt washing showers is basically the same as described previously for asbestos paper. The uses and flow patterns of water in millboard manufacturing operations are very similar to those in asbestos papermaking.

As with the asbestos products covered previously, most of the water in the millboard manufacturing process serves as an ingredient carrier and continually circulates in a loop through the millboard machine and the save-all. All water flowing out of the cylinder screen and that drawn by vacuum out of the wet millboard is pumped to the save-all. Solids that settle in the save-all are pumped to the stock chest or the beater. Save-all overflow water is used for beater make-up, dilution, deckle water, and occasionally shower water. Excess overflow water must be discharged from the plant or sent to a treatment facility for additional treatment before it can be reused.

When possible, trimmings from millboard sheets are returned to the beater and repulped for use in new millboard. Most millboards can accept from 5 to 10% reclaimed material. A typical asbestos millboard plant operates two or three shifts per day and five or six days a week.

There are seven known locations in the U.S. where asbestos millboard is manufactured. At all of these locations, the wastewaters are either discharged to municipal sewers or are combined with other asbestos manufacturing wastewaters. Consequently, there is almost no information from the industry about the quantity and quality of millboard wastewaters. The results presented below are based primarily upon the sampling program carried out for this study at two plants.

The water leaving the save-all systems at the two plants amounted to 41 and 136 cubic meters per metric ton (12,000 and 39,500 gallons per ton). One plant discharges its wastewaters to a large lagoon system and recycles all of the lagoon effluent into the plant. This is a multiproduct plant. The other plant normally recycles all of its save-all effluent. Surges due to upsets or shut-down are released to a municipal sewer. Since neither plant has any measurable effluent on a regular basis, the amounts of water used in the manufacturing process may not be representative of the amounts discharged by a plant that does not recycle its wastewater.

At the plant that discharges its wastewaters to the lagoon system, the constituents added to to the water were measured as follows:

	mg/l	kg/kkg	lb/ton
Suspended solids	35	1.8	3.5
BOD (5 day)	5	0.25	0.5

The total solids and COD levels in the water leaving the millboard save-alls were the same as those of the make-up water. The pH of the raw wastewater ranged from 8.3 to 9.2. Some millboard is manufactured with portland cement and the pH would be higher in such cases.

The effluent from the save-all system at the millboard plant that operates with a completely closed water system had the characteristics listed below. In such a plant, the waste constituents accumulate until a steady-state level is reached. The contribution of each manufacturing cycle cannot be determined directly and, consequently, raw waste loadings expressed in terms of production units are meaningless.

	Average (mg/l)	Range (mg/l)
Total solids	6,100	3,950 - 7,800
Suspended solids	5,100	3,060 - 6,270
BOD (5 day)	2	-
COD	60	10 - 145

The pH ranged from 11.8 to 12.1 and the alkalinity from 2,000 to 2,700 mg/l, mostly in the hydroxide form.

The temperatures of the raw wastewaters at the two sampled millboard plants were 12° and 26°C, with the higher temperature measured at the completely closed system. The highest reported summer temperature of the effluents at two other millboard plants was 31°C.

Small amounts of oil and grease, nitrogen, and phosphorus were detected in some of the samples collected in this study. No information is available from the millboard industry on the presence of plant nutrients, toxic constituents, or about the nature of the additive materials that are used in the many varieties of millboard.

No information is available by which to accurately estimate the degree of fluctuation in millboard wastewater characteristics. Judging from the differences in the two plants that were sampled and from the relatively broad range of raw materials used, the variability of wastewaters from millboard manufacture is high.

One plant that produces a wide variety of millboard products with a relatively small save-all system presently achieves almost complete recycle of the process water. The stimulus at this location was, at least in part, high costs for water and sewer services. The plant releases save-all overflow to the municipal sewer when upsets or product changes occur. With greater save-all capacity or a holding tank, this plant could accomplish zero discharge on a continuous basis.

In connection with this study, four of the seven known millboard plants in the country were visited. Since almost complete recirculation has been demonstrated in a typical plant, it is believed that zero discharge can be achieved soon by millboard manufacturing plants.

The applicable control measure for asbestos millboard plants is complete recycle of all process wastewaters. No end-of-pipe technology is specifically required if the plant's save-all capacity is adequate. Unlike settling tanks, save-alls can provide surge capacity.

Waste solids will normally be generated only when the plant is shut down. These will require dewatering and transportation to a land disposal site. Since asbestos millboard manufacturing operations are located in plants that make other asbestos products, the best means of solids handling and disposal will be dependent on the methods used for solids from the other product lines.

A variety of alternatives have been explored by EPA (13) for wastewater control in asbestos millboard manufacture as follows:

Alternative A — No Waste Treatment of Control

Effluent waste load is estimated to be 1.8 kg/kkg (3.6 lb/ton) of suspended solids and 0.25 kg/kkg (0.5 lb/ton) of BOD for the selected typical plant at this minimal control level. In-plant use of save-alls is assumed, as this is universally practiced in the industry.

Costs — None

Reduction Benefits — None

Alternative B — Sedimentation of Process Wastes

This alternative includes settling of all process wastewaters. Some form of sedimentation is applied at at least 40% of the plants. Costs include disposal of sludge. Effluent load estimated to be 0.8 kg/kkg (1.6 lb/ton) of suspended solids and 0.2 kg/kkg (0.4 lb/ton) of BOD.

Costs — Investment costs are approximately $40,000.

Reduction Benefits — Estimated reduction of effluent suspended solids of 55% and BOD of 20%.

Alternative C —Complete Recycle of Process Water

This alternative includes complete recycle of all process wastewater back into the manufacturing process and other in-plant uses. Fresh water taken into plant equals the quantity in wet product. Complete control of pollutant constituents without discharge is effected. One millboard plant is known to achieve complete recycle most of the time.

Costs — Incremental costs are approximately $12,000 over Alternative B; total costs are $52,000.

Reduction Benefits — Reduction of suspended solids, BOD, and all other pollutant constituents of 100%.

The annual costs and resulting effluent quality for the treatment of control technology alternatives for asbestos millboard are summarized in Table 12.

TABLE 12: WATER EFFLUENT TREATMENT COSTS FOR MANUFACTURING ASBESTOS MILLBOARD

Treatment or Control Technologies	Alternatives		
	A	B	C
Investment*	-	$40	$52
Annual Costs:*			
Capital Costs	-	3.2	4.2
Depreciation	-	2.0	2.6
Operating and Maintenance Costs (excluding energy and power costs)	-	31.0	24.3
Energy and Power Costs	-	5.0	7.0
Total Annual Costs*	-	41.2	38.1

*Costs in thousands of dollars

Effluent Quality:

Effluent Constituents Parameters (Units)	Raw Waste Load	Resulting Effluent Levels		
Suspended Solids - kg/MT	1.8	do	0.8	0
BOD (5-day) - kg/MT	0.25	do	0.2	0
Suspended Solids - mg/l	35	do	15	0
BOD (5-day) - mg/l	5	do	4	0

Source: EPA (13)

Asbestos Roofing

Asbestos roofing is made by saturating heavy grades of asbestos paper with asphalt or coal tar with the subsequent application of various surface treatments. The stock paper may be single or multiple layered and usually contains mineral wool, Kraft fibers, and starch as well as asbestos. Fiber glass filaments or strands of wire may be embedded between layers for reinforcement.

Figure 15 shows the major steps in the manufacture of asbestos roofing. Asbestos paper is pulled through a bath of hot coal tar or asphalt. After it is thoroughly saturated, the paper passes over a series of hot rollers to set the coal tar or asphalt in the paper. The paper then passes over cooling rollers that reduce the temperature of the paper and give it a smooth surface finish. At some plants, cooling water is sprayed directly on the surface of the saturated paper.

Roll roofing is coated with various materials to prevent adhesion between layers and then passed over a final series of cooling rollers. The roofing is then air dried and rolled up and packaged for marketing. The manufacture of asbestos roof shingles is similar from a wastewater point of view. A typical rool roofing plant operates one to two shifts a day on a five day per week schedule.

FIGURE 15: ASBESTOS ROOFING MANUFACTURING OPERATIONS

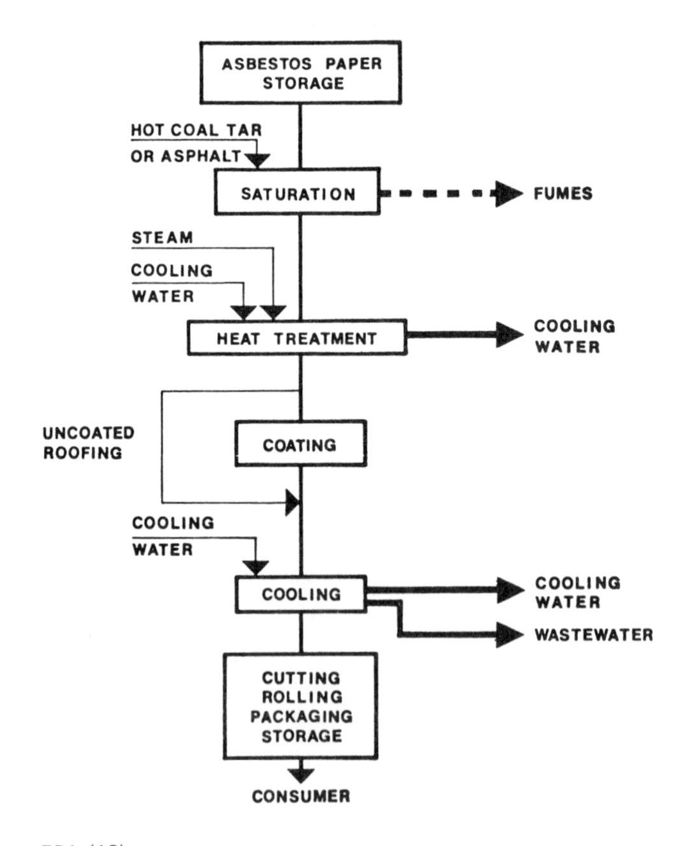

Source: EPA (13)

Water is used in two ways in the production of roofing. It is converted to steam to heat the saturating baths and hot rollers and for cooling the hot paper after it has been saturated. Condensate from the saturating bath coils and the hot rollers is collected and returned to the boilers. Fresh make-up water in small quantities is required to replace boiler blowdown water, steam, and condensate that escapes through leaks. Cooling water is used once and discharged unless cooling towers or other means of cooling the water are available. The only process wastewater associated with roofing manufacture is that originating in the spray cooling step. In many cases, this contaminated contact cooling water is discharged with the clean noncontact cooling water.

Unlike the asbestos products covered previously, water is not an integral part of roofing products. It is used, however, to cool the roofing after saturation. All plants use non-contact cooling and some use spray contact cooling. The roofing is largely, but not completely, inert to water and the contact cooling water becomes a process wastewater. This contaminated cooling water is discharged with the noncontact cooling water in some plants, resulting in a large volume of dilute process wastewater.

The discharge volumes vary widely among the few roofing plants that reported information on flows, ranging from 145 to 2,100 liters per metric ton (35 to over 500 gallons per ton) of product. The original temperature of the cooling water, whether it is once-through or recirculated, and whether noncontact water is included are factors influencing the reported amount of water discharged. The fluctuations in flow rate should be minimal at a given location.

The characteristics of spent cooling water from roofing manufacture are developed from sampling data taken at one plant. This plant employs surface sprays and discharges the contact and noncontact cooling water into a common sewer. The combined wastewater was sampled. At the time of sampling, the roofing was being made from organic (non-asbestos) paper. Since the water spray contacts only the outer bituminous surface and not the base paper, it is believed that the samples are representative of wastes from contact cooling of asbestos-based roofing.

The added quantities of the major constituents were as follows:

	mg/l	kg/kkg	lb/ton
Suspended solids	150	0.06	0.13
BOD (5 day)	6	0.003	0.005
COD	20	0.008	0.016

The pH of the wastewater averaged 8.2.

The temperature of the spent cooling water was 13°C, a 7 degree increase over the temperature of the intake water at a flow rate of about 1,420 cubic meters per day (0.375 mgd).

Information about the effluents from one other asbestos roofing plant was reported by the manufacturer. The wastewater is treated by settling, oil skimming, and passage through an adsorbant filter. The added quantities of materials are reported to be:

	mg/l	kg/kkg	lb/ton
Suspended solids	37	0.06	0.12
BOD (5 day)	38	0.07	0.13
COD	91	0.15	0.30

The average pH of the effluent is reported to be 6.8. Other constituents of interest were measured in this effluent with the following average results in terms of added quantities:

	mg/l	g/kkg
Total solids	93	0.16
Total organic carbon	1	0.00015

(continued)

	mg/l	g/kkg
Cyanide	0.00003	0.00005
Copper	0.019	0.03
Iron	0.031	0.05
Lead	0.001	0.0015
Nickel	0.003	0.005
Zinc	0.071	0.12
Oil and grease	1.6	0.0025
Phenols	0.003	0.005

Total nitrogen and phosphorus levels in the cooling water were each increased 0.5 mg/l by passage through the plant. Arsenic, cadmium, and chromium were analyzed for, but not detected in, the effluent.

The above information on treated roofing wastewaters is presented as supplemental data. It has not been verified, but it does provide an insight into the strength and character of the wastewaters from asbestos roofing manufacture.

There is insufficient information to describe variations in the characteristics within a plant or among plants in this category. Since the wastewater is spent cooling water, its characteristics should be unaffected by start-up and shut-down operations.

The asbestos roofing plants that practice contact cooling should evaluate the possibility of eliminating this source of process wastewater as one in-plant control measure. If this were done, and leaks and other losses of noncontact cooling were closed and dry cleaning practices instituted, the asbestos roofing industry would be able to operate without the discharge of process wastewaters.

In any case, noncontact cooling water and condensate should not be mixed with contact cooling water. This practice greatly increases the volume of process wastewater to be treated.

The applicable end-of-pipe technology for asbestos roofing wastewaters is sedimentation with skimming or filtration to remove insoluble materials. Properly designed and operated facilities should reduce the suspended solids levels to 15 mg/l and COD to 20 mg/l or less. If the organic materials are not adequately removed, further treatment, possibly by activated carbon adsorption, will be required. There is, at present, no information available by which to assess the suitability or efficiency of such treatment for these wastes. Information is lacking on the nature of the dissolved organics in wastewaters from asbestos roofing manufacture.

To completely eliminate the discharge of pollutant constituents will require that the contaminated cooling water that constitutes the process wastewater be treated, cooled, and reused. As noted above, the precise type and extent of treatment required is not known due to lack of information. An alternative solution would be the elimination of contact cooling and confinement of leaks so that the water remains uncontaminated.

A variety of alternatives have been explored by EPA (13) for wastewater control in asbestos roofing manufacture as follows:

Alternative A —No Waste Treatment or Control

Effluent waste load is estimated to be 0.06 kg/kkg (0.12 lb/ton) of suspended solids, 0.003 kg/kkg (0.006 lb/ton) of BOD, and 0.008 kg/kkg (0.016 lb/ton) of COD for the selected typical plant at this minimal control level.

Costs — None

Reduction Benefits — None

Alternative B — Sedimentation of Process Wastes (Contaminated Cooling Water)

This alternative includes settling of all process wastewaters (contaminated cooling water) with skimming or filtration as necessary to remove suspended matter. Effluent load estimated to be 0.006 kg/kkg (0.012 lb/ton) of suspended solids. BOD and COD waste loads remain the same as Alternative A.

Cost — Investment costs are approximately $24,000.

Reduction Benefits — Estimated reduction of effluent suspended solids of 90%.

Alternative C — Complete Recycle of Process Water

This alternative includes treatment, cooling, and reuse of process wastewater (contaminated cooling water). No process wastewaters are discharged and complete control of pollutant constituents is effected.

Costs — Incremental costs are approximately $24,000 over Alternative B; total costs are $48,000.

Reduction Benefits — Reduction of suspended solids, BOD, and COD and all other pollutant constituents of 100%.

The annual costs and effluent quality associated with each of the treatment or control alternatives for asbestos roofing are given in Table 13.

TABLE 13: WATER EFFLUENT TREATMENT COSTS FOR MANUFACTURING ASBESTOS ROOFING

Treatment or Control Technologies	Alternatives		
	A	B	C
Investment*	-	$24	$48
Annual Costs:*			
Capital Costs	-	2.0	4.0
Depreciation	-	1.2	2.4
Operating and Maintenance Costs (excluding energy and power costs)	-	6.0	0
Energy and Power Costs	-	1.3	2.0
Total Annual Costs*	-	10.5	8.4

*Costs in thousands of dollars

Effluent Quality:

Effluent Constituents Parameters (Units)	Raw Waste Load	Resulting Effluent Levels		
Suspended Solids - kg/MT	0.06	do	0.006	0
BOD (5-day) - kg/MT	0.003	do	0.003	0
COD - kg/MT	0.008	do	0.008	0
Suspended Solids - mg/l	150	do	15	0
BOD (5-day) - mg/l	6	do	6	0
COD - mg/l	20	do	20	0

Source: EPA (13)

Asbestos-Based Friction Materials

The manufacturing steps typically used in dry-mix molded brake lining manufacture are shown in Figure 16. The bonding agents, metallic constituents, asbestos fibers, and additives are weighed and mixed in a two-stage mixer. The mix is then hand-tamped into a metal mold. The mold is placed in a preforming press which partially cures the molded asbestos sheet. The asbestos sheet is taken from the preforming press, and put in a steam preheating mold to soften the resin in the molded sheet. The molded sheet is formed to the proper arc by a steam heated arc former, which resets the resin. The arc-formed sheets are then cut to the proper size. The lining is then baked in compression molds to retain the arc shape and convert the resin to a thermoset or permanent condition. The lining is then finished and, after inspection, is packaged.

The finishing steps include sanding and grinding of both sides to correct the thickness, edge grinding, and drilling of holes for rivets. Following drilling, the lining is vacuum-cleaned, inspected, branded, and packaged.

FIGURE 16: DRY-MIXED BRAKE LINING MANUFACTURING OPERATIONS

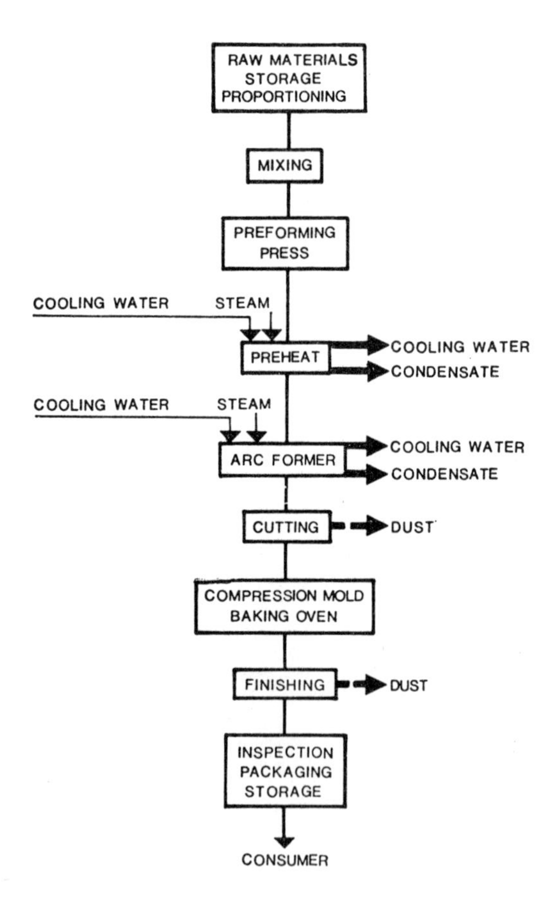

Source: EPA (12)

Figure 17 shows the major steps in the manufacture of wet-mixed molded brake linings. The name wet mix process is a misnomer and refers to the use of a solvent. The ingredients of the molded lining are actually relatively dry. After weighing, they are mixed in a sigma blade mixer. The mixed ingredients are then sent to grinding screens where the particle size of the mixture is corrected. The mixture is conveyed to a hopper and is forced from the hopper into the nip of two form rollers which compress the mixture into a continuous strip of friction material.

The strip is cut into the proper lengths and then arc-formed on a round press bar. The cutting and arc forming operations are done by separate units. The linings are then placed in racks and either air-dried or oven-dried to remove the solvent. An alternative process is to place the arc-formed linings in metal molds for baking in an oven. From the ovens, the linings are finished, inspected, and packaged.

Molded clutch facings are produced in a manner similar to the wet-mixed process. The rubber friction compound, solvent, and asbestos fibers are introduced into a mixer churn. After the churn mixes the ingredients, the mixture is conveyed to a sheeter mill which forms a sheet or slab of the materials.

FIGURE 17: WET-MIXED MOLDED BRAKE LINING MANUFACTURING OPERATIONS

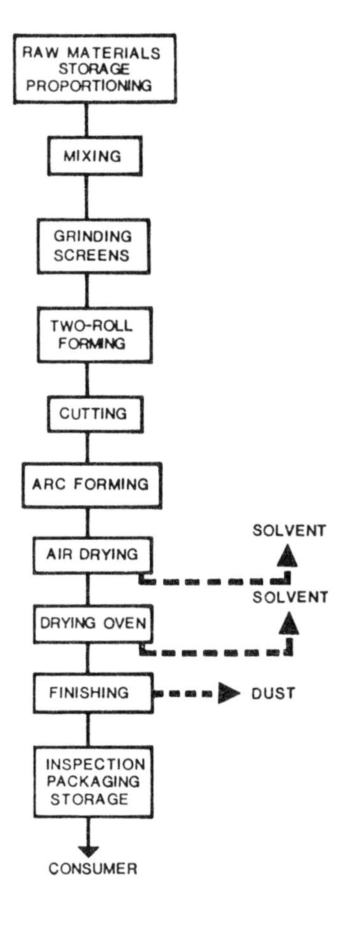

Source: EPA (12)

The sheet is then diced into small pieces by a rotary cutter. The pieces are placed in an extrusion machine which forms sheets of the diced material. The sheets are cut into the proper size and then punch-pressed into donut-shaped sheets. The scraps from the punch press are returned to the extrusion machine. The punched sheets are placed on racks and sent to a drying oven and then a baking oven for final curing and solvent evaporation. The oven dried sheets are finally sent to the finishing operations. Figure 18 illustrates the steps in the manufacture of molded clutch facings.

Woven clutch facings and brake linings are manufactured of high-strength asbestos fabric that is frequently reinforced with wire. The fabric is predried in an oven or by an autoclave to prepare it to be impregnated with resin.

FIGURE 18: MOLDED CLUTCH FACINGS MANUFACTURING OPERATIONS

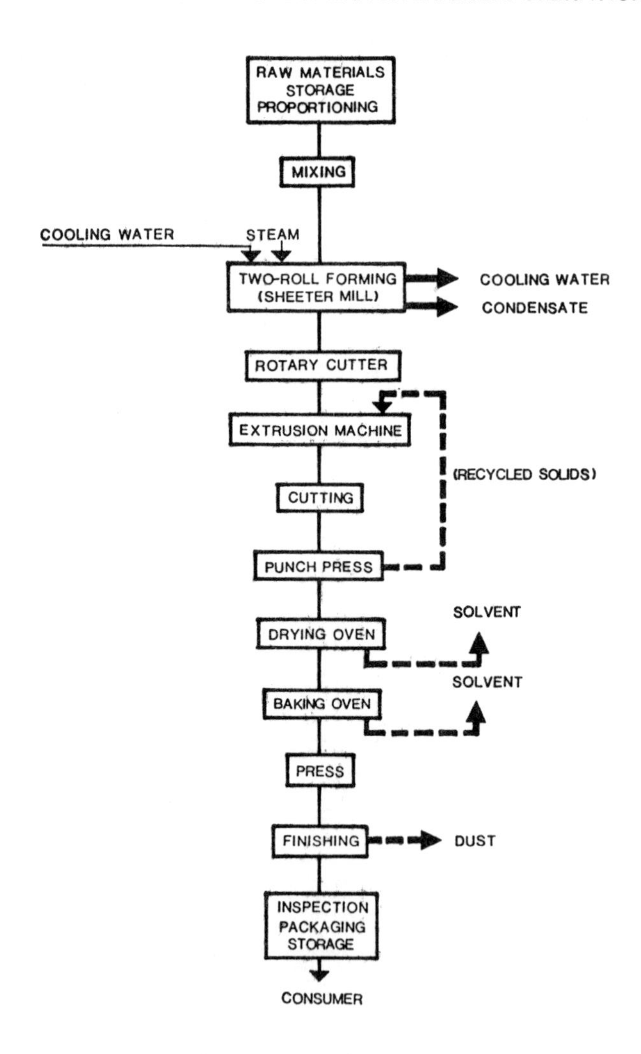

Source: EPA (12)

The fabric can be impregnated with resin by several techniques:

[1] immersion in a bath of resin,

[2] introducing the binder in an autoclave under pressure,

[3] introducing dry impregnating material into carded fiber before producing yarn, and

[4] imparting binder into the fabric from the surface of a roll.

After the solvents are evaporated from the fabric, it is made into brake linings or clutch facings. Brake linings are made by calendering or hot pressing the fabric in molds. The linings are then cut, rough ground, placed in molds, and placed in a baking oven for final curing. Following curing, the lining is finished, inspected, and packaged. The composition by weight of woven brake linings ranges from 40 to 60% asbestos, 10 to 20% cotton, 20 to 40% wire, and 5 to 20% binder.

Figure 19 illustrates the manufacture of woven clutch facings. The treated fabric is cut into tape-width strips by a slitting machine. The strips are wound around a mandrel to form a roll of the fabric. The roll is pressed in a steam-heated press and then baked in an oven to cure the resin in the clutch facing. Following curing, the clutch facing is finished, inspected, and packaged.

Water does not mix with the ingredients of friction materials and is not used in the manufacturing processes. Wastewaters are generated in a few friction materials plants in solvent recovery operations and in wet dust collection equipment used to control the quality of the air from the finishing areas. Most plants in this industry use dry dust collection equipment. Friction materials plants typically operate two or three shifts a day on a five or six day per week schedule.

FIGURE 19: WOVEN CLUTCH FACINGS MANUFACTURING OPERATIONS

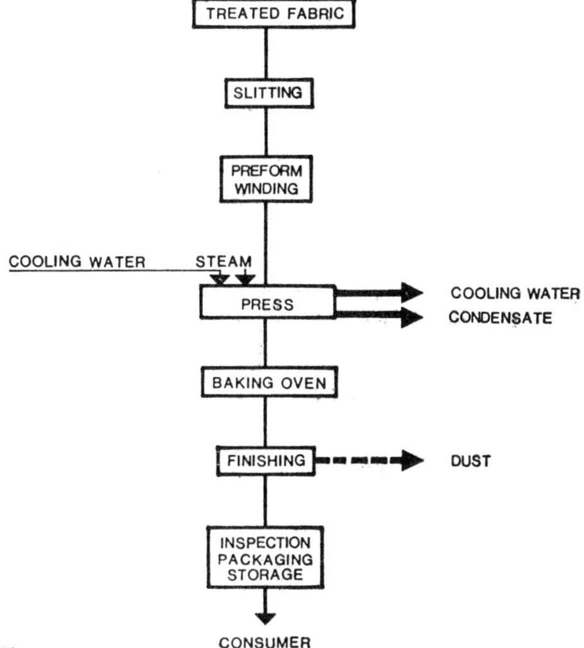

Source: EPA (12)

There are known to be four friction materials manufacturing plants that discharge wastewaters from wet dust collection equipment. It is estimated that the total number of such plants in the country is no more than eight. At all the known plants, the wastewaters are clarified before discharge to surface waters. At one, the wastes are combined with metal-finishing wastes in a physical-chemical treatment facility.

The water use rate in wet dust collectors varies from 0.06 to 1.3 l/sec/m^3/min of air scrubbed (0.5 to 10 gpm per 1,000 scfm). The plant air systems that are served by wet scrubbers that discharge wastewaters from the plant property range from 280 to 1,700 m^3/min (10,000 to 60,000 scfm), resulting in wastewater discharges of from 190 to 570 m^3 (50,000 to 150,000 gallons) per day. Units that incorporate recirculation discharge a settled slurry. The contents of the settling tank are dumped each week. The wastes are discharged to a settling lagoon.

The wastewaters from the wet dust collectors are slurries of dust emanating from grinding and drilling operations used in finishing the products. The principal parameter for characterizing the wastes is suspended solids. Because friction materials are designed to shed water, it is unlikely that the dust is solubilized. The COD test provides a means of detecting and monitoring this phenomenon. The quantity of friction material lost in the finishing operations may be 30%. Even with the relatively high price of asbestos fiber, it is not recovered. Once the resin has set up, it is not economical to break it down to salvage the fiber.

Wastewaters from wet dust collectors can be treated by sedimentation, with coagulation as needed. No data are available on the efficiency of plain sedimentation, but there is no reason it would not be effective. While the dust particles have a significant organic content, they are not treatable by biotreatment or activated carbon adsorption. If treatment beyond sedimentation is indicated, filtration would be the logical next step and complete removal could be accomplished. It would be more appropriate to substitute dry dust collectors, e.g., baghouses, for the wet scrubbers. This step, already being taken at some of the plants, eliminates discharge of wastewater. Detailed information about the engineering aspects of equipment for dry collection of particulates is available in the literature dealing with air pollution control.

A variety of alternatives have been explored by EPA (12) for wastewater control in asbestos-based friction material manufacture as follows:

Alternative A — No Waste Treatment or Control

Estimated effluent waste load is 380 m^3 (100,000 gallons) per day of concentrated dust slurry. The dissolved solids level is not significantly higher than that of the carriage water.

Costs — None

Reduction Benefits — None

Alternative B — Sedimentation

This alternative comprises sedimentation, with coagulation as necessary, to remove suspended solids. Sludge is dewatered for disposal in a controlled sanitary landfill. Daily effluent waste load is estimated to be 11 kg (25 lb) of suspended solids. All known plants use this alternative as a minimum level of control.

Costs — Investment cost is estimated to be $64,000.

Reduction Benefits — Reduction of suspended solids of over 95%.

Alternative C — Zero Discharge

This alternative comprises substitution of dry dust collection devices (baghouses) for the wet dust scrubbers. No wastewater is generated in using this control technology. Most of the friction materials plants now use such dry equipment.

Costs — Estimated investment cost is $94,000.

Reduction Benefits — Reduction of all pollutant constituents of 100%.

Tables 14 to 16 inclusive give the investment and annual costs and effluent quality associated with each of the above treatment or control alternatives.

TABLE 14: WATER EFFLUENT TREATMENT COSTS FOR ASBESTOS MANUFACTURING, WET DUST COLLECTION—SMALL PLANT

	(Costs in $1000)		
Treatment or Control Technologies:	A	B	C
Investment	-	44	43
Annual Costs:			
Capital Costs	-	3.5	3.4
Depreciation	-	1.8*	1.7**
Operating and Maintenance Costs (excluding energy and power costs)	-	7.7	4.3
Energy and Power Costs	-	4.0	-
Total Annual Cost	-	17.0	9.4

Effluent Quality:

Effluent Constituents	Raw Waste Load	Resulting Effluent Levels		
COD (Filtrate) - mg/l	Unknown	Unknown	50	Zero
Suspended Solids - mg/l	Variable	Variable	30	Zero
pH - units	6-9	6-9	6-9	-

* Expected lifetime - 25 years
** Expected lifetime - 20 years.

Source: EPA (12)

TABLE 15: WATER EFFLUENT TREATMENT COSTS FOR ASBESTOS MANUFACTURING, WET DUST COLLECTION—MEDIUM PLANT

	(Costs in $1000)		
Treatment or Control Technologies:	A	B	C
Investment	-	64	94
Annual Costs:			
Capital Costs	-	5.1	7.5
Depreciation	-	2.6*	4.7**
Operating and Maintenance Costs (excluding energy and power costs)	-	12.0	6.1
Energy and Power Costs	-	5.2	-
Total Annual Cost	-	24.9	18.3

(continued)

TABLE 15: (continued)

Effluent Quality:

Effluent Constituents	Raw Waste Load	Resulting Effluent Levels		
COD (Filtrate) - mg/l	Unknown	Unknown	50	Zero
Suspended Solids - mg/l	Variable	Variable	30	Zero
pH - units	6-9	6-9	6-9	-

* Expected lifetime - 25 years
** Expected lifetime - 20 years.

Source: EPA (12)

TABLE 16: WATER EFFLUENT TREATMENT COSTS FOR ASBESTOS MANUFACTURING, WET DUST COLLECTION—LARGE PLANT

Treatment or Control Technologies:	A	B	C
	(Costs in $1000)		
Investment	-	83	146
Annual Costs:			
Capital Costs	-	6.6	11.7
Depreciation	-	3.3*	7.3**
Operating and Maintenance Costs (excluding energy and power costs)	-	16.0	8.5
Energy and Power Costs	-	6.5	-
Total Annual Cost	-	32.4	27.5

Effluent Quality:

Effluent Constituents	Raw Waste Load	Resulting Effluent Levels		
COD (Filtrate) - mg/l	Unknown	Unknown	50	Zero
Suspended Solids - mg/l	Variable	Variable	30	Zero
pH - units	6-9	6-9	6-9	-

* Expected lifetime - 25 years
** Expected lifetime - 20 years.

Source: EPA (12)

Asbestos Textiles

The primary reasons for the use of asbestos fiber in textile products are its properties of durability and resistance to heat, fire, and acid. Asbestos is the only mineral that can be manufactured into textiles using looms and other textile equipment. The asbestos textile products are primarily used for friction materials, industrial packing, electrical insulation, and thermal insulation.

Figure 20 illustrates the steps in the manufacture of the various asbestos textile products. The textile plants receive the asbestos fiber by railcar in 100 pound bags. The bags are opened, and the fibers passed over vibrating or trommel screens for cleaning. The fibers are lifted from the screens by air suction and graded. The fiber is mixed and blended. Chrysolite is the predominant fiber used in textiles. Crocidolite and amosite asbestos fibers may also be added.

Small percentages of cotton, rayon, and other natural or synthetic fibers serve as carriers or supports for the shorter asbestos fibers, and they improve the spinnability of the fiber mixture. Typically, the organic fiber content is between 20 and 25%. The blending and mixing operations are primarily done during carding of the fibers, but can also be performed in multihopper blending units.

In the carding operation, the fibers are arranged by thousands of needle-pointed wires that cover the cylinders of the carding machine. The fibers are combed by passing between the carding machine main cylinder and the worker cylinders rotating in the opposite direction. The carding machine forms a continuous mat of material. The mat is divided into strips, or slivers, and mechanically compressed between oscillating surfaces into untwisted strands. The strands are wound on spools to form the roving. Roving is the asbestos textile product from which asbestos yarn is produced.

The roving is spun into yarn in a manner similar to that employed to manufacture cotton and wool yarns. The strands of roving are converted into a single yarn by the twisting and pulling operations of a spinning machine. The yarn produced by spinning and twisting is the basic component of several other asbestos textile products.

Asbestos twine or cord is produced by twisting together two or more yarns on a spinning frame similar to those used to manufacture cotton cord. Braided products are made by a series of yarn-carrying spindles, half traveling in one direction and half in the opposite direction to plait the yarn together and form a braided product.

Asbestos yarn can also be twisted or braided into various shapes to form packing and gaskets. The braided material can be impregnated with different compounds. Graphite is commonly used to impregnate braided packing material, the graphite serves to lower the frictional and binding properties of the packing.

Asbestos cloth is woven from yarn on looms that operate in a manner similar to those used for the manufacture of other textile products. The warp yarn is threaded through the heddles and the reed of the loom and the filler yarn is wound on quills and placed in a shuttle. The cloth is woven as the filler yarn in the shuttle interweaves the warp yarn transversely. Following weaving, the asbestos cloth is inspected for strength, weight, and asbestos content.

Asbestos yarn or cloth may be coated for fabrication into friction materials and special textile products. The material is drawn through one or more dip tanks and the coating material is spread by rollers, brushes, or doctor blades. The coated textile product then passes through a drying oven where the solvent is evaporated.

Water is not normally used in an asbestos textile manufacturing plant. Two exceptions are the addition of moisture during weaving or braiding and the coating operations. Wastewater is generated only in the latter process. A typical asbestos textile plant operates two or three shifts per day and five days per week.

Wastewaters result from the coating of asbestos textiles at two plants in the country at the present time. Where textile products are coated (impregnated) in the manufacture of friction materials and sealing devices, water is not used and no wastewater is generated.

FIGURE 20: ASBESTOS TEXTILE MANUFACTURING OPERATIONS

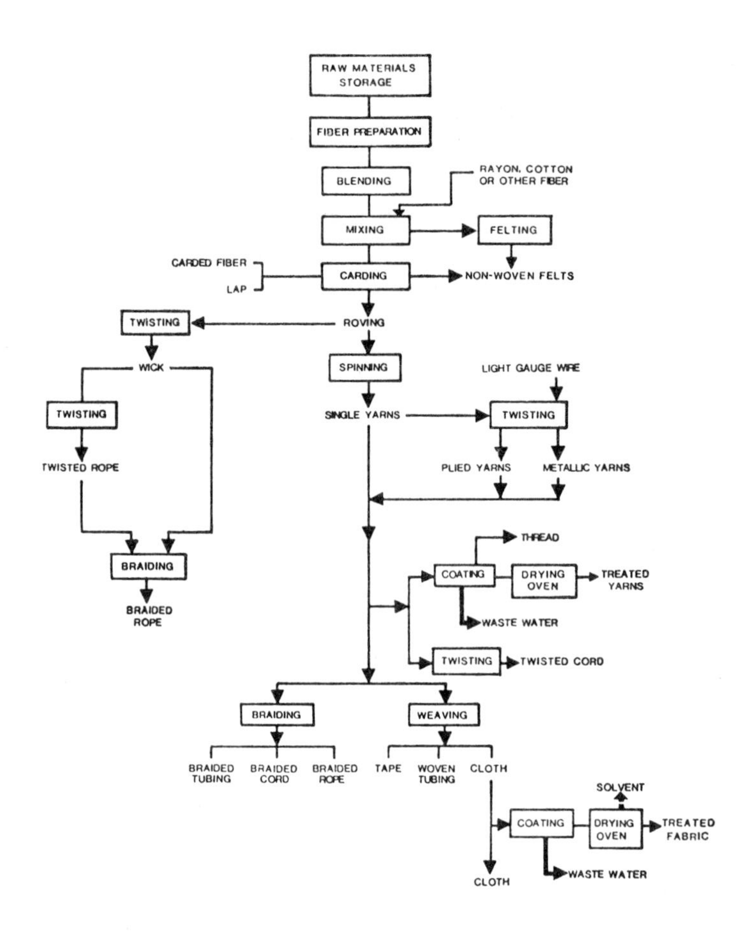

Source: EPA (12)

The volume of waste generated in the coating of asbestos textiles is estimated to be no more than 750 liters (200 gallons) per day. The coating of asbestos textiles is not presently a full-time operation at either of the two plants. The waste results from dumps and cleanup at the end of a run. The number and length of the runs varies on a typical day, making the quantity of waste largely independent of the quantity of textile treated, or the level of production.

Coated asbestos textiles are used in a variety of products; e.g., pipe lagging, paper machine felts, ironing board covers, etc. One of the purposes of coating is to encase the fibers, thereby reducing the potential health hazards in fabricating and using the final products. The coating has additional functions and its composition is normally specified by the fabricator. Consequently, the chemical constituents of the coating material, and subsequently those of the wastewater, vary at each of the plants. The ingredients include resins, elastomers, pigments, solvents, and fillers. The wastes are high in COD and suspended and dissolved solids. In addition to the organic components, trace quantities of heavy metals, phosphates, and fluorides may be present. At both of the plants that coat asbestos textiles, the wastewaters are discharged to municipal sewerage systems, one with pretreatment and the other without. Other than knowing the quantities of raw materials used, neither plant has information on the characteristics of its wastewaters.

The wastes from textile coating result from clean-up and dumping of unused coating material at the end of a run. The wastewaters are small in volume and relatively concentrated. Because of the high cost of treating this waste to make it suitable for discharge to a surface water, the recommended control measure is containment of the waste in undiluted form and containerization for salvage or land disposal. The required quantities of finishing material for each run should be estimated and prepared so that a minimal amount remains to be disposed of.

Dry cleaning techniques should be substituted for wet methods. Measures should be taken to eliminate or contain spills and dripped material. The waste would be placed in appropriate containers, e.g., steel drums, for salvage by a commercial waste handling firm or for disposal in a controlled sanitary land fill. If no commercial handling firm is available and state or local regulations prohibit disposal of solvents in sanitary landfills, it may be necessary to employ small batch incinerators for disposal of the reduced volumes of waste.

A number of alternatives have been explored by EPA (12) for wastewater control from coating operations in asbestos textile manufacture as follows:

Alternative A — No Waste Treatment or Control

Effluent waste load is a very small volume of concentrated organic material (COD) and suspended solids with potentially significant levels of heavy metals and plant nutrients. The waste is discharged on about half of the plant operating days.

Cost — None

Reduction Benefits — None

Alternative B — Zero Discharge

Discharge of wastewater is eliminated through in-plant control measures, including the use of dry cleaning methods and containment of dumped and spilled coating material. Waste is containerized for salvage by commercial waste salvage firm or for disposal in a controlled sanitary landfill. Some in-plant control measures are now in use, e.g., minimizing dumps, but no plant completely retains all waste.

Costs — Investment cost is approximately $2,000.

Reduction Benefits — Reduction of all pollutant constituents of 100%.

The estimated costs and associated pollution reduction benefits for the above alternative control technologies are given in Table 17.

At one of the two asbestos textile coating plants, a vapor absorption unit is used to scrub solvent from the drying oven exhaust. The fume scrubber at the single known installation in this industry is operated once or twice a month for a two-shift period each time. The water usage rate is about 3.8 liters per second (60 gallons per minute) for a total volume per period of approximately 220 cubic meters (58,000 gallons). The scrubber comprises four chambers, and water is recirculated within the unit.

The vapor absorption unit is charged with 22.7 kilograms (50 pounds) of sodium hydroxide in solution for each period of operation. The resulting wastewater, therefore, contains this caustic plus the absorbed solvent. The waste is pretreated in a two-stage lagoon prior to discharge to the municipal sewerage system. There are no records available that describe the characteristics of the raw wastewater resulting from the vapor absorption unit. It should have a somewhat elevated pH value and a significant COD content.

TABEE 17: WATER EFFLUENT TREATMENT COSTS FOR MANUFACTURING ASBESTOS TEXTILE COATING

Treatment or Control Technologies:	(Costs in $1000) A	B
Investment	-	2.0
Annual Costs:		
Capital Costs	-	0.2
Depreciation	-	0.2*
Operating and Maintenance Costs (excluding energy and power costs)	-	8.0
Energy and Power Costs	-	Zero
Total Annual Cost	-	8.4

Effluent Quality:

Effluent Constituents	Raw Waste Load	Resulting Effluent Levels	
COD - mg/l	Variable	Variable	Zero
Suspended Solids - mg/l	Variable	Variable	Zero
pH - units	Variable	Variable	-

* Expected Lifetime - 10 years.

Source: EPA (12)

The wastewater from vapor absorption operations resembles that from solvent recovery in that it contains organic material and has a negligible suspended solids content. In this industry, however, the vapor absorption operations are operated intermittently, and bio-treatment processes are not feasible. All biological facilities require a reasonably steady inflow of waste to function effectively. Carbon adsorption should be effective with these wastes, however. Adjustment of the pH to a lower level would probably be beneficial to increase the efficiency of the carbon.

Since recovery of the solvent is not a goal in vapor absorption, a fume incinerator could be substituted to remove the vapor from the exhaust air. Both direct-fired and catalytic types are available and either should be suitable for this application. Detailed information about the design, operation, costs, and applicability of various types of incinerators is beyond the scope of this report and is readily available in the technical literature on air pollution control. The use of an incinerator would eliminate the discharge of wastewater in this subcategory.

A variety of alternatives have been explored by EPA (12) for wastewater control from vapor absorption operations in asbestos textile manufacture as follows:

Alternative A — No Waste Treatment or Control

Daily effluent waste load is estimated to be 410 kg (900 lb) of COD at a pH level above 9.5. The suspended solids waste load is negligible. Discharge is presently intermittent in this subcategory.

Costs — None

Reduction Benefits — None

Alternative B — Carbon Adsorption

This alternative involves treatment of the raw wastewater in 2-stage granular activated carbon columns. The raw wastewater is acidulated as necessary, but does not require filtration. The carbon is regenerated off-site by the supplier. The daily effluent waste load is estimated to be about 10 kg (22 lb) of COD with the pH value in the neutral range, 6 to 9.

Costs — Investment cost is estimated to be $130,000.

Reduction Benefits — Reduction of COD of approximately 98% and neutralization of alkali in effluent.

Alternative C — Zero Discharge

Zero discharge is achieved by replacement of the vapor absorption unit with a fume incinerator. No wastewater is generated.

Costs — Estimated cost for this alternative is $152,000.

Reduction Benefits — Reduction of all pollutant constituents of 100%.

The estimated costs and associated pollution reduction benefits for the above alternative control technologies are given in Table 18.

TABLE 18: WATER EFFLUENT TREATMENT COSTS FOR ASBESTOS MANUFACTURING—VAPOR ABSORPTION

Treatment or Control Technologies:	(Costs in $1000) A	B	C
Investment	-	130	152
Annual Costs:			
Capital Costs	-	10.4	12.2
Depreciation	-	9.3*	15.2**
Operating and Maintenance Costs (excluding energy and power costs)	-	8.7	1.8
Energy and Power Costs	-	1.0***	16.8
Total Annual Cost	-	29.4	46.0

Effluent Quality:

Effluent Constituents	Raw Waste Load	Resulting Effluent Levels		
COD - mg/l	1800	1800	50	Zero
Suspended Solids - mg/l	30	30	30	Zero
pH - units	>9	>9	6-9	-

* Expected lifetime - 14 years
** Expected lifetime - 10 years
*** Not including carbon regeneration.

Source: EPA (12)

In addition to textile coating, it should be noted that there is a potential source of wastewater in this part of the asbestos industry: namely, the manufacture of yarn by the dispersion process. At the time of this study, two plants in the country have pilot-plant or experimental manufacturing operations using this process.

The level of production is extremely limited today, but it could increase in the future. While these operations are too limited to be considered in this study, it was determined that, even with in-plant controls, the associated wastewaters can be expected to contain both organic and inorganic pollutants. If this process becomes operational, separate effluent limitations guidelines should be developed.

The water use rate is in the order of 20 to 60 cubic meters (5,000 to 15,000 gallons) per day in these pilot-plant operations. Because these facilities are very small, water usage based on production cannot be realistically extrapolated to plant-scale operations. The water passes through save-alls in the process and there is at least a potential for recycle of water. Because of the waste characteristics, it is not feasible at this time to completely reuse all water in this process.

The wastewaters from the two plants that are developing the dispersion process differ significantly, in part because the processes are not exactly the same. It is possible that the wastes will change significantly as the processes are refined and developed. Some of the parameters that should be measured are total and suspended solids; COD and BOD; hexane extractables: MBAS; zinc and other metals; and the plant nutrients, nitrogen and phosphorus.

Asbestos Packing and Gaskets

The gaskets, packings, and sealing devices group includes a wide variety of products, many of which contain metallic components. The asbestos content of these products varies widely from one type to another. The typical plant making these products is a fabricator rather than a manufacturer, purchasing materials that are ready for cutting and assembly. There are many specialized hand operations in some plants in this category. Gaskets and packings may be made from asbestos paper, felt, and millboard; yarn, cloth, wick, and rope; and sheet gasket material. The wastewaters associated with asbestos paper, felt, and millboard were covered above.

The variety of materials and forms comprising this group of products is so wide that it precludes general descriptions of typical manufacturing processes. In this study, no plant was found that used water in the manufacture of gaskets, packing, and/or sealing devices. The manufacture of sheet gasket material may involve cooling and solvent recovery operations that produce wastewaters. Among the plants contacted in this study, only one was found that generated wastewater from a sheet gasket production facility, and this was from the solvent recovery operations.

In summary, the fabrication of asbestos-containing gaskets, packings, and sealing devices does not normally result in process wastewaters, although the manufacture of some of the raw materials may result in process-associated wastes. Sealant manufacturing plants normally operate one or two shifts for five days a week.

An operation which is significant in water pollution is solvent recovery which is involved in both gasketing and friction materials manufacture but which will be discussed here under gasketing.

The quantity of wastewater from solvent recovery operations varies, depending upon the type and the size of the equipment. A typical value is 38,000 liters (10,000 gallons) per day for this industry. The discharge is normally steady and, since it is a function of the activated carbon regeneration process, it cannot be directly related to the level of production in the plant.

The wastewaters from solvent recovery units may contain residual solvent and/or other organic materials that are either evaporated from the product or generated during the recovery operations. The suspended solids level is normally very low, and the wastewater may have an elevated tempeature. Typical wastewater characteristics from one solvent recovery operation are as follows: 1,125 mg/l BOD (5 day); 1,930 mg/l COD; and 0 mg/l suspended solids.

The wastewaters from this plant are discharged with the sanitary wastes to the municipal sewerage system. The wastewaters from the other known plant that recovers solvent are combined with larger volumes of industrial wastewaters for treatment prior to discharge to a surface water. The BOD of the combined, treated effluent from the plant is less than 20 milligrams per liter. There are plans at this plant to completely recycle all process-related wastewaters.

At least one plant in this industry recovers solvent without generating wastewater. It is not known if this technique is applicable at other plants using different solvents. The solvent recovery wastewaters may contain significant organic loads and may have an elevated temperature.

If the organic material is not refractory, biotreatment by the activated sludge process after cooling, as necessary, would be suitable for meeting the Level I limitations. For the scale of operations encountered in this industry, i.e., approximately 40 cubic meters (10,000 gallons) per day, the extended aeration variation would be appropriate. Excess sludge could be removed by a commercial hauler for disposal at a municipal treatment plant.

In order to meet the Level II limitations, or if the waste is refractory to biotreatment, adsorption on activated carbon is recommended. If properly designed and operated, this process should reduce the concentrations of organic materials to acceptable levels. Because of the relatively small volume to be handled, carbon regeneration by the supplier would probably be more economical than on-site thermal regeneration.

In preparing to apply either of the treatment technologies described above, their suitabilities for a particular waste stream must be evaluated. There are standardized testing procedures to measure both the biodegradability and sorptive characteristics of wastewaters. In the event that neither of these technologies is feasible, more sophisticated processes, such as reverse osmosis, are available to achieve the desired results.

A variety of alternatives have been explored by EPA (12) for wastewater control from solvent recovery in asbestos packing and gasket manufacture as follows:

Alternative A — No Waste Treatment or Control

Daily effluent waste load is estimated to be 75 kg (165 lb) of COD and 45 kg (100 lb) of BOD for the typical plant at this minimal control level. The suspended solids waste load is negligible. All known plants in the industry provide only this level of control.

Costs — None

Reduction Benefits — None

Alternative B — Biological Treatment

This alternative involves using the extended aeration variation of the activated sludge process with removal of excess sludge to a municipal sewage treatment plant. The daily effluent waste load is estimated to be about 2 kg (5 lb) of COD and 1.1 kg (2.5 lb) of BOD with this alternative.

Costs — Investment costs are approximately $73,000.

Reduction Benefits — Estimated reduction of effluent COD and BOD of 97%.

Alternative C —Carbon Adsorption

This alternative involves treating the effluent from the biotreatment process in 2-stage granular activated carbon columns. The carbon is regenerated off-site by the supplier. Costs for filtration of the biotreatment process effluent are not included. The daily effluent waste load is estimated to be less than 0.2 kg (0.4 lb) for both COD and BOD.

Costs — The estimated incremental cost for this alternative is $146,000. Total costs are $219,000.

Reduction Benefits — Reduction of COD and BOD of more than 99.8%.

The estimated costs and associated pollution reduction benefits for the above alternative control technologies are given in Table 19.

TABLE 19: WATER EFFLUENT TREATMENT COSTS FOR ASBESTOS MANUFACTURING—SOLVENT RECOVERY

Treatment or Control Technologies:	(Costs in $1000) A	B	C
Investment	-	73	219
Annual Costs:			
Capital Costs	-	5.9	11.7
Depreciation	-	2.9*	10.5**
Operating and Maintenance Costs (excluding energy and power costs)	-	12.5	20.6
Energy and Power Costs	-	11.0	1.0***
Total Annual Cost	-	32.3	43.8

Effluent Quality:

Effluent Constituents	Raw Waste Load	Resulting Effluent Levels		
BOD (5-day) - mg/l	1200	1200	30	5
COD - mg/l	2000	2000	50	5
Suspended Solids - mg/l	30	30	30	5
pH - units	6-9	6-9	6-9	6-9

* Expected lifetime - 25 years
** Expected lifetime - 14 years
*** Not including carbon regeneration.

Source: EPA (12)

SOLID WASTE DISPOSAL

Solid waste control must be considered. The waterborne wastes from the asbestos industry may contain a considerable volume of asbestos particles as a part of the suspended solids pollutant. Best practicable control technology and best available control technology as they are known today, require disposal of the pollutants removed from wastewaters in this industry in the form of solid wastes and liquid concentrates. In some cases these are non-hazardous substances requiring only minimal custodial care. However, some constituents may be hazardous and may require special consideration. In order to ensure long term protection of the environment from these hazardous or harmful constituents, special consideration of disposal sites must be made.

All landfill sites where such hazardous wastes are disposed of should be selected so as to prevent horizontal and vertical migration of these contaminants to ground or surface waters. In cases where geologic conditions may not reasonably ensure this, adequate legal and mechanical precautions (e.g., impervious liners) should be taken to ensure long term protection to the environment from hazardous materials. Where appropriate the location of solid hazardous materials disposal sites should be permanently recorded in the appropriate office of legal jurisdiction.

Consideration should also be given to the manner in which the solid waste is transferred to an industries waste disposal area. Solids collected in clarifiers, save-alls or other sedimentation basins should first be dewatered to sludge consistency. Transportation of this asbestos containing sludge should be in a closed container or truck in the damp state so as to minimize air dispersal due to blowing. Precautions should also be taken to minimize air dispersal when the sludge is deposited at the industries waste disposal areas.

The quantities of solids associated with treatment and control of wastewaters from paper, millboard, roofing, and floor tile manufacturing are extremely small. For example, the reported volume of dewatered waste solids from a paper plant is 1.5 cubic meters (2 cubic yards) per month. Solids are wasted only when elastomeric binders are being used, which is 25 to 35% of the time.

Another example is that provided by one of the larger floor tile plants in the country. The sludge and skimmings from the sedimentation unit amount to about 625 liters (165 gallons) per week. Unlike other asbestos manufacturing wastes, this material is highly organic and is disposed of by a commercial firm that incinerates it. The treatment facility at this plant is not highly efficient, but is believed to capture at least 50% of the waste solids.

Contrary to the above categories, the waste solids associated with asbestos-cement product manufacture are significant in volume. The reported losses at one pipe plant are in the order of 5 to 10% of the weight of the raw materials. The losses of asbestos fibers are kept to a minimum in this industry, to 1% or less, and the fiber content of the waste solids is low. The solids have no salvage or recovery value.

In summary, the solid wastes disposal associated with the application of treatment and control technologies in the asbestos manufacturing industry does not present any serious technical problems. The wastes are amenable to proper landfill disposal. Full application of control measures and treatment technology will not result in major increases at most plants. In many cases, complete recycle will result in lower losses of solids.

The volumes of solid wastes resulting from application of the control technologies to asbestos textile, friction material and gasket manufacture will not be large compared to many industries. The wastes do not present any unusual problems in handling or in disposal. A properly planned, designed, and operated sanitary landfill with capability for receiving industrial solid waste will be adequate. The disposal of dust is already practiced at all known friction materials plants and implementation of the control technologies will not create any unusual problems. Transportation of dust should be in closed vehicles or the dust should be heavily dampened to eliminate air emissions. The containerized waste from textile coating does not pose a health or environmental hazard if properly disposed of at a licensed landfill site. There is no known recovery value in any of the residues from this industry with with the possible exception of use as fuel substitute. No data are available by which to evaluate this possibility.

A process developed by H.B. Johnson (16) is one in which the solid wastes of combustible and noncombustible roofing material are comminuted in an attrition mill, fed to a surge bin, and then used as fuel in a solid waste and fume incinerator which also receives and ecologically incinerates the exhaust fumes from a saturator. A fume incinerator also receives saturator exhaust fumes along with effluent gas from the oxidizer of the factory, and ecologically incinerates the same with a minimum consumption of conventional fuel.

The fume ducts from the saturator exhausts to the incinerators include a cross duct and switching valves for selectively controlling the flow of the saturator exhaust fumes from one or both saturators to one or both incinerators for optimum operational efficiency of the system in accordance with current production of the roofing factory. Granules resulting from the solid scrap incineration are washed and recovered for reuse.

BRICK INDUSTRY

Brick manufacture dates back thousands of years. Bricks were formed by hand or in crude molds and baked in the sun. The art of baking or burning brick to produce a hard, durable product was developed prior to 500 BC. The two basic operations in the manufacture of brick or tile, the forming of the ware, and firing, persist to this day. The basic raw material is, as it was in the earliest times, naturally occurring clay. The properties of clay products depend upon the shape into which they are formed, and to a very large extent upon the nature of the clay from which they are produced.

Clays comprise natural earth materials which form plastic self-adherent masses when wet, and after drying form hard, brittle structures. All clays are the result of decomposition of rock, and consist of very fine, water-insoluble particles which have been carried in suspension in groundwater and deposited in geologic basins according to their specific gravity and degree of fineness. Chemically, the clays are hydrates of alumino-silicates with various impurities such as powdered feldspar, quartz, sand, limestone, carbonaceous materials such as coal, and pyrites.

The manufacture of brick and related products such as clay pipe, pottery, and some types of refractory brick involves the mining, grinding, screening, and blending of the raw materials, and the forming, cutting or shaping, drying or curing, and firing of the final product. Surface clays and shales are mined in open pits; most fine clays are found underground. After mining, the material is crushed to remove stones and stirred before it passes onto screens that are used to segregate the particles by size.

At the start of the forming process, clay is mixed with water, usually in a pug mill. The three principal processes for forming brick are: stiff mud, soft mud, and dry process. In the stiff mud process, sufficient water is added to give the clay plasticity; bricks are then formed by forcing the clay through a die and using cutter wire to separate the bricks. All structural tile and most brick are formed by this process.

The soft mud process is usually used when the clay contains too much water for the stiff mud process. The clay is mixed with water until the moisture content reaches 20 to 30%, and the bricks are formed in molds.

In the dry press process, clay is mixed with a small amount of water and formed in steel molds by applying a pressure of 500 to 1,500 psi. The brick manufacturing process is shown in Figure 21. Before firing, the wet clay units that have been formed are almost completely dried in dryers that are usually heated by waste heat from the kilns.

FIGURE 21: BASIC FLOW DIAGRAM OF BRICK MANUFACTURING PROCESS.

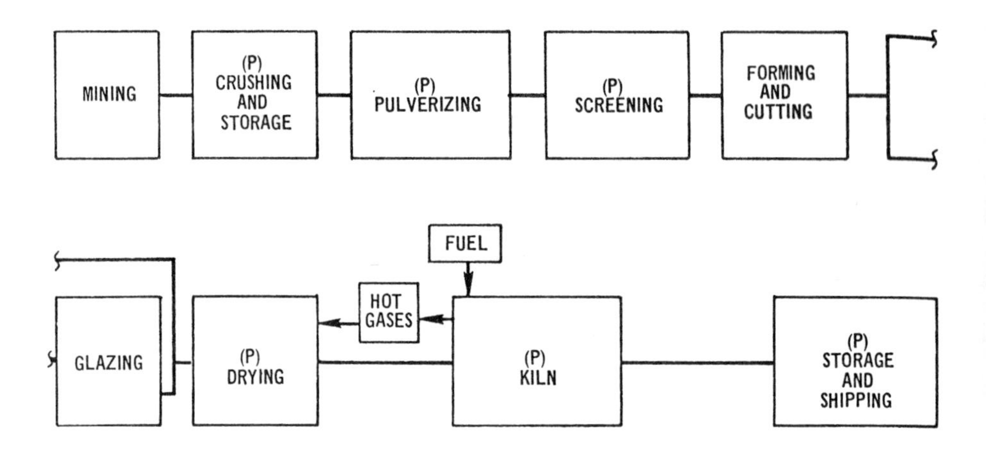

(P) denotes a major source of particulate emissions.

Source: EPA (15)

Many types of kilns are used for firing brick, however, the most common are the tunnel kiln and the periodic kiln. The downdraft periodic kiln is a permanent brick structure that has a number of fireholes where fuel is fired into the furnace. The hot gases from the fuel are drawn up over the bricks, down through them by underground flues, and out of the oven to the chimney.

Although fuel efficiency is not as high as that of a tunnel kiln because of lower heat recovery, the uniform temperature distribution through the kiln leads to a good quality product. In most tunnel kilns, cars carrying about 1,200 bricks each travel on rails through the kiln at the rate of one six foot car per hour. The fire zone is located near the middle of the kiln and remains stationary.

In all kilns, firing takes place in six steps: evaporation of free water, dehydration, oxidation, vitrification, flashing, and cooling. Normally, gas or residual oil is used for heating, but coal may be used. Total heating time varies with the type of product, for example, 9 inch refractory bricks usually require 50 to 100 hours of firing. Maximum temperatures of about 2000°F (1090°C) are used in firing common brick.

There are about 520 manufacturers of brick products in the United States. Most brick plants also produce other clay products like tile. Since clay is the principal raw material used, and clay is rather abundant in all of the United States, no geographic preference is found for location of these plants (18).

Other types of kilns include beehive, scove and shuttle kilns, but the tunnel kiln is the most popular. Typical size of a tunnel kiln is 12 feet wide by 10 feet high and 800 to 900 feet long for the newer kilns and 400 to 500 feet long for older tunnel kilns.

The temperature in the tunnel ranges from 1400° to about 2200°F and varies with the structural properties of the brick desired. The exhaust gases from the tunnel are used as a dryer for the fresh extruded brick before it enters the high temperature tunnel.

Typical plant capacity is about 100 million bricks per year. There is no stack used at the end of a tunnel kiln to exhaust emissions at an elevation. There is no particulate emission associated with the tunnel kilns other than products of combustion. The few beehive kilns still in use may use coal as a heat fuel which produces a distinct plume.

Seventy-five percent of the brick manufacturers use natural gas as their heat source for the tunnel kiln. The remaining manufacturers use coal or oil. The major air pollution emission sources are the grinding, screening and mixing operations which emit dust. Most plants will control emission with baghouses from the grinding, screening and mixing operations.

The building brick is made by mixing one or two clays with water (10 to 25% water and 75 to 90% clays). About one ton of clay will yield one ton of finished brick. Operational time for these plants are usually one or two shifts per day and five days a week.

Practically all modern brick and tile plants use a tunnel kiln to fire their ware as noted above. The configuration of a typical tunnel kiln is shown in Figure 22. The ware is placed on cars and charged to the left end of the kiln and moves continuously to the right. As it moves it is gradually heated, reaching a maximum temperature in the hot zone between the furnaces.

FIGURE 22: PLAN SECTION OF TUNNEL KILN

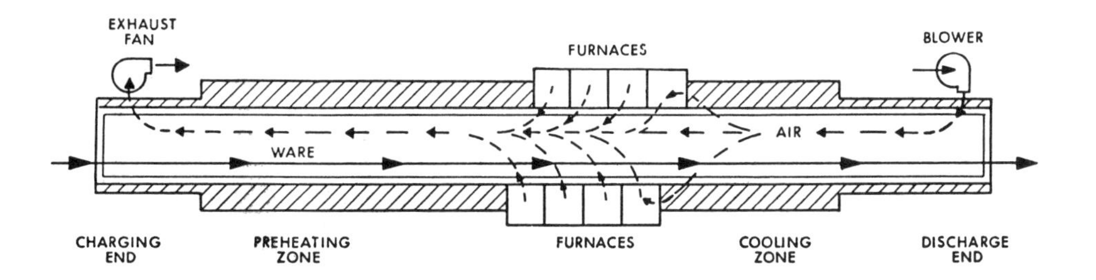

Source: L.C. Hardison and C.A. Greathouse (17)

The charge is then cooled as it passes out of the kiln. Air is passed through the kiln countercurrent to the direction of movement of the ware. Cold air is forced in the right end of the kiln and passes through the charge, cooling it by exchanging heat. Some air is withdrawn from this section for use as the primary air for combustion in the burners. The remaining air continues to the left into the combustion zone, mixes with the combustion gases, and then passes through the incoming charge, losing heat to it.

The temperature of the flue gases ranges from 150° to 300°C, depending on the length of the preheating zone and the amount of air recirculated. Air is drawn out the left end of the kiln with a suction fan. Air locks are used at both ends so that the flow conditions in the kiln will not be disturbed by the entrance or exit of cars. The output of tunnel kilns varies from 100 to 250 tons per day with an airflow of 15,000 to 37,000 acfm. The exact operating parameters of a kiln are determined by the raw material used and the nature of the product desired.

AIR POLLUTION

Particulate matter is the primary emission in the manufacture of bricks. The main source of dust is the materials handling procedure, which includes drying, grinding, screening, and storing the raw material. Combustion products are emitted from the fuel consumed in the curing, drying, and firing portion of the process. Fluorides, largely in gaseous form, are also emitted from brick manufacturing operations. Sulfur dioxide may be emitted from the bricks when temperatures reach 2500°F (1370°C) or greater, however, no data on such emissions are available.

A variety of control systems may be used to reduce both particulate and gaseous emissions. Almost any type of particulate control system will reduce emissions from the material handling process, but good plant design and hooding are also required to keep emissions to a minimum.

The emissions of fluorides can be reduced by operating the kiln at temperatures below 2000°F (1090°C) and by choosing clays with low fluoride content. Satisfactory control can be achieved by scrubbing kiln gases with water; wet cyclonic scrubbers are available that can remove fluorides with an efficiency of 95% or higher. Further details of fluoride emissions from brick manufacture and their control are available (19) from a study by Resources Research, Inc. and TRW Systems Group.

Emission factors for brick manufacturing are presented in Table 20. Insufficient data are available to present particle size information. Data on total particulate emissions for various types of brick manufacture from a study by Midwest Research Institute (20) are given in Table 21.

TABLE 20: EMISSION FACTORS FOR BRICK MANUFACTURING WITHOUT CONTROLS[a]

Type of process	Particulates		Sulfur oxides (SO_x)		Carbon monoxide (CO)		Hydrocarbons (HC)		Nitrogen oxides (NO_x)		Fluorides (HF)	
	lb/ton	kg/MT	lb/ton	kg/MT	lb/ton	kg/MT	lb/ton	kg/MT	lb/ton	kg/MT	lb/ton	kg/MT
Raw material handling[c]												
Dryers, grinders, etc.	96	48	–	–	–	–	–	–	–	–	–	–
Storage	34	17	–	–	–	–	–	–	–	–	–	–
Curing and firing[d]												
Tunnel kilns												
Gas-fired	0.04	0.02	Neg[e]	Neg	0.04	0.02	0.02	0.01	0.15	0.08	1.0	0.5
Oil-fired	0.6	0.3	4.0S[f]	2.0S	Neg	Neg	0.1	0.05	1.1	0.55	1.0	0.5
Coal-fired	1.0A	0.5A[g]	7.2S	3.6S	1.9	0.95	0.6	0.3	0.9	0.45	1.0	0.5
Periodic kilns												
Gas-fired	0.11	0.05	Neg	Neg	0.11	0.05	0.04	0.02	0.42	0.21	1.0	0.5
Oil-fired	0.9	0.45	5.9S	2.95S	Neg	Neg	0.1	0.05	1.7	0.85	1.0	0.5
Coal-fired	1.6A	0.8A	12.0S	6.0S	3.2	1.6	0.9	0.45	1.4	0.70	1.0	0.5

[a]One brick weighs about 6.5 pounds (2.95 kg). Emission factors expressed as units per unit weight of brick produced.
[b]Based on data from the literature.
[c]Based on data from sections on ceramic clays and cement manufacturing in this publication. Because of process variation, some steps may be omitted. Storage losses apply only to that quantity of material stored.
[d]Based on data from the literature and emission factors for fuellcombustion.
[e]Negligible.
[f]S is the percent sulfur in the fuel.
[g]A is the percent ash in the coal.

Source: EPA (15)

TABLE 21: PARTICULATE EMISSIONS

Source	Quantity	Emission Factor	Efficiency of Control C_c	Application of Control C_t	Net Control $C_c \cdot C_t$	Emissions tons/yr
Heavy Clay Products	23,700,000 tons					
A. Grinding	20% of heavy clay	76 lb/ton prod.	--	--	0.60	72,000
B. Drying	30% of heavy clay	70 lb/ton prod.	--	--	0.60	99,000

(continued)

TABLE 21: (continued)

Source	Quantity	Emission Factor	Efficiency of Control Cc	Application of Control Ct	Net Control Cc·Ct	Emissions tons/yr
Refractories	3,440,000 tons					
A. Kiln-Fired						
1. Calcining	20% of kiln-fired	200 lb/ton prod.	--	--	0.64	25,000
2. Drying	30% of kiln-fired	70 lb/ton prod.	--	--	0.64	13,000
3. Grinding	100% of kiln-fired	76 lb/ton prod.	--	--	0.64	47,000
B. Castable Refracts.	550,000 tons	225 lb/ton prod.	--	--	0.77	14,000
C. Dead-Burned Magnesite	125,000 tons	250 lb/ton prod.	--	--	0.56	7,000
D. Mortars	120,000 tons					
1. Grinding		76 lb/ton prod.	--	--	0.60	2,000
2. Drying		70 lb/ton prod.	--	--	0.60	2,000
E. Gunning Mixes	250,000 tons	76 lb/ton prod.	--	--	0.60	4,000

Source: A.E. Vandegrift, L.J. Shannon, E.W. Lawless, P.G. Gorman, E.E. Sallee and M. Reichel (20)

Due to the diverse nature of the raw material and its effect on the emission from the kiln, three types of operation will be discussed:

(1) where the clay contains no sulfur or fluorine-containing material;

(2) where the raw material does contain sulfur and fluorine; and

(3) where the clay contains organic matter such as lignite or sawdust.

In the first case, the contaminants are derived only from the fuel used. Where natural gas is used, there should be no problems. High sulfur fuel oil or coal will produce both SO_2 and fly ash emissions. There is a possibility of CO emissions from passing the hot gases over the incoming bricks, but the concentration should be negligible. In the second case, the fuel will produce contaminants as it does in the first case. Fluorides and additional SO_2 will be emitted from the impurities in the clay. One common fluorine-containing impurity is fluorite or fluorspar, CaF_2, which can react as follows (17):

$$CaF_2 + \tfrac{3}{2}SiO_2 \rightarrow CaSiO_3 + \tfrac{1}{2}SiF_4$$

$$CaF_2 + \tfrac{1}{2}CaSiO_3 \rightarrow \tfrac{3}{2}CaO + \tfrac{1}{2}SiF_4$$

$$CaF_2 + H_2O \rightarrow CaO + 2HF$$

$$CaF_2 + H_2O + SiO_2 \rightarrow CaSiO_3 + 2HF$$

In addition, silicon tetrafluoride can react with water vapor as follows:

$$SiF_4 + 2H_2O \rightarrow SiO_2 + 4HF$$

The equilibrium constants for these reactions at 1200°C are, respectively: 0.13; 1.6×10^{-6}; 2.0×10^{-4}; 0.36; and 16.4. It can be seen from the above that essentially all SiF_4 formed in the presence of the water vapor from the combined and free water in the clay should be hydrolyzed to HF. Therefore, the fluorine is emitted in the form of HF rather than SiF_4.

With a fuel containing 15% ash and 2% sulfur, the flue gas of a kiln using 150 lb of fuel per ton of ware fired and 600% excess air will contain about 0.74 gr/acf fly ash and 125 parts per million SO_2. If the raw material contains 0.1% sulfur and 300 to 500 parts per million fluorine which is 30 to 90% volatilized, the flue gas will contain about 290 parts per million SO_2 and from 25 to 125 parts per million HF. The HF probably hydrolyzes to form hydrofluoric acid mist at the flue gas condition. The third case, involves the generation of air pollutants when organic matter such as sawdust or powdered coal is added to the clay with the objective of burning it out in the kiln and leaving a porous, low density brick. Such bricks have improved insulating qualities as well as being light in weight. In this case, and also when there is a high percentage of naturally occurring organic material such as coal in the clay, there may be a partial volatilization of the organic matter in the kiln followed by condensation and partial oxidation.

One result of this sequence is the production of a black organic smoke consisting of very tiny carbon particles. Unlike the sulfur oxides or hydrofluoric acid, the carbonaceous smoke may be decomposed to some extent in the furnace. However, there is likely to be sufficient emission to cause violation of visible smoke ordinances in circumstances where a substantial amount of organic matter is included in the clay.

For example, if a clay is blended with sawdust to form a 1% organic matter mixture, the total amount of carbonaceous material present in the clay would be sufficient to produce a grain loading of 0.85 gr/acf at the kiln discharge. However, only a fraction of the total carbonaceous matter is likely to be vaporized and survive as black particulate matter. It is apparent that air pollution abatement equipment must be tailored to the specific contaminants generated from impurities in the clay or in the fuel. These may be divided into:

Gaseous Contaminants	Particulates
SO_2	fly ash
HF	smoke

The gaseous contaminants can be removed by either absorption in a solvent or adsorption on a solid material. Of the two, absorption using water as the scrubbing medium is the method accepted in practice. Wet scrubbers are suitable for removal of both gaseous impurities.

Gaseous absorption is carried out in a variety of scrubbing devices, most of which involve countercurrent contacting of the gas and liquid. Where gases are absorbed into liquid streams free of solids, fixed beds of packing material are most frequently used. The presence of solids in either the liquid or gas phases tends to cause plugging problems and requires the use of nonplugging scrubbers. These may be cocurrent Venturi scrubbers, crossflow packed scrubbers, or a variety of proprietary devices utilizing moving packings or self-cleaning impingement surfaces.

Where collection of particulate matter and absorption are required, Venturi scrubbers, mobile packing devices and self-cleaning scrubbers are necessary. This case was chosen for the specification of a hypothetical kiln in which sulfur-bearing coal is burned and both SO_2 and HF are generated by decomposition of the clay impurities.

HF is readily absorbed in water until the pH becomes quite low. However, fluoride-containing effluent water cannot ordinarily be discharged into natural bodies of water, so it is necessary to add some reagent which will precipitate the fluoride as a solid. Typically lime or limestone is used for this purpose and insoluble CaF_2 is produced. This material is most frequently deposited in a pond in which the scrubber effluent is impounded and from which water is recycled.

Where SO_2 is present in the gas, it may be removed by absorption, but the pH requirement is higher than for HF absorption. For this reason, addition of lime to the scrubber system rather than to the pond may be chosen for a system specification. The removal of fly ash can be accomplished by wet scrubbing, electrostatic precipitation or fabric filtration. However, the fly ash problem is relatively limited in scope because of the predominance of gas fired kilns and because of the low ratio of coal to total ventilating air. The fly ash collection has been limited, for purposes of this report, to wet scrubbing with the concurrent removal of HF and SO_2. Special circumstances at a given plant might indicate the use of an electrostatic precipitator or fabric collector for fly ash collection where no gaseous contaminants are involved.

Smoke produced by volatilization of organic material present in the clay or added to modify the properties of the ware presents a somewhat different problem. Here the conventional particulate collection devices such as fabric collectors and precipitators may operate satisfactorily or may be subject to a variety of operating problems because of the nature of the particles. These can vary from droplets of liquid oil to dry, solid carbon particles.

Where there is a possibility of caking or of wetting the collecting elements, both filters and precipitators present special design problems. In particular, fabric collectors are prone to blinding of the cloth, which restricts the gas volume sharply. This would interfere with or prevent the normal operation of the kiln. Precipitators have difficulty in handling solids with a caking tendency, and are also subject to fire hazards when operating with combustible particulate in oxygen-rich gas streams.

Scrubbers have difficulty collecting particulate smokes which are formed by volatilization and carbonization of organic materials. This is due to the small particle size rather than to the hydrophobic nature of the particulate matter, and high pressure drop across a Venturi scrubber contributes toward improved operation. The application of a high energy scrubber for smoke abatement usually requires careful measurement with a pilot unit to determine the pressure drop and horsepower requirement.

Incineration is an acceptable method of abatement for smokes generated by volatilization of organic material in ovens. There are two limiting cases which have different requirements, however. Where the volatile material is vaporized at relatively low temperature and passes through the oven without oxidation, the result is usually a white or blue-white plume similar in appearance to a light steam plume. This material is generally in the vapor phase at temperatures above 500°F and can be oxidized by passing it over a catalyst, or by thermal incineration. Typical operating conditions for catalytic and thermal incinerators on volatile hydrocarbons which tend to produce white smoke are:

	Catalytic	Thermal
Temperature, °F	700	1250
Residence time, sec	0.05	0.5

The second condition involves a partial incineration or oxidation of the organic vapors in the furnace at a high temperature, and frequently in the absence of sufficient oxygen to produce complete combustion. The resultant material is a carbonaceous solid similar to lampblack. The appearance of a plume of this material is gray to black.

This material must be treated as a solid in the incineration equipment. Catalytic incineration is not suitable, in that only materials reaching the surface of the catalyst as vapors are subject to the rate-increasing action of the catalyst. Thermal incineration is suitable but requires a much more severe combination of time and temperature to provide time for complete burning of the carbon particles. Reasonable conditions for incineration of the black smoke are in the range of 1400° to 2000°F and one to two seconds residence time.

The smoke produced by brick kilns is relatively low in concentration and is likely to require no more than one second residence time at 1500°F. Because of the possibility that both types of organic emissions can exist in a kiln firing clay to which organic materials have been added, a thermal incinerator was specified for the hypothetical plants covered by the specifications in this section. Thermal incinerators have a substantial fuel requirement and some form of heat recovery equipment is usually included. In this case, a self-regenerative heat exchanger was prescribed for the incinerator.

The choice between this kind of heat recovery and using the heat to preheat furnace makeup air is purely an economic one and will be specific to each application. Because emissions from brick and tile kilns are limited to those cases where impurities in the clay are present, it is difficult to describe a general case which covers all of the possibilities involved in economic analysis and cost estimation. The alternatives considered are: no air pollution control required; inorganic gaseous pollutants generated by fluorides and sulfur in the raw materials; organic emissions from vegetable matter or oil in the clay; and both inorganic and organic impurities.

To cover these possibilities, two specifications were written. The first specifies the installation of a wet scrubbing system for limiting fluoride and SO_2 emissions. This was based on the presumption of a high level of natural fluoride minerals in the clay and emission requirements of the same order of magnitude as those currently imposed by the state of Florida. In addition, sulfur and fly ash from combustion of high sulfur coal are included. The second specification covers the installation of thermal incineration equipment for the removal of carbonaceous smoke produced in the kiln by incomplete burning of sawdust inclusions in the clay. The first cost for the scrubber installations varies considerably because these systems are not common and there is no stereotype which can be followed. It might be expected that the costs for commercial installations solicited without a preliminary process design might vary over a wide range.

The thermal incineration system quotations were received from two companies who furnish this type of equipment. Of these, only one quoted the complete turnkey installation, while the other supplied only the cost of the incineration equipment. There are few operating systems using either incineration or scrubbing equipment. It is unlikely that any single instance exists where both of the problems described are present in the same operation. If there is such a situation, it would be necessary to install the two systems in tandem, and the costs would approach the sum of the individual system costs. Figures 23 and 24 show capital and annual costs respectively for the application of wet scrubbers to brick and tile kilns from data prepared by the Industrial Gas Cleaning Institute Inc. of Stamford, Conn. (17). Similarly, Figures 25 and 26 show capital and annual costs respectively for the application of thermal incinerators to brick and tile kiln effluent gases (17).

FIGURE 23: CAPITAL COSTS FOR WET SCRUBBERS FOR BRICK AND TILE KILNS

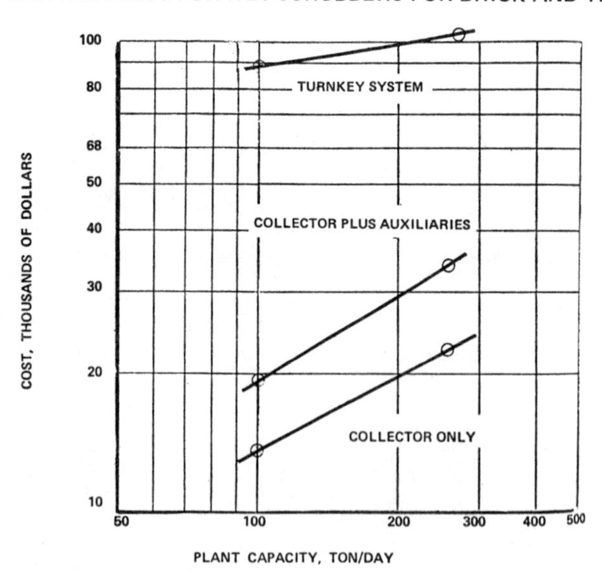

Source: L.C. Hardison and C.A. Greathouse (17)

FIGURE 24: ANNUAL COSTS FOR WET SCRUBBERS FOR BRICK AND TILE KILNS

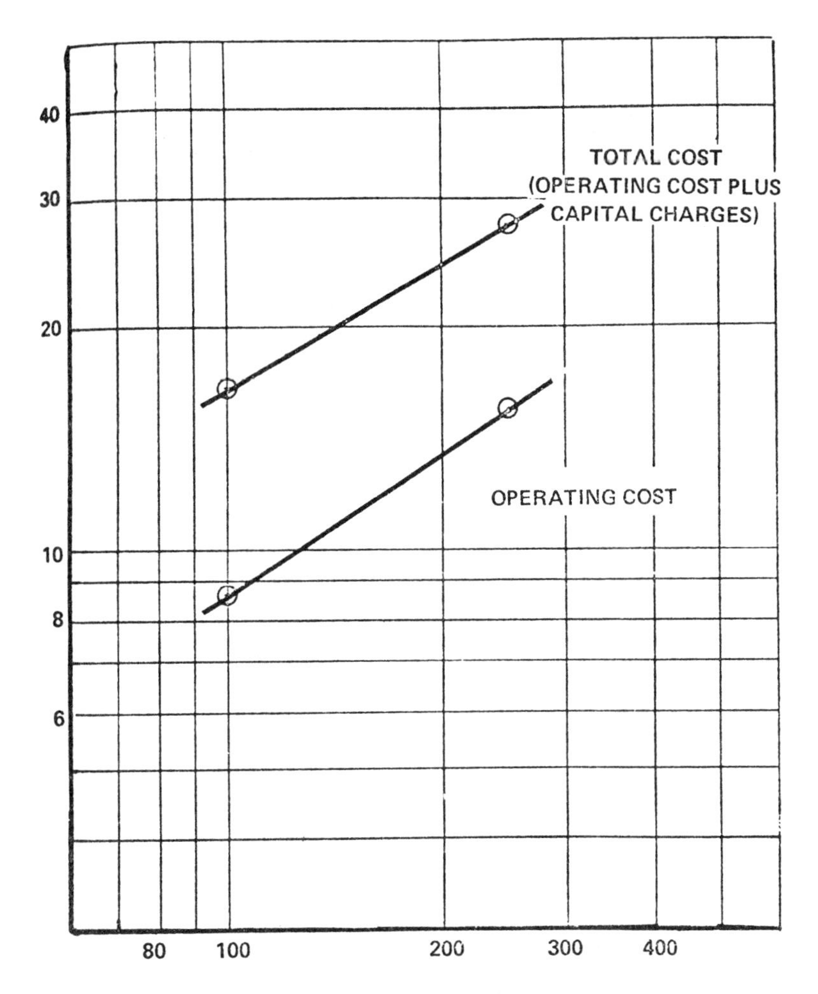

Source: L.C. Hardison and C.A. Greathouse (17)

FIGURE 25: CAPITAL COSTS FOR THERMAL INCINERATORS FOR BRICK AND TILE KILNS

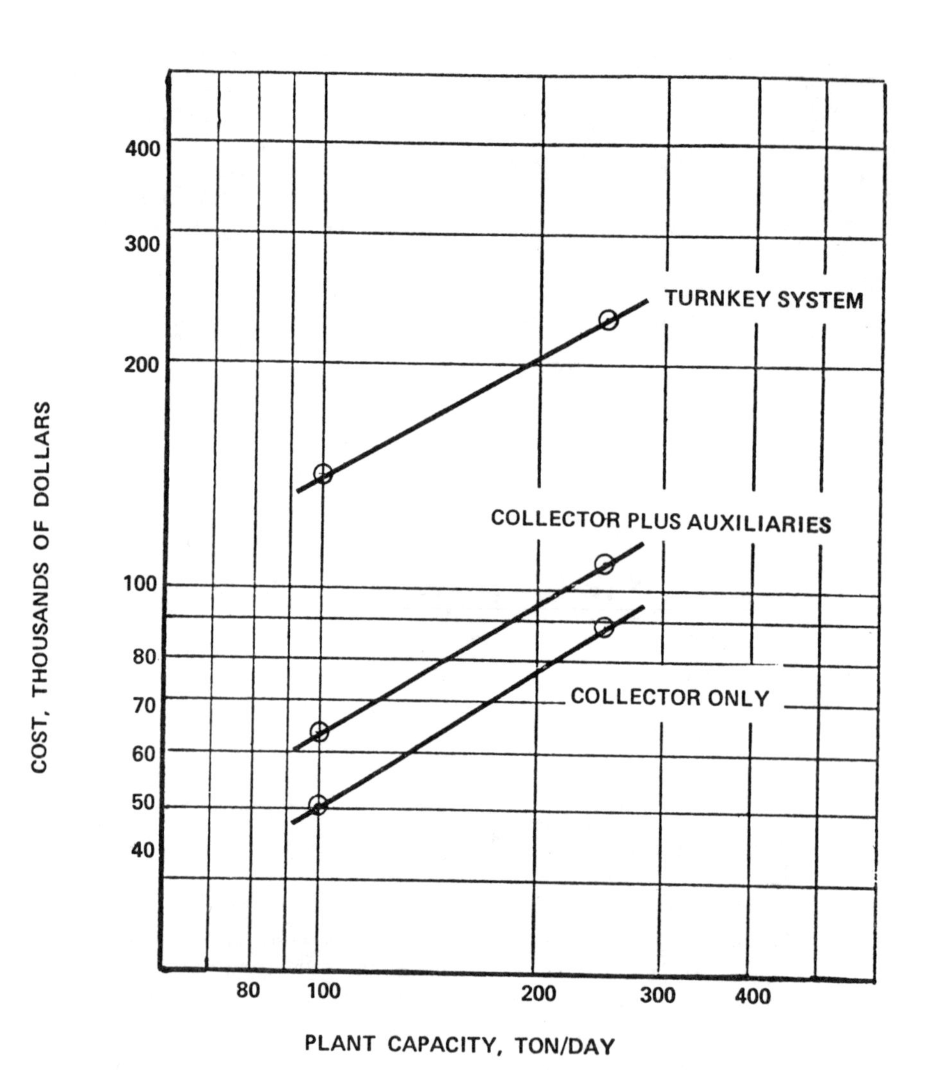

Source: L.C. Hardison and C.A. Greathouse (17)

FIGURE 26: ANNUAL COSTS FOR THERMAL INCINERATORS FOR BRICK AND TILE KILNS

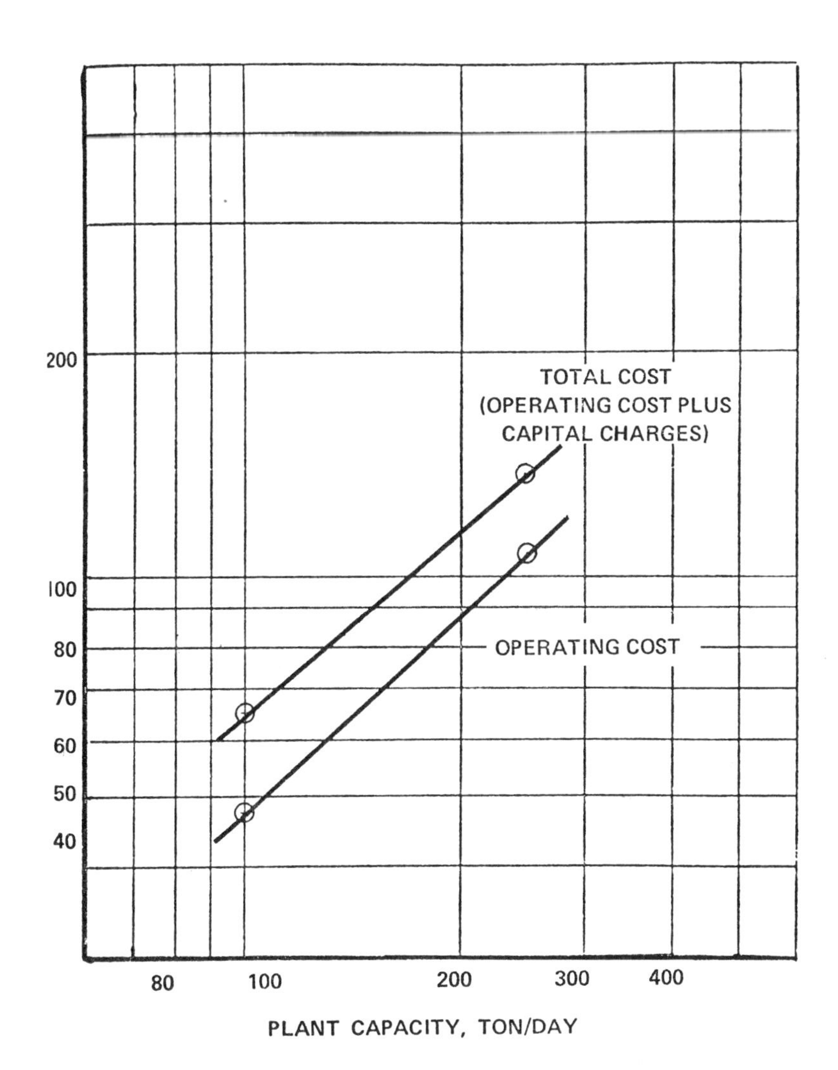

Source: L.C. Hardison and C.A. Greathouse (17)

CEMENT INDUSTRY

Portland cement is a powder which when mixed with water will bind sand and stone into a hardened mass called concrete. Portland cement concrete is a widely used construction product because of its low cost, high compressive strength, and durability. It is used, in varying amounts, in all major forms of construction, i.e., highway, residential, industrial commercial, and public works. Grey Portland cement, the most common form, is a low priced, relatively undifferentiated commodity selling for approximately one cent per pound. Its low value-to-weight ratio has meant that land transportation, storage, and distribution costs are high relative to most other manufactured products.

The cement manufacturing industry is classified by the Department of Commerce as SIC group 3241 (hydraulic cement). The products produced by the industry are various types of Portland cement, manufactured to meet different requirements.

There are 51 companies with 166 plants currently in operation in the United States and Puerto Rico. These plants are widely distributed, being located in all but nine states, in areas close to sources of raw materials, transportation routes, and local markets.

The number of plants in operation has declined from a high of 188 in 1967 to the estimated 166 plants at the end of 1972. In addition to these, about five plants are presently shut down for modernization, and five new plants are under construction. Expansion programs are also underway or planned at about 20 existing plants. The annual capacity of these plants ranges from 0.18 to 2.7 million metric tons (0.2 to 3.0 million short tons).

In 1971 the production of clinker by domestic plants was about 68 million kkg (75 million tons). According to U.S. Department of Commerce, the value of cement shipments will grow from $1.6 billion in 1971 to $2.2 billion by 1975 and $3.1 billion by 1980.

Excess capacity has existed in the industry since a major expansion in the early sixties. In 1971, the capacity utilization was about 88%, and is estimated at 90% for 1972, the highest in over ten years. Expansion programs currently underway should increase capacity about 2% in 1973.

Most cement, about 60% of the total, is shipped to ready-mix concrete firms who supply construction projects in the above categories in a regional metropolitan market. Lesser amounts of cement are sold to manufacturers of concrete building products such as block, pipe, precast panels, etc.; to highway contractors; and to building material dealers. Most shipments are now in bulk by truck, which has largely replaced rail shipment. Cement is also shipped in bags (of 94 pounds), and water transportation by barges has been increasingly used.

Overall demand growth has been low; about 2.8% per annum in volume for 1953-1969. A projected higher growth in construction activity will accelerate the growth somewhat to 3.4 to 4.1% per annum in the 1971-1980 period. Shifts in end use and customer type will be very moderate, although the trend to increased shipments to ready-mix dealers will continue.

Cement manufacture is highly capital intensive, with a typical sales-to-fixed-assets ratio for a new plant of about 1:2. Construction of a major new cement plant today could easily cost $50 million to $100 million. Investment time horizons are long, with two to three years of design and construction time and possibly five to eight years for a new plant to show an operating profit. The practical operating life of a cement plant may be thirty to fifty years, with obsolescence due to size being the chief factor leading to most plant closings. The need for costly pollution control devices has caused a number of older, small plants to close in the past few years, and will cause additional closings in 1975.

The growing national concern about the effects of air and water pollution has led cement companies to increasing levels of spending to hold down emission of pollutants. The main pollutants from cement plants are dust, which consists of very fine particles of cement-like material carried aloft from plant stacks, and waterborne alkalies which are leached from the dust. Pollution control devices generally consist of fabric bag dust collectors (known generally as "bag houses") or electrostatic precipitators. These systems trap the cement dust particles, which arise primarily from the kiln and from the clinker cooler associated with the kiln.

In the ten years through 1971, the industry estimates that it has spent approximately $216 million on capital equipment, or about 44 cents per barrel of installed capacity to improve air and water quality (21).

While these expenditures have done much to improve the overall level of emissions industry-wide, and some plants are now virtually emission-free, more spending is required to clean up existing plants. The Environmental Protection Agency has estimated that the U.S. cement industry will have to make capital expenditures of an additional $96.6 million for air pollution control, and $25.3 million for water pollution control, to further upgrade facilities built prior to 1967. These amounts are over and above the sums which the EPA projects the industry would have spent anyway, to meet standards existing before those promulgated by the EPA in November of 1971. Moreover, these amounts do not include costs of operation.

Plants built subsequent to 1967 were assumed by the EPA to meet all existing and newly-issued pollution control standards. New plants built after 1971 must meet federal emission standards, which are generally more stringent than previous state standards. The EPA has estimated that the increased cost of meeting these more stringent standards will be about 30 cents per barrel of new capacity installed, again over and above the amounts which would have been spent for pollution controls in any event.

The industry has disputed this argument, generally on the following grounds:

[1] Industry leaders do not recognize the distinction between sunk and still-to-be-expended capital costs, and think in terms of "total costs to depollute."

[2] The industry projects harsher future state-by-state regulation of existing plants than is contemplated by the EPA, and consequently higher investment costs.

[3] They recognize that improved air pollution control increases the disposal problem substantially, especially of particulate

material high in alkalies which lead to water pollution, and will add significantly to total control costs.

As a result, many leading cement company executives estimate total capital costs of approximately 88 cents per barrel to depollute. Individual estimates naturally span a fairly wide range, but the average is close to this figure. Of that amount, half or about 44 cents per barrel will have to be expended in the next few years if all existing plants are required to be brought up to projected standards.

Annual operating costs, including depreciation, of air pollution control facilities are estimated by the EPA at 8 to 10 cents per barrel; industry estimates would raise the upper figure to perhaps 20 cents per barrel annually.

Cement is manufactured by a continuous process, normally interrupted only to reline the kilns. There are 3 major steps in the production process: grinding and blending of raw materials; clinker production; and finish grinding. These steps are illustrated in Figure 27.

The ordinary ingredients for the production of cement include lime (calcium oxide), silica, alumina, and iron. Lime which constitutes the largest single ingredient, is most commonly obtained from limestone, cement rock, oyster shell marl, or chalk, all of which are principally calcium carbonate. Materials such as sand, clay, shale, iron ore and blast furnace slag are added to obtain the proper proportions of the other ingredients. At some plants it is necessary to beneficiate the raw materials before they can be used. For example, if the most economical supply of clay contains too much sand, the mixture can be separated by washing with water.

Two types of processes are used in the manufacture of cement, "wet" and "dry." At wet plants, the raw materials are ground with water and fed to the kiln in a slurry. At dry plants the raw materials are dried before grinding, and are ground and fed to the kiln in a dry state. The moisture content of the raw materials available at a given location frequently determines which process a plant will use. For example, if clay and marl with a high water content are available, the wet process may prove more economical.

After the raw material has been finely ground it is placed in storage containers, silos for dry process and slurry tanks for wet process. The material is analyzed and the composition is adjusted as necessary to obtain the correct formulation for the type of cement being produced.

The ground raw materials are fed to a kiln consisting of a large rotating metal tube, usually 3.7 m (12 ft) or more in diameter and 75 to 150 m (250 to 500 ft) long, lined with refractory brick on the inside. The kiln is inclined slightly so that the contents are transferred forward as the kiln rotates. The raw materials are fed into the elevated end, and the kiln is heated by a flame at the lower end. An array of heavy steel chains near the entrance is used sometimes and serves to transfer heat from the gas stream to the raw materials.

The fuel for the kiln may be coal, gas or oil. Most cement plants are equipped to burn more than one type of fuel, and the fuel used at any particular time is dictated by availability and cost. When available, natural gas is usually the least expensive fuel, but in order to obtain gas at the most favorable price, the manufacturer must agree to curtail its use when supplies are limited, and must, therefore, use coal or oil as a standby fuel.

The amount of fuel used to manufacture cement varies with the efficiency of the kiln, the composition of raw materials, the process used, and many other operational factors. In 1963, on the average, the production of one metric ton of cement required about 246 kg (541 lb) of coal, or 187 m^3 (6,670 ft^3) of natural gas which is equivalent to approximately 1.5 million kg cal (5.8 million Btu). (29) Newer plants would be expected to consume about 20% less fuel. Although the wet-process kiln has a higher heat requirement than the dry-process kiln, the fuel consumption difference, in many cases, is partially

FIGURE 27: FLOW SHEET FOR THE MANUFACTURE OF PORTLAND CEMENT

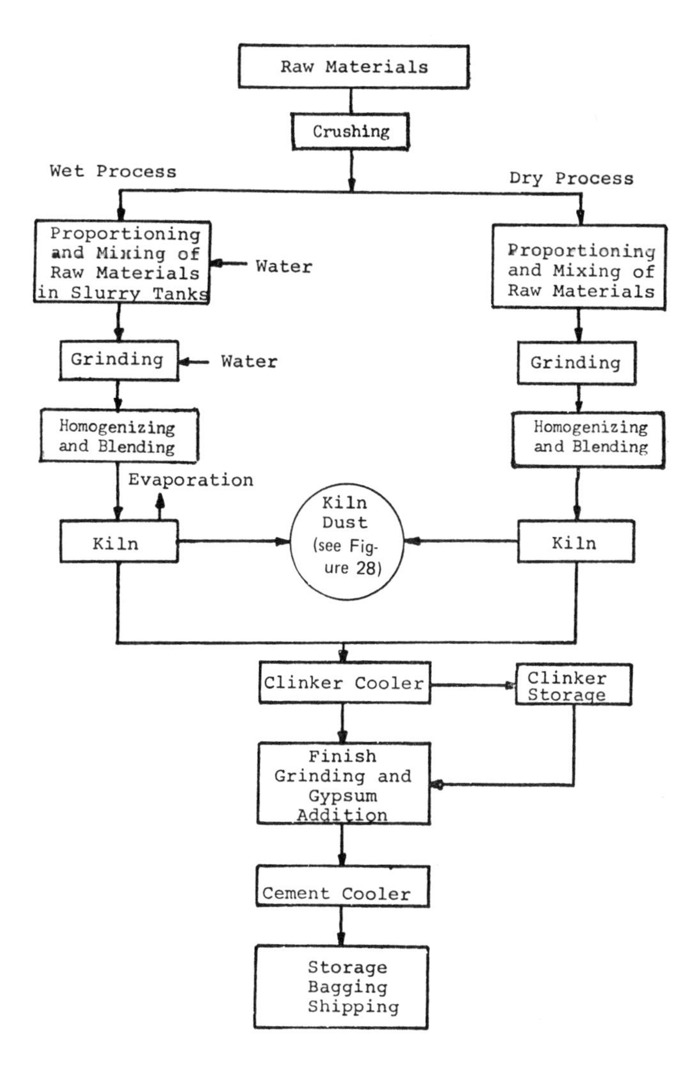

Source: EPA (22)

offset by the heat consumed in those dry-process plants in which dryers precede the raw materials grinding.

As the raw materials proceed down the kiln their temperature increases to about 1600°C (2900°F). At this temperature the raw materials reach a point of incipient fusion and hard, marble-sized balls, called clinker, are formed as the clinker comes from the kiln it is rapidly cooled by air (part of which is subsequently used as combustion air in the kiln).

The clinker along with a small amount of gypsum added to regulate the setting time is ground into a fine powder. The grinding energy is dissipated as heat in the product, and the cement is cooled before being bagged or shipped in bulk to the user. One type of cement cooler consists of a large, vertical cylinder with a rotating screw that pushes the cement through the cooler. The heat is removed by water, which flows through an enclosed jacket around the cooler or cascades in the open down the outside.

The finely ground cement is transported within the plant by pneumatic pumps. The air is supplied by water cooled compressors. After the air has been used to convey the cement it is cleaned with bag filters, and the dust removed is returned to the product stream. In dry-process plants much dust is associated with the grinding and pneumatic pumping of raw materials. This dust can also be collected in bag houses and returned to the process.

AIR POLLUTION

The cement dust which presents the air pollution problem is nontoxic, noncorrosive, nonflammable and nonexplosive. So far as is known it is not hazardous to health, except, as any dust, in extreme concentrations. In fact, because of the alkali content, the dust makes excellent liming products for agricultural purposes. The dust is very fine, however, and is emitted in suspension with a high temperature, high volume air stream. The dust is carried aloft by the heated air and can be scattered over a wide area downwind of the plant. If unsuppressed, it has the effect of producing a greyish coating on plants and buildings.

Little is known about the potential health hazard, if any, of cement dust. There have been two suits brought against cement companies for health damage done to workers, but both were decided in favor of the companies. It is known that most dust particles larger than one micron are removed in the nasopharengeal tract.

A large portion of particles smaller than one micron are deposited in the bronchial and pulmonary regions, where their effect is unknown. These very fine particles are known to have the highest concentration of alkalies, but these alkalies are converted into inert salts by the action of carbon dioxide. As cement plants improve their control of dust emissions, the small amounts released are naturally composed primarily of the very finest particles. A great deal of research must be done before it can be said with any certainty whether these very diffuse concentrations of submicron particles create any form of health hazard, or to what extent they may pose a hazard (23).

Temperature, velocity and flow quantity figures for gaseous emissions from cement manufacture have been presented in a report by Engineering Science, Inc. of Washington, D.C. (18).

Atmospheric emissions from the manufacture of Portland cement have been reviewed by Kreichelt et al (24).

A bibliography with abstracts on the air pollution aspects of cement manufacturing has been published by EPA (25).

The greatest source of dust at most cement plants is from the kiln. The rotation of the kiln plus the rapid flow of gases (from the evolution of carbon dioxide from the raw materials) and the motion of the chains cause a large amount of the finely ground material to become airborne. The high-velocity gases flowing through the kiln carry large quantities of this dust (typically 10 to 20% of the kiln feed) out of the feed end of the kiln. The large dust particles can be removed from the gases by mechanical collectors (cyclones), but the smaller particles require more expensive dust collectors (electrostatic precipitators, bag filters, or wet scrubbers).

Reuse of collected dust, if compatible with the process, is advantageous from three points of view: conservation of raw materials, reduction of disposal costs, and reduction in accumulation of solid wastes.

There are two ways to return collected dust to the kiln. In some plants the dust is mixed with the raw feed. In other plants the dust is blown in through a pipe in the hot end of the kiln, a technique known as insufflation. A portion of the dust is often wasted to prevent buildup of a large amount of fine particulate matter containing alkali salts that continuously cycles between dust collector and kiln.

The dust that is removed from the kiln gases by the dust collectors is a mixture of particles of raw material, clinker, and materials of intermediate composition. The gases also contain alkalies from raw materials and fuel that are volatilized in the hottest portion of the kiln and condensed into a fume as the gases pass through the kiln. The alkalies in the raw material are insoluble because they are tightly bound in a mineralogical matrix. The high temperature in the kiln alters the matrix sufficiently to free a large portion of the alkalies. The free alkali is volatile at high temperatures, and it is also water soluble.

ASTM and Federal specifications require that the alkali content of certain cement products not exceed 0.6%. The low-alkali specification is only necessary in cases of known or suspected alkali reactions with the aggregate being used, but many building and construction contractors routinely specify low alkali cement regardless of the characteristics of the aggregate.

Therefore, since many manufacturers have difficulty marketing high-alkali cement, they strive to make a low alkali cement as a standard product. For plants that use raw materials with a high alkali content, the dust cannot be returned directly to the kiln, and its reuse and disposal constitute a serious problem in the industry.

As air pollution control regulations have become more stringent, the amount of high-alkali dust that is collected has increased, and as more manufacturers install dust collectors that remove more than 99% of the particulate load from the stack gases, the problem of disposal of high-alkali dust will increase. Measures to minimize water pollution stemming from increased amounts of high-alkali dust are described elsewhere.

Figure 28 shows a schematic of the kiln-dust collection and handling systems currently employed in the industry.

Dust can be adequately arrested in the cement industry by proper selection of dust control equipment. Dust emissions as low as 0.03 to 0.05 grain/scf have been obtained in newly designed well-controlled plants. Table 22 gives ranges of dust emissions for various combinations of control devices. An emission level of 0.1 grain/scf is probably the value needed to preclude nuisance complaints from nearby residents.

The hot kiln gases are the main source of emission and they present a major problem because gas volumes are large; they contain acid gases such as H_2S and SO_2, varying amounts of H_2O, and a temperature range usually above 500° or 600°F. A kiln producing 20 tons per hour of cement clinker will produce about 240,000 lb/hr of exit gases, or about 92,000 acfm.

Multicyclones: Although a number of types of dust collectors are used in the cement industry, only the high-efficiency collectors such as the electrostatic precipitator and fabric filter, sometimes used in series with inertial collectors, effectively collect fine dust. The multicyclones alone are not an acceptable means of reducing dust emission from the kiln to the atmosphere.

Multicyclones, when preceding other control equipment, can be expected to scalp off about 70 weight percent or all of the coarser particles.

Electrostatic Precipitators: In a wet-process plant the performance of an electrostatic precipitator is greatly enhanced by the extra water vapor present in the exhaust gases from the slurry. Dry-process kilns do not have this water in the feed and often it is necessary to add it as an aid to precipitator operation.

FIGURE 28: KILN DUST COLLECTION AND HANDLING

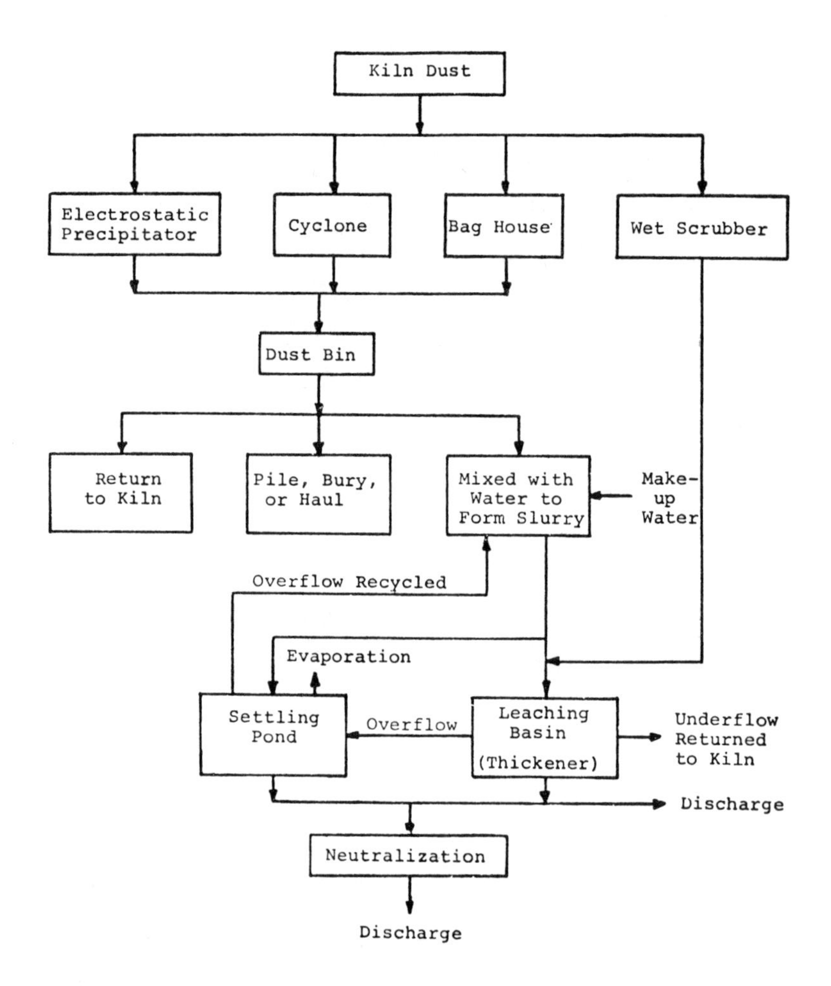

Source: EPA (22)

The operation of electrostatic precipitators has not been entirely satisfactory in the past because of decreasing efficiency over extended periods due to the effects of the cement dust on the high voltage components. Also, when kilns have been shut down and then restarted, it may be necessary to by-pass the electrostatic precipitator for periods up to 24 hr because of the danger of explosion from the presence of combustible gas or coal dust.

The total installation cost of an electrostatic precipitator is shown to be as high as 400% of the purchase cost in the HEW publication AP-51 (27). It is reported that this figure should be somewhat higher based on costs frequently experienced in the cement industry. Recent cost data for electrostatic precipitators used in the cement industry indicate that the installed cost for precipitators having efficiencies of 99.0 to 99.9% ranges from

TABLE 22: RANGES OF DUST EMISSIONS FROM CONTROL SYSTEMS SERVING DRY- AND WET-TYPE CEMENT KILNS

Source	Type of Dust Collector	Range of Dust Emissions from Collector grains/scf*	lb/ton of Cement
Kiln-dry type	Multicyclones	1.55 - 3.06	26.2 - 68.6
	Electrical precipitators	0.04 - 0.15	1.7 - 5.7
	Multicyclone and electrical precipitators	0.03 - 1.3	0.6 - 29.4
	Multicyclone and cloth filter	0.039	0.7
Kiln-wet type	Electrical precipitators	0.03 - 0.73	0.52 - 9.9
	Multicyclone and electrical precipitators	0.04 - 0.06	4.3 - 24.2
	Cloth filter	0.015	0.35

*Grains/scf - Grains/standard cubic foot of gas corrected to 60°F and 1 atm pressure.

Source: A.E. Vandegrift, L.J. Shannon, E.W. Lawless, P.G. Gorman, E.E. Sallee and M. Reichel (20)

$1.00 to $3.50/acfm with an average of $1.80/acfm (20). Additional details of the application of electrostatic precipitators to the cement industry have been presented by S. Oglesby, Jr. and G.B. Nichols of Southern Research Institute (26).

Fabric Filters: Fiber glass baghouse filters have had much success in controlling kiln emissions. Bag life averages 2 years or more. A big plus in baghouse installations is the fact that duct designs are simple and uncomplicated, requiring little study for the flow of gases when compared with the frequently complicated model studies necessary for good gas-flow patterns in the electrostatic-type dust collector.

Moisture condensation in glass-fabric filters can present problems. However, dew point temperatures are normally avoided by proper application of insulation to ducting, etc., and by proper operation to avoid condensation.

The simplicity of design and operation of the fiber glass filter system, which lowers the cost, is balanced to some extent by increased fan power needed to overcome pressure drop across the baghouse. Many baghouses operate with a pressure drop of 3 to 7 inches of water.

The total installation cost of fabric filters is shown to be as much as 400% of the purchase cost in HEW Publication AP-51 (27). This figure is claimed to be in line with cement industry experience (26).

Table 23 summarizes the methods employed to dispose of kiln dust as reported by 80 plants. As shown in the Table, only 27 (34%) of these plants are able to return all of the collected dust to the kiln. These data are derived from RAPP applications, questionnaires, and plant visits.

Presently most manufacturers are wasting the collected kiln dust that cannot be returned to the kiln. The dust is hauled or slurried either to an unused part of the quarry or to vacant land near the plant. The presence of the dust limits the future use of the dumping

TABLE 23: SUMMARY OF METHODS OF DUST UTILIZATION AND DISPOSAL

Method	Number of Plants Reporting	% of 80 Plants Reporting*
All dust returned to kiln	27	34
Surface piling (dry)	29	36
Returned to quarry (dry)	11	14
Leached	9	11
Slurried and discarded	7	9
Some sold or hauled away by contractor	8	10

*Percentage total is greater than 100 because some plants report more than one method.

Source: EPA (22).

site. Moreover, leaching of the dust piles by rainwater overflow from slurrying can cause pollution of streams and ground water.

To avoid wasting high-alkali dust, some manufacturers have installed kiln-dust leaching systems. The dry dust is mixed with water in a pug mill to make a slurry containing about 10% solids. The soluble alkalies, usually at least half of the alkali content, immediately dissolve. The slurry flows into a clarifier where the solid material falls to the bottom. The underflow from the clarifier which contains 40 to 60% solids is returned to the kiln. The overflow, which contains the alkalies is discharged. This discharge constitutes the most severe water pollution problem in the industry.

Another alternative is to use only raw materials of low alkali content. Many cement manufacturers do not have a dust disposal problem because their quarries contain low-alkali raw materials. However, the alkali content of the raw materials is only one of the many factors that must be considered in selecting a plant site, and many of the present cement plants were constructed long before alkali problems became significant.

Cohrs (20) made a survey of 30 plants built since 1960 and found that only ten had anticipated dust disposal problems prior to construction and had made plans to handle it. In some cases plants have hauled in low-alkali raw materials to avoid a dust disposal problem, but most plants would find this solution economically prohibitive.

Since waste kiln dust has a high potassium content and considerable capacity for neutralizing acids, suitable uses for the material have been proposed. Some of the applications that have been considered are fertilization, soil stabilization, and neutralization of acidic wastes from metal pickling operations and mine drainage. Although such uses for waste dust have been pursued for many years, most of the dust now being collected is discarded.

Table 24 summarizes emission factors for cement manufacturing and also includes typical control efficiencies of particulate emissions.

Table 25 indicates the particle size distribution for particulate emissions from kilns and cement plants before control systems are applied.

Sulfur dioxide may be generated from the sulfur compounds in the ores as well as from combustion of fuel. The sulfur content of both ores and fuels will vary from plant to plant and with geographic location. The alkaline nature of the cement, however, provides for direct absorption of SO_2 into the product. The overall control inherent in the process is approximately 75% or greater of the available sulfur in ore and fuel if a baghouse that allows the SO_2 to come in contact with the cement dust is used. Control, of course, will vary according to the alkali and sulfur content of the raw materials and fuel.

TABLE 24: EMISSION FACTORS FOR CEMENT MANUFACTURING WITHOUT CONTROLS[a,b,c]

Pollutant	Dry Process		Wet process	
	Kilns	Dryers, grinders, etc.	Kilns	Dryers, grinders, etc.
Particulate[d]				
lb/ton	245.0	96.0	228.0	32.0
kg/MT	122.0	48.0	114.0	16.0
Sulfur dioxide[e]				
Mineral source[f]				
lb/ton	10.2	–	10.2	–
kg/MT	5.1	–	5.1	–
Gas combustion				
lb/ton	Neg[g]	–	Neg	–
kg/MT	Neg	–	Neg	–
Oil combustion				
lb/ton	4.2S[h]	–	4.2S	–
kg/MT	2.1S	–	2.1S	–
Coal combustion				
lb/ton	6.8S	–	6.8S	–
kg/MT	3.4S	–	3.4S	–
Nitrogen oxides				
lb/ton	2.6	–	2.6	–
kg/MT	1.3	–	1.3	–

[a]One barrel of cement weighs 376 pounds (171 kg).

[b]These emission factors include emissions from fuel combustion, which should not be calculated separately.

[c]PHS and EPA data.

[d]Typical collection efficiencies for kilns, dryers, grinders, etc., are: multicyclones, 80%; electrostatic precipitators, 95%; electrostatic precipitators with multicyclones, 97.5%; and fabric filter units, 99.8%.

[e]The sulfur dioxide factors presented take into account the reactions with the alkaline dusts when no baghouses are used. With baghouses, approximately 50% more SO_2 is removed because of reactions with the alkaline particulate filter cake. Also note that the total SO_2 from the kiln is determined by summing emission contributions from the mineral source and the appropriate fuel.

[f]These emissions are the result of sulfur being present in the raw materials and are thus dependent upon source of the raw materials used. The 10.2 lb/ton (5.1 kg/MT) factors account for part of the available sulfur remaining behind in the product because of its alkaline nature and affinity for SO_2.

[g]Negligible.

[h]S is the percent sulfur in fuel.

TABLE 25: SIZE DISTRIBUTION OF DUST EMITTED FROM KILN OPERATIONS WITHOUT CONTROLS

Particle size, μm	Kiln dust finer than corresponding particle size, %
60	93
50	90
40	84
30	74
20	58
10	38
5	23
1	3

Source: EPA (15).

The major raw materials contain calcium fluoride; during the direct-fired kiln calcination process, the calcium fluoride serves as a source material for the evolution of gaseous HF.

Cement production is of special interest since it involves evolution of fluorides in the presence of limestone. Information concerning the fate of the fluoride in this circumstance may be used to infer conclusions for similar situations, e.g., iron and steel processes or dry limestone process SO_2 control.

The kilns are the major points of fluoride evolution in the cement process. At equilibrium, 100% of the charge fluoride would be evolved as gaseous HF at 2700°F. At the same time as the gaseous HF is evolved, very large quantities of limestone and high free-lime-content particulate material are dispersed into the combustion product stream. The active alkaline surface area thus made available for adsorption and reaction with the evolved HF is enormous, and much of the evolved gaseous HF is removed from the gas stream.

Unfortunately, normal operating data are not available defining fluoride emission factors for cement production. The cement industry has concentrated on the particulate problem. Limited experience with fluorides added to the feed does indicate that [1] 70 to 80% of the evolved fluoride can be collected in an electrostatic precipitator, and [2] gaseous and water-soluble fluorides were only about 10% of the total fluoride emissions. This would appear to verify the effectiveness of limestone, and possibly other particulate matter, in adsorbing or reacting with gaseous fluorides. The result is that alleviation of the particulate emission problem, which is being actively pursued, will also alleviate fluoride emission problems.

Assuming average fluoride content of limestone and shale (650 ppm F^-) with no fluorspar addition to feed, soluble fluoride emissions will grow from 270 tons (as F^-) in 1964 to 800 tons in 2000 if current control levels (no gaseous fluoride control) are maintained. If 99% efficient control systems are utilized, the emissions will decrease to less than 10 tons (as F^-) in 2000. Table 26 summarizes the emission data.

TABLE 26: SOLUBLE FLUORIDE EMITTED FROM THE CEMENT INDUSTRY

	1964	2000
Cement Production (10^6 ton/year)	68	200
Soluble Fluoride Evolution Factor (lb F^-/ton cement)	0.008	0.008
Soluble Fluoride Emission Factor with Current Practice (lb F^-/ton cement)	0.008	0.008
Soluble Fluoride Emission Factor with 99% Control (lb F^-/ton cement)	-	0.00008
Soluble Fluoride Evolution (10^3 ton F^-/year)	0.27	0.80
Soluble Fluoride Emission with Current Practice (10^3 ton F^-/year)	0.27	0.80
Soluble Fluoride Emission with 99% Control (10^3 ton F^-/year)	-	0.008

Source: J.M. Robinson, G.I. Gruber, W.D. Lusk, and M.J. Santy (19)

Equipment used in handling cement includes hoppers, bins, screw conveyors, elevators, and pneumatic conveying equipment. The equipment to be discussed in this section is that involved in the operation of a bulk cement plant, which receives, stores, transships, or bags cement. Its main purpose is usually to transfer cement from one type of carrier to another, such as from railway cars to trucks or ships (14).

In the handling of cement, a dust problem can occur if the proper equipment or hooding is not used. A well-designed system should create little air pollution. Sources of emissions include the storage and receiving bins, elevators, screw conveyors, and the mobile conveyances.

Railway cars are usually unloaded into an underground hopper similar to the one described for trucks in the preceding section. The canvas tube is usually, however, permanently attached to the receiving hopper and is attached by a flange to the discharge spout of the hopper car. When flanges fit properly, emissions from equipment such as this are usually negligible.

Bins filled by bucket elevators must be ventilated at a rate equal to the maximum volumetric filling rate plus 200 fpm indraft at all openings. The area of openings is usually very small. Since most bulk plants have a number of bins, a regular exhaust system with a dust collector provides a more practical solution than the silo filter vents do that were described for concrete batch plants. Bins filled by pneumatic conveyors must, of course, use a dust collector to filter the conveying air. Gravity-fed bins and bins filled by bucket elevators can use individual filter vents if desired.

Bucket elevators used for cement service are always totally enclosed. Ventillation must be provided for the bin into which it discharges. Since elevators are nearly always fed by a screw conveyor that makes a dust-tight fit at the feed end, no additional ventilation is usually required. Another type of conveyor used for cement service is a vertical screw conveyor. These, of course, cause no dust emissions as long as they have no leaks. Horizontal screw conveyors are frequently fed or discharged through canvas tubes or shrouds. These must be checked regularly for tears or leaks.

Hopper trucks and railraoad cars are usually filled from overhead bins and silos. The amount of dust emitted is sufficient to cause a nuisance in almost any location. Figure 29 shows a type of hood and loading spout that permits these emissions to be collected with a minimum amount of air. The ventilation rate is the same as for bins, the displaced air rate plus 200 fpm through all openings.

If the hood is designed to make a close fit with the hatch opening, the open spaces are very small and the required exhaust volume is small. The hood is attached to the telescoping cement discharge spout in such a way that it can be raised and lowered when hopper trucks are changed.

A baghouse has been found to be the most satisfactory dust collector for handling the ventilation points described. All sources are normally ducted to a single baghouse. Cotton sateen cloth with a filtering velocity of 3 fpm is adequate. Dacron cloth, which provides longer wearing qualities but is more expensive, can also be used.

Measurement of emission levels poses a difficult problem (23). Many states have stack plume opacity standards as one index of pollution. Since stack gas contains a good deal of water vapor, it may have considerable opacity under certain atmospheric conditions, though the plume is virtually devoid of particulates. Public opinion, too, may consider a plant a polluter because of a stack plume even though the plume may be only vapor condensation.

A more serious measurement problem is that of measuring bag collector output. Many baghouses are built such that a large volume of surrounding air is drawn into the baghouse on the exhaust, or clean, side of the bags. This contributes to efficient operation and cools

FIGURE 29: HOOD FOR TRUCK-LOADING STATION

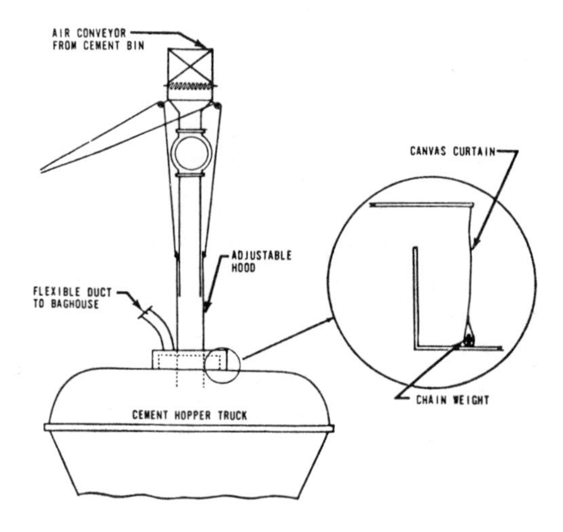

Source: J.A. Danielson (14)

the exhaust gases before they are released to the atmosphere. This mixed volume of air is discharged through openings under the eaves of the structure. This opening may be 100 feet long and a foot or so high, on both sides of the building. There is no established technique for sampling such a stream to provide a reliable index of particulate content.

In addition to establishing firm and reasonable standards, it is important that sound and reliable measurement techniques be specified. In the absence of measurement technology, companies otherwise willing to comply may delay their investment.

In determining the impact which pollution legislation will have on the cement industry, one of the most significant factors will be the amount of capital expenditures required at the outset of an intensified control program, and over what period these funds will be expended. In discussing the magnitude of expenditures required with industry representatives, it quickly became apparent that substantial differences exist between industry and government estimates.

The source of the differences in viewpoint is the definition of what is included in pollution control expenditures. This has several elements. From the point of view of most industry spokesmen, all expenditures for pollution control look the same. This is regardless of whether they have been made in the past, are currently planned or required by state or local codes, or whether they will be specifically required by amendments to the Clean Air Act.

The EPA acknowledges that the aggregate of all pollution control spending is a large figure. However, it has been specifically concerned with the economic impact of incremental spending resulting exclusively from the most recent legislation. In preparing cost estimates, the EPA has deducted from the cost of controls in new plants the estimated amount of pollution control equipment that would have been incorporated in any event, for economic reasons, plant and public health motivations, or existing codes. It has reduced the estimated cost of cleaning up existing plants by the projected amounts which would have been spent on those plants without current legislation.

The EPA has provided an estimate of the "direct, incremental investment required to attain environmental standards." This figure represents the investment "required to attain standards in addition to investment that had already been made prior to and during 1971." The EPA estimate for the entire industry over the next five years (1972-1976) totals $96.6 million in capital costs for existing plants, $63 million in operating costs, and $25 million in interest. This amounts to about 20 cents per barrel capital cost. If all current plants are assumed to be operating in 1976, the operating cost amounts to 5 cents per barrel per year at that time.

The Boston Consulting Group (23) asked a large number of cement manufacturers to estimate their past and projected expenditures for pollution control on a plant-by-plant basis. The replies have been consolidated, and found to have a high degree of uniformity of cost figures among a wide variety of companies. Their figures, based on the total costs of pollution control rather than the incremental costs of current legislation, are naturally substantially higher than the EPA figures.

For purposes of analysis, the results were divided between "new" and "old" facilities (see Tables 27 through 29). Plants constructed or substantially remodeled during or after 1960 are considered new. All others are considered old. Plants are also divided into "large" and "small" categories, with plants in excess of 4.0 million barrels being considered large. The plants were also broken down between single and multiplant companies. The plants owned by the participants are classified in Table 27.

TABLE 27: CLASSIFICATION BY PLANT TYPE (millions of barrels capacity)

	Small	Large	Total
Old	114	36	150
New	102	72	174
Total	216	108	324

TABLE 28: BARRELS BY PLANT TYPE (percent)

	Small	Large	Total
Old	35	11	46
New	31	23	54
Total	66	34	100

TABLE 29: PLANTS – SINGLE AND MULTIPLANT COMPANIES

	Single Plant	Multi-Plant	Total
Million Barrels	41	283	324
Percent	13	87	100

Note: Data on 115 different plants of the 180 in the U.S. were suitable for inclusion (23) and were used to prepare the data in Tables 30 and 31.

Source: Boston Consulting Group (23)

The results appear consistent with the exception that new small plants (Table 30) below have a slightly higher combined cost ($0.87) than the old small plants ($0.83). The projected incremental expenditures on new small plants ($0.35) are not as large as those for old small plants ($0.50), however.

TABLE 30: CONTROL EXPENDITURES

Plant Type	$/Barrel Already Expended*	$/Barrel of Projected Expenditure	Total Required Expenditure
New large	$0.48	$0.43	$0.91
New small	0.52	0.35	0.87
Old large	0.39	0.45	0.84
Old small**	0.33	0.50	0.83
All new	0.48	0.36	0.84
All old	0.37	0.52	0.89
All large	0.45	0.44	0.89
All small	0.42	0.43	0.85

*Amounts spent on capital equipment to control pollution 1962-1971.

**Two companies, representing a substantial amount of capacity were significantly higher than the rest of the industry in their estimates of the amount to be spent for emission control on small old plants. Exact capacities must be concealed to avoid disclosure of confidential information. Their results are shown separately to avoid the distortion of the industry figures (Table 31).

TABLE 31: CONTROL EXPENDITURES – OLD SMALL PLANTS

	$/Barrel Already Expended	$/Barrel of Projected Expenditure	Total Required Expenditure*
Company A	$0.75	$1.26	$2.01
Company B	0.20	1.18	1.38
All others	0.33	0.50	0.83

*Either these firms have overestimated the cost of meeting the control requirements or the other 34 firms who provided cost information have substantially underestimated the cost of control. The results of the two firms are excluded from the "old small", "all small" and "all old" categories.

Source: Boston Consulting Group (23)

Costs for the entire industry may be estimated by extrapolating from these results. The total projected expenditure for all plants responding is $140 million spread over 324 million barrels. (This excludes the two divergent companies.) Projecting this figure over the current industry capacity of approximately 500 million barrels gives a total projected future cost of $216 million, which is $119 million above, or 2.2 times, the government estimate.

In other words, the EPA figures seem to imply that about 55% of the industry's anticipated spending would have been required regardless of the amendments to the Clean Air Act.

Another way of estimating total costs is to divide total existing capacity into large and small facilities (Table 32), and use the previous figures to estimate costs. The result, shown on the following page, is substantially the same when computed this way. Table 32 shows the estimated total incremental costs of the entire industry for emission control, extrapolated from Table 30.

TABLE 32: ESTIMATED TOTAL INCREMENTAL COSTS

	No. Plants	Capacity (Million Barrels)	Incremental Cost Per Barrel	Incremental Cost ($MM
Large	33	193	$0.44	$ 85
Small	147	307	0.43	132
Total	180	500		$217

Source: Boston Consulting Group (23).

Additional data on the economic impact of air pollution control costs on the cement industry have been published by T.A. Le Sourd and F.L. Bunyard of Research Triangle Institute (28).

On June 29, 1973, the U.S. Court of Appeals for the District of Columbia remanded to EPA the standards of performance for Portland cement plants promulgated under Section 111 of the Clean Air Act, directing the Agency to reconsider the standards and provide additional explanation of the action taken. EPA has reviewed the standards for these plants pursuant to the remand and has concluded that the standards other than the opacity standard should not be revised. This document (29) is a detailed technical justification and explanation of the actions taken by the Administrator in promulgating the standards for Portland cement plants. Included is the justification for revising the opacity standard. Obviously, final definition of control costs will depend on settlement of this dispute.

Proprietary Control Processes

A process developed by R.E. Burton (30) makes use of natural fibers such as redwood or other bark fibers to separate airborne solids from gases. It is desirable to prewet the fibers to insure the electrostatic attraction between the fibers and colloidal particles present in the gases.

Figure 30 shows such a scheme as applied to a typical cement plant operation. In such system the exhaust air from the kiln carrying the colloidal cement dust is introduced at **60**. Bales of redwood bark fibers from commercial lumber-mill operations simultaneously enter the system at **62**, and are reduced to substantially individualized fibers by the apparatus represented by **64**. Such apparatus can comprise a pair of superposed belts which flex and break up the bales, coupled with a flail to reduce the fibrous mass to substantially individualized fibers. Alternatively, such apparatus can be of the type described in U.S. Patent 3,042,977 up to the point of discharge from the fiber distributing station.

The fibers discharged at **66** are formed into a loosely felted mat **68**, employing vibration to obtain fiber orientation and a uniform density. These mats can be formed in a batch process or preferably in a continuous process as illustrated.

In the latter process, the mats are deposited on an endless belt **69** and passed under a liquid spray **70** to effect a prewetting of the mat prior to delivery to a zone of contact with the entering gases at **72**. Suitable means such as the moving endless screens **74** can be employed to support the mat as it passes through the contact zone **72**.

As indicated previously, the cement-laden gas enters the system at **60**, and is initially cooled from an inlet temperature of approximately 750° to about 220°F. In the illustrated apparatus, cooling is effected by means of a blower or other suitable means **76**, which circulates cooling medium about the gas passing through the conduits **78**. If desired, the heat which is removed at this stage can be utilized for preheating the limestone or other materials used in the making of cement to increase the efficiency of the entire operation.

FIGURE 30: GAS FILTER FOR CEMENT PLANT GASES USING REDWOOD BARK FIBER FILTER MEDIUM

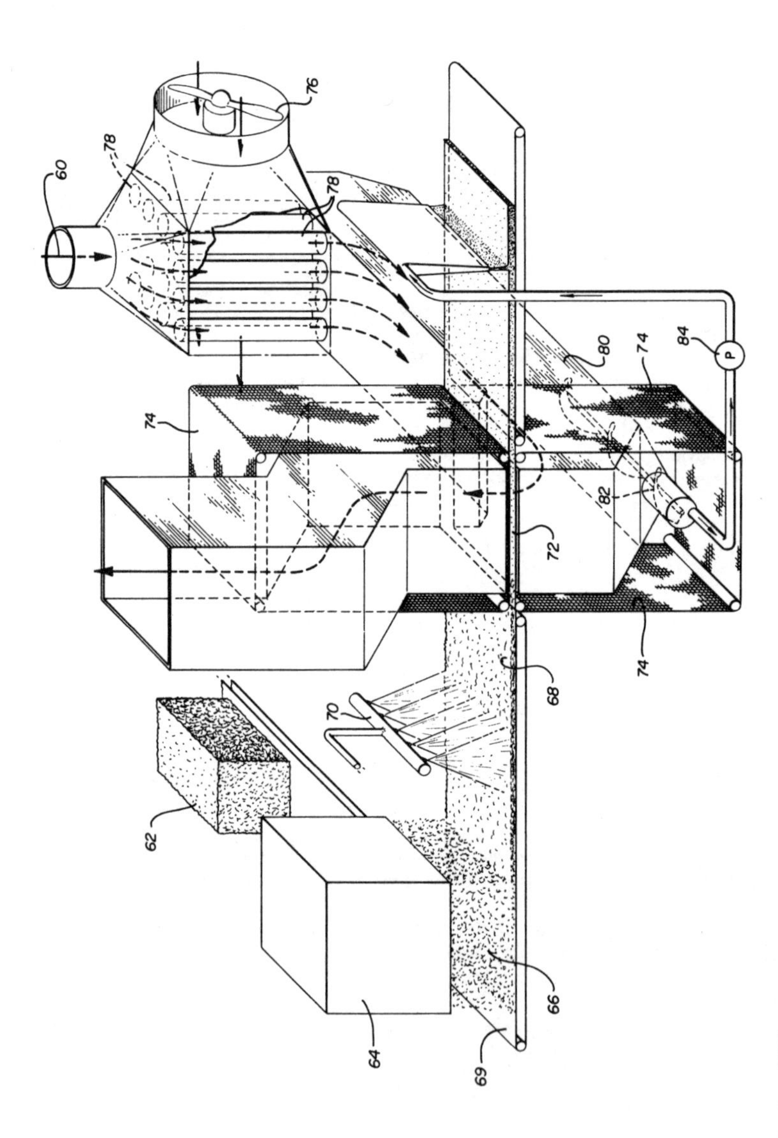

Source: R.E. Burton (30)

Since the exhaust gases from the kiln may constitute as much as 40% water vapor, the indicated reduction in the temperature of the exhaust gases causes the formation of a colloidal suspension of water vapor droplets, which attract the cement dust particles to produce agglomerates or micelles of colloidal sizes. These agglomerated particles then pass with the gas to the contact zone at **72**. Some of the larger particles however may fall out of the air stream, in which case they can be collected as a moist powder or slurry in the bottom **80** of the plenum chamber for discharge by the conveying mechanism **82** and pump **84**.

In the contact zone **72**, the agglomerated particles of cement dust are attracted to the fibers in the mat **68** and generally are removed from the exhaust gases in a single pass. Almost complete removal of cement dust and similar contaminants can be obtained with a layer of redwood bark in the contact zone **72** no more than 1 to 2 inches in thickness. While this unexpected efficiency of dust removal is not entirely understood, it is believed to result from a phenomenon of electrostatic attraction or cataphoresis.

Initially, water vapor molecules are also dispersed in the air, however, as cooling of the stack gases commences, the water vapor molecules condense into tiny droplets and flocculation takes place as Van Der Waals forces become stronger than the repelling force of the electrical charge. Although gravity causes the larger micelles to fall, producing a moist dust in the bottom trough **80**, the bulk of the dust micelles are carried along with the air stream to the contact zone **72**. As the water droplets and smaller micelles of the colloidal dust pass into contact with the bark fiber pad **68** in the zone **72**, the hydroxyl ions of the water are adsorbed on the bark fiber.

This process continues until the spaces around the individual fibers are full of the colloidal dust micelles. In the gaseous system, the benefits of an electrostatic precipitation system (i.e., as in a Cottrell system) are obtained without recourse to external sources of electricity.

It is to be noted that the condensed moisture in the circulating gas serves to replace the moisture initially adhering to the fibers as a result of the prewetting operation induced by the spray **70**. In this way the exposed surfaces of the fibers are continually wetted, and maintained in the desired prewetted condition.

In the cement dust system described, the rate of advance of the bark fibers is such that sufficient fiber surface is presented to the exhaust gases to insure removal of substantially all of the dust from the air. The cement impregnated mat is then advanced continuously from the contact zone **72** for further processing. If desired, the slurry of moist dust discharged by the means **82, 84** can be deposited on the top of the discharged fibrous mat so that additional cement dust is incorporated within the interstices between the fibers.

This processing is particularly desirable where the impregnated mat is to be used in fiberboard manufacture, to produce a heavier board. Following the above described processing, the mat can be subjected to pressure (e.g., about 250 psi) to compress it to a desired thickness for use as cement board. Thereafter the material can be cured (e.g., approximately two weeks at room temperature) and the resultant board trimmed to size for use as conventional cement board (i.e., it can be sawed, will receive and hold nails, etc.). It is also substantially fireproof.

In test runs on airborne colloidal wastes, employing apparatus of the type illustrated, cement kiln gases at about 750°F and containing approximately ½ pound of cement dust per 1,000 cubic feet of gas are cooled to 275°F, and passed through a mat of individualized redwood fibers of the type produced by the process of U.S. Patent 3,042,977 (approximate thickness 1 inch, approximate density 0.06 pound per cubic feet). The average rate of advance of the mat is about 6 to 18 inches per hour.

A white powdery dust slurry collects on the bottom of the plenum chamber on the inlet side of the fiber mat indicating the presence of condensed water vapor and larger agglomerates of cement particles due to cooling of the kiln gases. A high speed blower is employed (circulating 50,000 cubic feet of gases per minute, per 100 ft^2 of filter area) requiring the use of a screen (¾ by 2 inch mesh) on the downstream side of the moving fiber pad. No measurable amount of cement dust is observed in the air discharged from the fiber pad, whereas the fibers of the pad are coated with the colloidal dust, indicating dust removal of the order of 95 to 98%. The temperature of the purified exhaust air is approximately 130°F.

A process developed by G.W. Barr (31) provides an improved baghouse construction designed to operate continuously at high efficiency and capable of filtering large quantities of undesirable contaminants from the gases discharged by industrial processing equipment as for example a continuously operating cement kiln.

A process developed by S. Matsuda (32) provides a wet system electrostatic dust precipitator which is particularly applicable to the treatment of effluent gases from cement kilns.

A process developed by H. Deussner (33) for removing dust from the exhaust gases of a cement manufacturing installation is one in which the dust laden exhaust gases, after leaving a raw-powder-preheater, are conducted into the lower portion of a vertically extending moistening compartment. Water in an excess amount is sprayed into the upwardly moving gas stream and the excess water loaded with dust collects in a sump at the bottom of the compartment. A pump withdraws the muddy water from the sump and discharges it firstly into nozzles tangentially extending into the sump to circulate its content, and secondly into the upper portion of the compartment to wet the inner wall of the same. Another pump conveys a portion of the muddy water from the sump to the raw-powder-preheater at a point before the one where the raw-powder is introduced.

Figure 31 shows a suitable form of apparatus for the conduct of this process. A rotating furnace **1** for the production of cement is traversed in direction of the arrow by the furnace gases. The gases in a rising main flow direction pass through a group of consecutively connected cyclones **2**, which serve for the preheating of the raw-powder-cement.

The raw-powder-cement flows through this aggregate in known manner against the main direction of flow of the gases. A blower **3** withdraws the gases from the cyclones and conveys the gases to the lower portion of a vertically extending cooling- and moistening-compartment **4**, through which the gases flow in rising direction. By means of a fresh water inlet, for example, represented by a pump **5**, water is fed into the compartment **4**, and control members **6** adjust the fresh water supply for example in relation to the water level above the bottom of the compartment **4**, the so-called muddy water-sump **8**.

The fresh water is sprayed finely divided by means of nozzles **7** into the compartment **4**. Insofar as the fresh water does not evaporate in the rising gases, it precipitates in the form of dust-laden drops on the floor or on the inner wall **12** of the compartment **4** and collects in the sump **8** as muddy excess water. By means of a water-pipe **9** and a pump **10**, the excess-muddy water is forced into the overflow-ring-trap **11** from where it uniformly moistens the inner wall **12** of the compartment.

Furthermore spray pipes **13** are likewise connected with the muddy water pump **10** and extend tangentially into the sump **8**. Through the kinetic energy of the muddy water jets, the contents of the sump is caused to perform a circular movement, so that it does not come to any stagnant deposits. With a constant number of revolutions of the pump **10**, the built up fluid pressure is dependent on the weight and on the viscosity of the muddy water.

A manometer, not shown in the drawing, may accordingly furnish an indication, under the mentioned prerequisities, as to whether the muddy water has the desired consistency, or whether more or less fresh water should be supplied or whether more or less excess water is to be sprayed into the raw powder-preheater **2**. Other members **6a** of a type known per se, may, however, also be employed for determination of the specific weight and/or viscosity of the excess water, for example in operative connection with an adjusting member for the

FIGURE 31: SCRUBBER FOR REMOVAL OF DUST FROM CEMENT KILN EXHAUST GASES

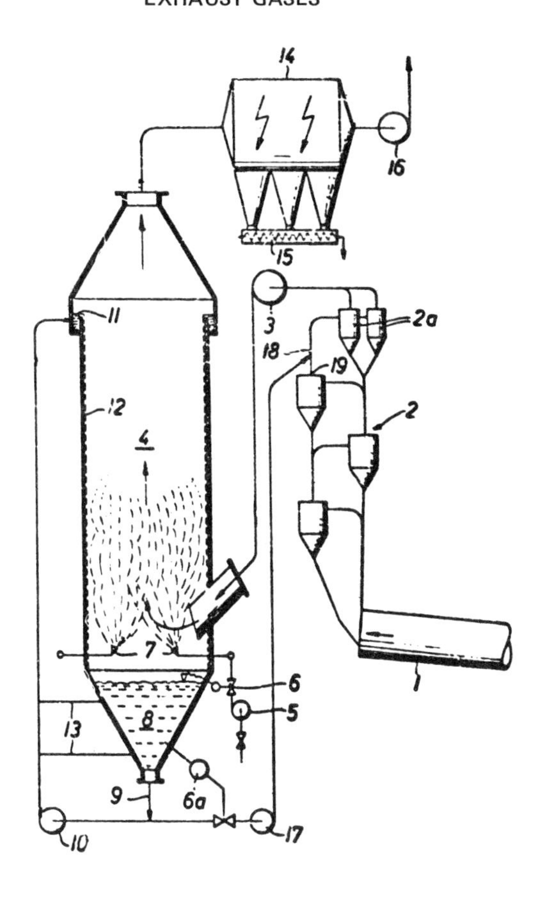

Source: H. Deussner (33)

feed quantity of the pump **17**. The greatly cooled and moistened gases in the compartment **4**, are conveyed to the electrofilter **14** and after the electro-dust-removal are conveyed through the blower **16** into the atmosphere. The dust dropping down in the dust discharge **15** may in known manner again be used for cement production.

The muddy excess water loaded with dust particles is withdrawn from the sump **8** by means of the pump **17** and by means of turbulence nozzles **18** is sprayed into the gas conduit leading to step **2a** of the cyclone group **2** mainly carried out as double-cyclone, last viewed in direction of the gas flow. At suitable distance in front of the turbulence nozzles **18**, likewise viewed relatively to the direction of gas flow, is disposed the raw-powder delivery point **19** constructed according to known art.

From the gas, which in this area still has a temperature of 500°C, a considerable quantity of heat will be removed or extracted by the introduced cool raw powder. Nevertheless, the spraying in of muddy water taking place shortly thereafter causes a further cooling off of the gases through removal of evaporation heat and heating of the dust precipitated by the evaporation. By means of the spraying of the muddy water in the gases loaded with raw powder, a substantial part of the muddy water agglomerizes with the raw powder particles, so that only relatively little fine-grain results, which in the last cyclone step **2a** is not separated and accordingly once more passes through the blower **3** and into the compartment **4**.

With the method and devices according to the process, a highest possible cooling of the exhaust gases is attained at a highest possible dewpoint, without that in this way loss of heat is incurred. Owing to the quality of the exhaust gases attained, the filter may be dimensioned smaller than heretofore. Incrustations in the moistening compartment, in the cyclone group as well as depositions of mud are effectively prevented, furthermore no waste mud is produced whose removal is always problematic and only possible with appreciable expenditures.

Due to the fact that muddy water is sprayed into the raw-powder-preheater, not only the solids contained therein are again supplied to the manufacturing method, but also they themselves effect a cooling of the gases. This is advantageous by supplying as small amount of water as possible to the raw-powder-preheater itself.

A process developed by G. Deynat (34) extracts alkali from exhaust gases of a cement kiln and includes a duct including a curtain of endless chain elements with chain cleaning arrangements for the chains.

A process developed by L. Kraszewski et al (35) involves a cement clinker burning apparatus having a suspension preheater which delivers preheated raw meal to a rotary kiln. A high percentage of the alkali content of the raw meal is removed by the following steps:

[a] a portion of the hot exhaust kiln gases is withdrawn from a point adjacent the gas exhaust of the rotary kiln;

[b] the portion is passed through a wet scrubber to remove alkali and solid material therefrom and produce a thin slurry;

[c] the gaseous portion obtained from step (b) is passed to the electrostatic precipitator; and

[d] the slurry then is discharged from the scrubber to waste.

A process developed by J.G. Hoad (36) utilizes a combined wet scrubber and clarifier or settling basin for removing dust and other particulate matter from exhaust gases associated with cement production.

WATER POLLUTION

As described above, there are two basic processes for the manufacture of Portland cement: the wet process in which the raw materials are slurried with water before being fed to the kiln and the dry process in which the raw materials are ground and fed to the kiln without use of water. A review of the characteristics of the wastewater and inspections of both types of processes, indicate that the type of process need not have a direct effect on the quality of the wastewater.

Table 33 shows the average loading of several selected parameters for wet- and dry process plants and the percentage of plants of each type that report less than 0.005 kg per metric ton (0.01 lb/ton) of cement produced. The average loadings for wet-process plants are slightly greater, due to the high loadings of the leaching plants, almost all of which are wet, but the average is still relatively low. Moreover, a significant number of plants in both groups report very low loadings.

The two different processes offer basically different options for water management and reuse. However, acceptable options are available for both types of processes. Any difference that may exist in the cost of implementing these options is likely to vary as much among plants of the same type of process as among plants of different types of process. Therefore, wet- and dry-process plants may be included in either subcategory.

TABLE 33: COMPARISON OF LOADINGS OF SELECTED PARAMETERS FOR WET- AND DRY-PROCESS PLANTS

	Wet-Process Plants[a]		Dry-Process Plants[b]	
Parameter	Average, kg/kkg (lb/ton) of product	Percent of Total Reporting Less Than 0.005 kg/kkg product	Average, kg/kkg (lb/ton) of product	Percent of Total Reporting Less Than 0.005 kg/kkg product
Alkalinity	0.394 (0.79)	50	0.096 (0.19)	75
Total Dissolved Solids	1.723 (3.45)	36	0.611 (1.22)	32
Total Suspended Solids	0 (0)	38	0 (0)	74
Sulfate	0.535 (1.07)	50	0 (0)	67
Potassium	1.075 (2.15)	46	0.040 (0.08)	50

a. Includes 9 leaching plants.
b. Includes 1 leaching plant.

Source: EPA (22)

Portland cement plants range in age from 2 years to more than 75 years since initial plant start-up. Analysis of the reported start-up dates for plants representing 75% of the establishments in the industry indicates that 16% of the plants are less than 10 years old while 37% of the plants are less than 20 years old, and about 32% of the plants are more than 50 years old.

Analysis of the quantity of water used and the wastewater constituents with respect to plant age shows no correlation between plant age and either the volume of water used or the wastewater characteristics. There are probably two basic reasons for this lack of correlation: first, the basic process for the manufacture of Portland cement has changed little in the last 50 years; and second, cement plants in general are constantly undergoing updating and modification.

Thus, a plant that was constructed in 1906 may be operating with kilns and other equipment that are identical to those in a recently constructed plant. Therefore, plants of different ages may be included in either subcategory.

Analysis of the available data and inspection of plants of various sizes indicate that there is no correlation between plant size and the quality of wastewaters as shown in Table 34. The lowest and highest average values for alkalinity and total solids are within one standard deviation.

Also shown in the table are the gross water discharged and the water discharged per ton of product, which vary widely among the large and small plants with no obvious relationship to plant size. While a smaller plant may, through water conservation and good management practices, consume and discharge far less water, this is not necessarily the case.

Differences in the amount of water discharged and possibly requiring treatment may be reflected in higher costs of control and treatment technology; however, since such differences are not directly relatable to plant size, plants of all sizes may be included in either subcategory.

TABLE 34: COMPARISON OF AVERAGE LOADINGS AND WATER DISCHARGED FOR PLANTS OF DIFFERENT CAPACITY

	Akalinity			Total Solids		
Rated Annual Capacity, 1000 kkg (Thousand tons)	Number of Plants Reporting	Average kg/kkg (lb/ton of product	Standard deviation kg/kkg (lb/ton) of product	Number of Plants Reporting	Average kg/kkg (lb/ton) of product	Standard deviation kg/kkg (lb/ton) of product
All plants	75	0.283 (0.57)	0.879 (1.76)	76	1.491 (2.98)	3.363 (6.73)
0–270 (0–300)	10	0.244 (0.49)	0.392 (0.78)	10	1.456 (2.91)	2.086 (4.17)
270–450 (300–500)	26	0.263 (0.53)	0.930 (1.86)	26	1.515 (3.03)	3.425 (6.85)
450–900 (500–1000)	33	0.361 (0.72)	1.045 (2.09)	34	1.569 (3.14)	3.662 (7.32)
over 900 (over 1000)	6	0.013 (0.03)	0.147 (0.29)	6	1.568 (3.14)	3.856 (7.71)

	Water Discharged				
		10^6 1/day (mgpd)		1/kkg (gal/ton) of product	
Rated Annual Capacity, 1000 kkg (thousand tons)	Number of Plants Reporting	Average	Standard Deviation	Average	Standard Deviation
All plants	117	7.9 (2.1)	27 (7.2)	5,103 (1,760)	12,268 (4,220)
0–270 (0–300	18	2.7 (0.7)	7.3 (1.9)	4,075 (1,400)	11,638 (4,000)
270–450 (300–500)	38	3.3 (0.9)	8.8 (2.3)	3,807 (1,310)	9,244 (3,180)
450–900 (500–1000)	53	8.5 (2.2)	18.3 (4.8)	6,076 (2,090)	14,115 (4,850)
over 900 (over 1000)	8	36.4 (9.6)	9 (2.4)	7,116 (2,450)	14,474 (5,070)

Data derived from 88 RAPP applications and 29 questionnaires.

Source: EPA (22)

The results of an analysis by EPA shown in Table 35, indicated that at least 20% of the plants in the industry are currently achieving essentially no discharge of pollutants, that is, they are discharging less than 0.005 kg/kkg (0.01 lb/ton) of product which corresponds to about 1 mg/l, the minimum measurable concentration at the flow rates common in this industry. The reliability of the reported data was verified by sampling and analysis at ten plants. The average measured and reported loadings of seven nonleaching plants and three leaching plants are shown in Table 36.

TABLE 35: DISTRIBUTION OF PLANTS BY REPORTED LOADING FOR 18 PARAMETERS

		Waste Load, kg/kkg				
	Number of Plants Reporting	Less than .005	.005 to .049	.05 to .49	0.5 to 4.9	Greater than 5
Alkalinity	78	44	8	15	11	0
BOD	74	59	14	1	0	0
COD	69	40	17	12	0	0
Total solids	79	28	15	11	13	12
Dissolved solids	77	27	11	19	8	12
Suspended solids	75	35	13	18	8	1
Volatile solids	73	34	13	15	11	0
Ammonia	69	69	0	0	0	0
Kjeldahl nitrogen	67	65	1	1	0	0
Nitrate as N	69	66	3	0	0	0
Phosphorus	71	71	0	0	0	0
Oil and grease	56	51	3	2	0	0
Chloride	67	48	6	9	4	0
Sulfate	68	36	11	10	10	1
Sulfide	50	50	0	0	0	0
Phenols*	56	52	1	3	0	0
Chromium*	62	55	2	4	1	0
Potassium	15	7	1	3	3	1

*Load expressed in g/kkg.

Source: EPA (22)

The raw materials required for the manufacture of Portland cement are chemically similar, including the oyster shell used at a small number of plants located along the Gulf Coast. Analysis of the available data and on-site studies of exemplary plants indicate that with the exception of alkali content, which will be discussed below, only minor differences in the quantity or quality of wastewater may be related to the type of raw materials used.

The raw materials that are available to some plants, especially limestone and clay, may contain higher-than-average amounts of potassium and sodium. These differences will be reflected in the wastewater streams only at those plants where the kiln dust comes in contact with the waste stream. Plants where such contact is purposeful rather than incidental are considered as a separate subcategory.

TABLE 36: COMPARISONS OF REPORTED AND MEASURED WASTE LOADS AT PLANTS VISITED

	Nonleaching Plants (7) Average Waste Loads, kg/kkg (lb/ton) of Product				Leaching Plants (3) Average Waste Loads, kg/kkg (lb/ton) of Product			
Parameter	Reported by Plants	Measured by SRI staff	Mean of Reported and Measured Average	Deviation from Mean kg/kkg of Product	Reported by Plants	Measured by SRI staff	Mean of Reported and Measured Average	Deviation from Mean kg/kkg of Product
Alkalinity	0.001 (0.002)	0.001 (0.002)	0.001 (0.002)	±0.000	1.09 (2.18)	1.21 (2.42)	1.15 (2.30)	±0.006
Dissolved Solids	0.029 (0.058)	0.032 (0.064)	0.030 (0.061)	±0.002	5.65 (11.30)	2.98 (5.96)	4.32 (8.63)	±1.34
Suspended Solids	0.009 (0.018)	0.022 (0.044)	0.015 (0.031)	± 0.006	0.045 (0.09)	0.045 (0.09)	0.045 (0.09)	±0.000
Sulfate	0.001 (0.002)	0.006 (0.012)	0.003 (0.007)	±0.002	- -	1.06 (2.12)	- -	-
Potassium	- -	0.001	- -	-	- -	0.885 (1.77)	- -	-

Data derived from visits to and RAPP applications for 10 plants.

Source: EPA (22)

Based on the significant differences in waste loadings and treatability of the wastewaters, two subcategories are defined:

leaching plants, in which the kiln dust comes into direct contact with water in the leaching of kiln dust for reuse or in the wet scrubbing of dust to control stack emissions.

nonleaching plants, in which contamination of water is not inherently associated with the water usage.

Few plants use only one type of fuel. The type of fuel burned may affect the amount of water-soluble constituents in the kiln dust; and minor differences may be found in the wastewater characteristics of plants using different fuels, if the kiln-dust comes in contact with the water. These differences are considered in the defined subcategories. Leaching of coal piles by rainfall and subsequent runoff may be a problem at some coal-burning plants, however, adequate methods for controlling such runoff are available in other industries that have large coal storage piles. Such methods include spraying the piles with latex films that prevent moisture from entering the piles, and diking the coal pile combined with lime or limestone neutralization to prevent discharge of acidic run-off water.

Water usage for the cement industry is summarized in Table 37 and in the flow diagram in Figure 32. These uses and the characteristics of the associated discharges are discussed below.

FIGURE 32: DIAGRAM OF WATER USAGE IN CEMENT MANUFACTURING

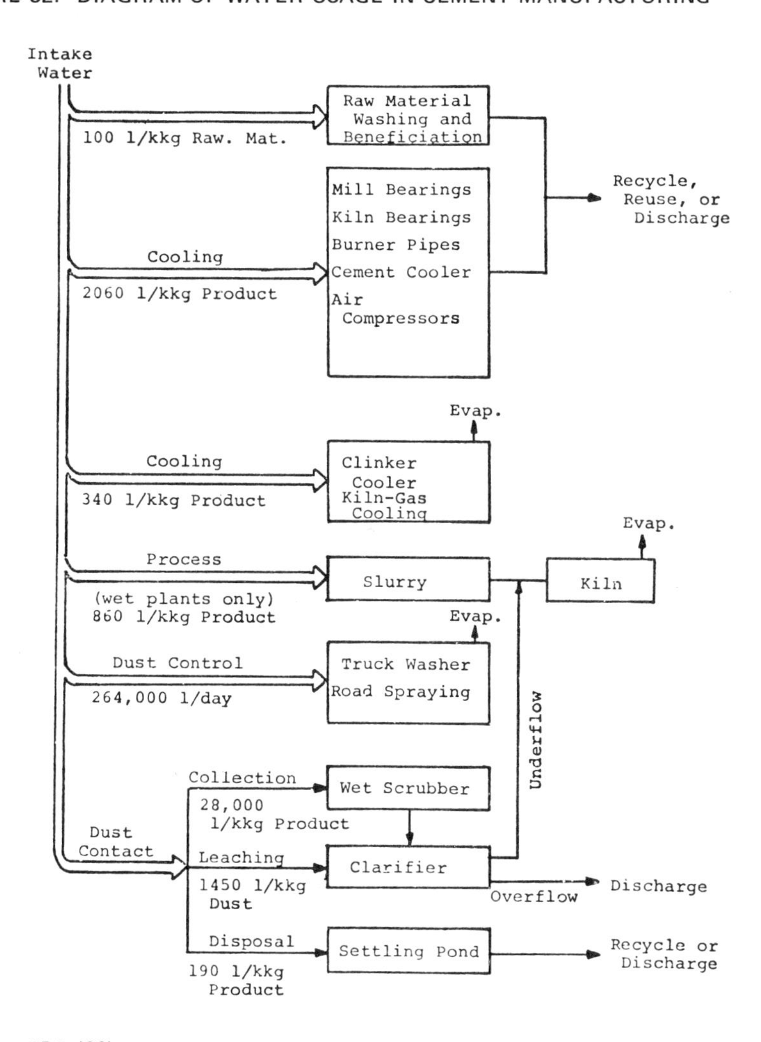

Source: EPA (22)

TABLE 37: SUMMARY OF WATER USAGE FOR THE CEMENT INDUSTRY

Use	Number of Plants	Average	Reported Flow Minimum	Maximum	Units
Cooling	117	1,550	17	72,000	l/kkg of Product
Raw Material Washing and Beneficiation	4	100	2.1	405	l/kkg of Raw Material
Process	78	860	246	1,740	l/kkg of Product
Dust Control	13	264,000	1,890	600,000	l/day
Dust Leaching	11	2410	2150	2650	l/kkg of Dust
Dust Disposal	5	190	7.9	490	l/kkg of Product
Wet Scrubber	3	28,000	4,150	42,500	l/kkg of Product

Data derived from 88 RAPP applications and 29 questionnaires.

Source: EPA (22)

The major use of water at most cement plants is for cooling. This water is used to cool bearings on the kiln and grinding equipment, air compressors, burner pipes and the finished cement prior to storage or shipment. A summary of average volumes of cooling water used for specific purposes is given in Table 38.

While cooling water is mostly noncontact, it can become contaminated to some extent through poor water management practices. This contamination may include oil and grease, suspended solids, and even some dissolved solids. If cooling towers are used, blow down discharges may contain residual algicides.

For purposes of this discussion, process water is defined as the slurry water used at wet plants to feed the raw material to the kiln. This water is subsequently evaporated in the kiln and, therefore, does not constitute a discharge.

The relatively constant volume of water used in the preparation of slurry averages 860 l/kkg (260 gal/ton). At a few plants, excess water containing a high concentration of suspended solids is discharged from the slurry thickeners. This practice constitutes a nonessential discharge and is easily avoided by recycling this water for making the slurry. Other losses of slurry may occur due to poor maintenance of pumps, which become worn and develop leaky seals. The resulting spillage may result in a waste discharge with high solids if not controlled.

There are three operations in which water contacts collected kiln dust. The wastewater generated by plants with these operations constitutes the highest loadings of pollutants within the industry.

The most significant of these operations is the leaching (removal) of soluble alkalies from the collected dust so that the dust may be returned to the kiln as recovered raw material. This operation occurs at about nine plants. In all cases the overflow (leachate) from this operation is discharged, sometimes without treatment. The wastewaters from this operation are essentially identical for all plants, varying to some extent in the concentration of individual constituents because of differences in raw materials at each plant. These constituents include high pH, alkalinity, suspended solids, potassium, and sulfate.

TABLE 38: REPORTED COOLING WATER USAGE IN CEMENT PLANTS

Use	Average Flow, l/kkg (gal/ton) Product	Number of Plants	Range	
			Minimum	Maximum
Bearing cooling	1,080 (284)	39	3.8 (1.0)	5,800 (1,530)
Cement Cooling	760 (200)	22	1.9 (0.5)	3,750 (985)
Clinker cooling	60 (23)	12	2.1 (0.6)	242 (64)
Kiln-gas cooling	322 (85)	4	92 (24)	770 (203)
Burner-pipe cooling	265 (70)	2	258 (68)	272 (72)

Data derived from 39 questionnaires.

Source: EPA (22).

The second most common operation is the wet disposal of dust. In this operation a slurry is also made of the collected kiln dust and fed to a pond, where the solids settle out. The settled solids are not recovered for return to the kiln, and the overflow (leachate) may be discharged. The constituents of this discharge are essentially the same as those from the leaching operation. At least five plants use this wet method to dispose of collected kiln dust and the volume of water used ranges from 70,000 to 760,000 l/day (18,000 to 200,000 gal/day).

The use of wet scrubbers for air pollution control constitutes the third example of water in direct contact with the kiln dust. At least three plants in the industry use wet scrubbers to collect kiln dust from effluent gases.

All cement plants have some accumulation of settled dust on the plant property and this dust may show up in the wastewater in a number of ways. Many plants spray water on the roads to prevent the dust from becoming airborne by truck traffic. Most plants also routinely wash accumulated dust off the trucks. At some plants, certain parts of the plant areas are also washed down to remove accumulated dust. The amount of water used for these purposes varies widely, ranging from 950 to 9,500 l/day (250 to 2,500 gal/day) as reported in a sample of 12 plants. Some of this water undoubtedly evaporates, but depending on the topography of the plants, some of this water may drain into storm sewers or natural waterways.

Water from surface runoff after rain may also be laden with the dust that accumulates on the plant site. Runoff from dust piles, coal piles, and raw material piles may also become contaminated. Plants with boilers, cooling towers, and intake water-treatment facilities, have blowdown and backwash discharged associated with these operations.

At some plants, raw materials are washed and at others the raw materials are enriched by a beneficiation process; these processes may result in wastewater discharges containing suspended solids.

Where an active or abandoned quarry is used as a receiving basin for dust disposal or plant wastewater, the discharge from the quarry may be contaminated with wastes associated with cement manufacturing. However, where a quarry is used exclusively for the production of raw material, discharge of any accumulated water (dewatering) is not considered in this report and is intended to be considered in the Mineral Mining Industry Effluent Guidelines Study. For nonleaching plants the average net loading of suspended solids is less than zero, indicating that more solids are removed from the intake water used in the plant than are added by the process.

However, 4 of the 58 plants of this group report over 1 kg/kkg (2 lb/ton) of product indicating a moderate level of suspended solids is possible if not properly controlled. For leaching plants the average discharge of suspended solids is 0.9 kg/kkg (1.8 lb/ton) of product.

There are relatively few operations in cement manufacturing where the addition of pollutants to the water used is inherently associated with the use of that water. For most of the plants in the industry, pollution results from practices that allow materials to come in contact with the water. Pollutant levels at these plants can be greatly reduced or eliminated by suitable in-plant control measures that prevent wastes from entering the water or by more extensive reuse and recycling of water that may become contaminated. (In these discussions recycle means using again for the same purpose; whereas, reuse means to recover for another use.)

For the plants in the leaching subcategory, wastes are necessarily introduced into the water and recycling is not feasible. Thus, for these plants, treatment is required to reduce the pollutant loading. Only a limited improvement can be expected from the application of available control technology. The main control and treatment methods for the cement industry involve recycle and reuse of wastewater. The devices employed include cooling towers or ponds, settling ponds, containment ponds, and clarifiers. Cooling towers or ponds are used to reduce the temperature of waters used to cool process equipment. Settling ponds are used primarily to reduce the concentration of suspended solids. Containment ponds are used to dispose of waste kiln dust. Clarifiers are mainly used to separate solids in dust-leaching operations.

With the exception of plants in the dust-contact subcategory, both wet-process and dry-process plants can achieve virtually complete reuse of wastewater with existing state-of-the-art technology.

With respect to wastewater management, wet-process cement plants have features that distinguish them from dry-process cement plants. In all wet-process plants, except for those that leach collected dust, the wastewaters from subprocesses (e.g., plant clean-up, truck washing, and cooling) and storm runoff waters, can be used in the raw mills to prepare the slurry fed to the kiln. In the kiln the water is evaporated, any inorganic matter in the water enters the product, and any organic matter in the water is burned.

Thus, for wet-process plants complete reuse of wastewaters is possible, although in some existing plants installation of cooling towers or ponds may be necessary to permit recycling of excess cooling water.

In contrast to the practices possible in wet-process plants, for dry-process plants disposal of wastewaters from subprocesses in the kilns is not possible. Nevertheless, a number of dry-process plants have achieved virtually complete recycle of wastewaters by the employment of cooling towers or ponds. The only discharge from these plants is the small volume of blowdown or bleed water from cooling towers that is required to prevent buildup of dissolved solids in the cooling water, and where contaminated, these small volumes can be evaporated at low cost.

Even without recycling, control measures can be taken to prevent introduction of contaminants into the water effluent from the plant. Cooling water streams can be segregated from other streams, and precautions can be taken to avoid entry of dust into the cooling water circuit.

In-Plant Control Measures

In-plant measures are primarily limited to the control of nondust-contact streams. For plants within the leaching subcategory, control technology consists of segregation of the leaching streams from other plant discharge streams and conservation of water to minimize the volume of water requiring treatment.

Control technology applicable to noncontact streams is discussed below for the major water uses and potential sources of wastewater. Several plants are discussed in detail in Table 39.

In either wet- or dry-process cement plants, water is used to cool process equipment such as bearings, compressors, burner tubes, and cement coolers by noncontact heat exchange. The wastewaters from these cooling operations are hotter than the entering water. The temperature rise in waters used to cool bearings is normally small, and desirably low temperatures can often be achieved by a simple recycle system in which heat is lost to the atmosphere from a small amount of pipe or a package recycle system as is the practice at Plant A.

In waters used to cool compressors, burner tubes, or cement coolers, the temperature rise is larger. However, if the temperature of cooling wastewater is reduced, the waters may be recycled. Temperature reduction has been accomplished in cooling towers (Plants B and E) and in spray ponds, or by simply recycling to a storage pond of sufficient area so that surface evaporation maintains a stable temperature.

The suspended solids concentration in recycled waters used in cement coolers can increase because the cooling stream in many cement coolers is open to a dust-laden atmosphere. If a cooling pond is used to cool the water before recycling, the pond can also serve as a settling pond. However, if cooling towers are used, a small-volume bleed or blowdown stream from the recycle stream is normally provided to maintain suspended and precipitable dissolved solids at a low concentration.

At a few plants, waste cooling waters from bearings or compressors may contain lubricants. Such cooling waters can be segregated to prevent dilution and treated to remove lubricants if necessary. Flotation and skimming usually suffice for removal of lubricants, but emulsion breaking may be needed in extreme cases.

Process water as defined in this report refers only to the slurry water used in wet plants. Since this water is evaporated, no direct discharge is associated with it. However, precautions need to be taken to insure that overflow from slurry tanks, leaks from slurry lines, and tank clean-up is prevented from entering the discharge from the plant or is adequately treated before discharge.

As discussed above, at many wet-process plants the slurry mix itself can represent a convenient control measure for handling at least some wastewater generated in the plant. Unless these wastewaters are highly alkaline, they can be used to prepare the slurry, as is done at plants A, B and C; the water is evaporated in the kiln, and the wastes that would otherwise have to be treated or eliminated by other control measures are consumed in the product.

For plants collecting a high-alkali dust not returnable to the process, surface dumping on the plant site or in an adjacent quarry is most common. Disposed of in this way, the dust could affect the quality of the plant effluent through runoff or quarry dewatering. Therefore, adequate precautions must be taken to enclose the dust disposal area with dikes to contain runoff and to use areas of the quarry not subject to flooding by ground water.

Another technique for disposal of dust is mixing it with water to make a slurry that is pumped into a lagoon. In some cases the overflow from the lagoon is discharged. However, in the past few years, at least three plants that slurry their discarded dust have eliminated the overflow from the lagoons by following the practice of recycling this water for slurry disposal.

Contaminants, primarily in the form of suspended solids, can enter wastewaters in other ways; such as, in plant clean-up and truck washing, and by pick-up of dust by storm runoff waters. The amounts of solids introduced into wastewaters by plant clean-up can be minimized by good maintenance and operating procedures to minimize solid spillage and to return dry dust to the process, and the solids introduced into storm runoff waters can be minimized by paving areas for vehicular traffic, by providing good ground cover (e.g., grass) in other open areas, and by removing accumulations of dust from roofs and buildings for return to the process. Implementation of more stringent air pollution controls is expected to result in a significant reduction in suspended solids in runoff water.

If introduction of solids into wastewaters cannot be prevented, settling ponds can be provided for the wastewaters that are affected by suspended solids build-ups (e.g., the waters from floor-drainage sumps, waters from raw-mill cleaning and slurry-pump leakage in wet-process plants, and storm water runoff).

Treatment Technology

With the exception of settling ponds for the removal of suspended solids, treatment of wastewater in the cement industry is practiced primarily at leaching plants.

Leachate water contains as pollutants: pH, alkalinity, suspended solids, and total dissolved solids (principally potassium and sulfate). The treatment technology currently practiced can adequately control pH, alkalinity, and suspended solids, but not dissolved solids. Neutralization by the addition of mineral acids such as sulfuric acid has the following effects: it lowers the pH to any desired level; it eliminates alkalinity by neutralization of hydroxyl, carbonate, and bicarbonate ions if it is followed by aeration to remove carbon dioxide; and it dissolves acid-soluble particulate matter such as lime that is present as suspended solids in the leachate overflow. However, it adds to the total dissolved solids content because the sulfate ions are heavier than any of the anions that are removed by neutralization.

Carbonation lowers the pH by replacing hydroxyl ions with carbonate ions. Additional carbonation converts carbonate ions to bicarbonate ions. Total alkalinity is not reduced by carbonation, because the carbon dioxide escapes when the bicarbonate solution is acidified or aerated. However, carbonation can be used to reduce the hardness of the leachate. The solubility of calcium reaches a minimum value of 16 ppm (at 16°C) when the pH has been lowered to 9.5 by carbonation. Any subsequent addition of carbon dioxide to lower the pH raises the solubility of calcium because calcium bicarbonate has nearly the same solubility as calcium oxide.

The above discussion suggests that carbonation might be advantageous as a treatment for leachate. Overflow from the primary clarifier could be carbonated with stack gas to lower the pH to 9.0, near the pH required for minimum solubility and an acceptable pH for discharge. This would cause precipitation of calcium carbonate which could be removed in a secondary clarifier or settling pond.

Carbonation may reduce the dissolved solids by converting dissolved calcium oxide to less soluble calcium carbonate which appears as fine suspended solids that must be removed by settling. Suspended solids may be controlled to less than 50 mg/l by proper design and operation of the clarifiers.

The degree of clarification is determined by several factors including the length of time the leachate remains in the clarifier, the turbulence in the clarifier, and the characteristics of the dust. The residence time and the degree of turbulence in the clarifier are fixed design parameters. However, the characteristics of the dust can be controlled to some extent.

One way of controlling the dust characteristics is by selecting what dust is to be leached. Maximum flexibility of selection is achieved when electrostatic precipitators are used to collect the dust from the kiln exhaust gases. In electrostatic precipitators the larger particles are more easily removed from the gas stream, so they are recovered in the first stages of the precipitator. The smallest particles are collected in the last stage.

Precipitators are designed so that these fractions of dust are segregated in several hoppers. The fine particles in the last hopper have significantly higher alkali content than the coarse particles in the first hopper. By leaching only the dust from the last hopper, the load of the leaching system can be significantly reduced. However, in some plants, all the collected dust is leached because the coarse particles make the slurry easier to handle.

The settling characteristics of the dust can also be controlled by the addition of flocculating agents to the water used for leaching the dust. Although none of the leaching plants use a treatment process to remove dissolved solids from the leachate effluent, there are methods and technologies that are potentially applicable. Several processes that might be employed include evaporation, precipitation, ion exchange, reverse osmosis, electrodialysis, and combinations of these. Each process must be considered in relation to the problem of disposal of the removed salts.

Some of these processes have technical limitations associated with their use in this application. For example, in ion exchange large amounts of acid and base are required to regenerate the resins. The amount of waste material would be approximately twice as great as for other separation processes. Similarly, although reverse osmosis is useful for desalination of dilute solutions, the dissolved solids content of the leachate is too high for this process to be practical.

Evaporation of the leachate could potentially eliminate the effluent. Although solar evaporation would have low operating cost, it could be used only in arid climates and where a large amount of land is available for evaporation ponds. Evaporation by submerged combustion or heat exchangers involves considerable cost for fuel and equipment. Waste heat from the kiln might be employed for evaporation of leachate, however, the economic feasibility of this practice is uncertain in the absence of industry experience. Reduction of the quantity of water to be evaporated by concentrating the leachate in some other process may be desirable.

A technology that appears promising for concentration of leachate is electrodialysis (ED), which has been successfully applied to the concentration of seawater for the recovery of salt (30). If ED were used, the concentrated stream would be more easily evaporated and the concentration of salts in the dilute stream would be low enough to allow it to be recycled to the leaching system. ED could be transferred directly to the concentration of leachate with two variations.

First, calcium ions must be removed to prevent precipitation and fouling of membranes. Reducing the pH to 9.5 by carbonation with stack gas will reduce the concentration of calcium ions to a minimum as was discussed above.

Second, reduction of the concentration of salts to a point where the water could be recycled in the leaching process will raise the cell resistance. Thus, more power must be provided than is needed for recovery of salt from seawater. A third desirable feature is additional carbonation to reduce the pH of the clarified leachate from 9.5 to about 8.0.

A flow diagram of a conceptual design for electrodialytic concentration of leachate is shown in Figure 33. At a typical leaching plant, about 6.5 kg/kkg (13 lb/ton) of dissolved solids are generated in the leachate stream, of which potassium salts are a major component. If the typical daily production of clinker is 1,600 metric tons (1,750 tons), the plant will generate about 10 metric tons (11 tons) of salts per day or about 3,300 metric tons (3,650 tons) per year. The costs of operating such a facility would amount to about $350/day.

Conventional electrodialytic equipment may be used. The only major change from the practices used in electrodialysis for desalination is that the concentrating compartments are not fed any water: the water that overflows the concentrating compartments and is withdrawn as brine is transferred through the membranes by electroosmosis and osmosis.

A diagram of an electrodialytic stack for concentrating electrolytes is shown in Figure 34. The stack consists of many (up to 2,000) cation- and anion-exchange membranes arranged alternately to form solution compartments, as indicated, between a cathode and an anode. The solution to be concentrated is circulated through alternate compartments, as shown.

The other set of compartments are closed at the bottoms. No solution is fed to them but they are filled with solution. When electrical current flows through the stack, cations and anions transfer from the circulating solution through the ion-exchange membranes into the closed compartments. Simultaneously, water transfers from the circulating solution through the membranes as a result of electroosmosis and osmosis. The water, so transferred, overflows from the tops of the closed compartments along with the transferred ions and is withdrawn as concentrated brine. It should be reemphasized that although only a few membranes and solution compartments are shown in Figure 34, commercial stacks may have as many as 2,000 membranes and 1,000 solution compartments.

The usual mode of operation for electrodialytic concentration stacks is known as feed-and-bleed operation. In this mode of operation only a small portion of the circulating solution is bled from a recycle line and returned to the cement process for reuse in slurrying dust. Most (perhaps 80%) of the solution is mixed with a volume of fresh leachate equal to the amount bled from the system and recycled to the feed side of the electrodialysis stacks. With this feed-and-bleed mode of operation, it is possible to transfer ions through the membranes at a high rate, without decreasing the concentration of ions in the circulating solution

FIGURE 33: FLOW DIAGRAM SHOWING STEPS IN ELECTRODIALYTIC CONCENTRATION OF LEACHATE

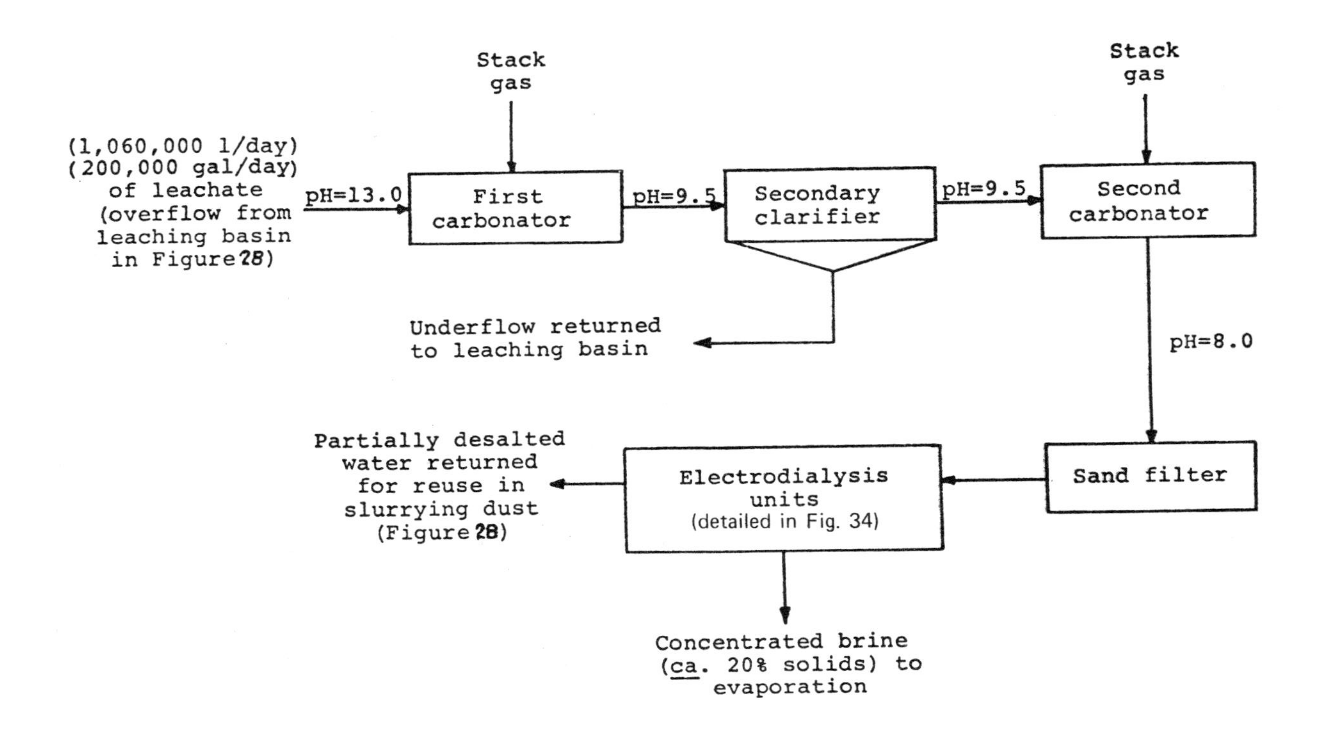

Source: EPA (22).

FIGURE 34: DIAGRAM OF ELECTRODIALYTIC CONCENTRATION STACK

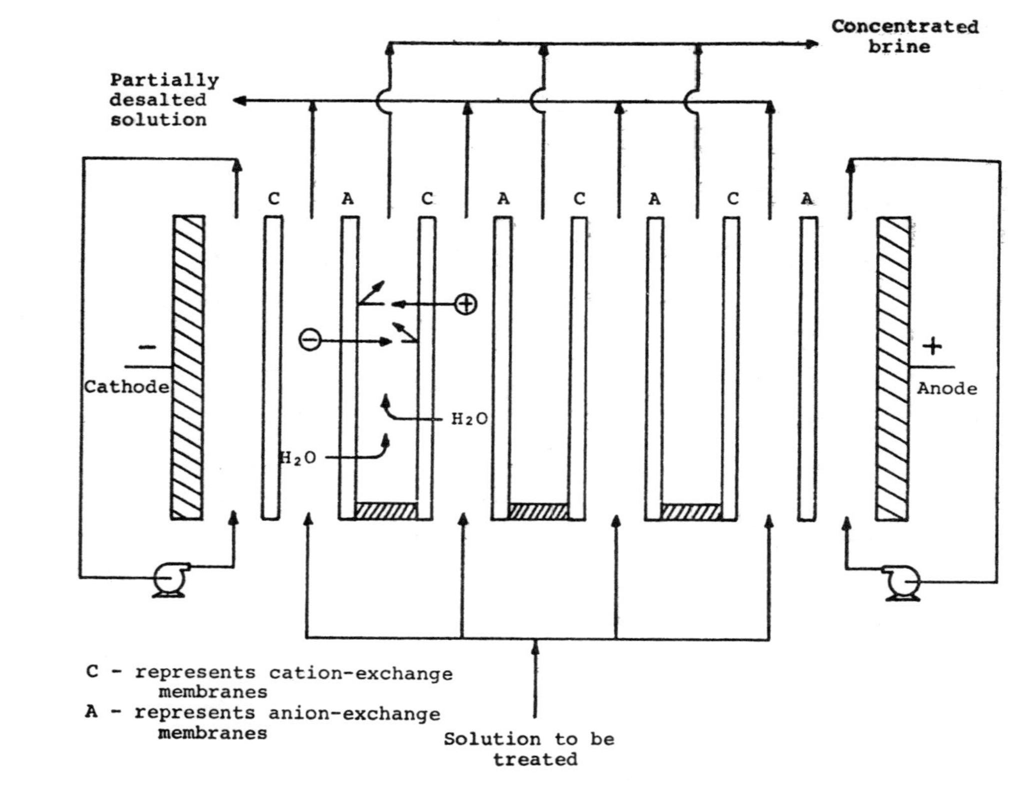

Source: EPA (22).

appreciably in any one passage through the stack. It is desirable to maintain a relatively concentrated circulating solution because with very dilute solutions the resistance of the stack would be high. Therefore, the energy requirements, which depend on resistance, would be high.

In the conceptual design, shown in Figure 33, leachate from the primary clarifier would be carbonated with stack gas in two turbo-agitated tanks arranged in series to reduce the pH to 9.5 so that $CaCO_3$ will precipitate. The liquid will be pumped to a secondary clarifier in which $CaCO_3$ can deposit on existing $CaCO_3$ particles carried within the clarifier as inventory. The underflow from this clarifier would be pumped back to the primary classifier; the overflow would be transferred to two secondary carbonators of the same type as the primary ones.

In the secondary carbonators the pH is reduced to 8.0 to convert the $CaCO_3$ remaining in solution to $Ca(HCO_3)_2$. This step is expected to prevent precipitation of calcium ions as the carbonate, since calcium bicarbonate is more soluble than calcium carbonate. As an added precaution against precipitation of calcium as either the bicarbonate or the sulfate, univalent selective cation-exchange membranes should be used.

No pretreatment of the feed other than that described above and filtration is expected to be needed. Iron and manganese, which have caused troubles with ED units for desalination, should not be present in this feed because any iron or manganese present in the dust should be fully oxidized, and should not leach from the dust at the high values of pH in the leaching section. If silica leaches from the dust, it could present a problem with silica slimes building up on the membranes. The extent to which silica might be leached is not clearly evident.

The solution from the secondary carbonators would be pumped through sand filters and into the ED stacks. As discussed previously, the ED stacks would be operated by a feed-and-bleed method. The partially desalted solution bled from the feed-and-bleed system would be returned to the primary clarifier for reuse in slurrying dust. The concentrated brine that overflows from the closed compartments of the stacks would be sent to an evaporation step. The evaporation could be performed in a solar pond in arid climates, or by other means in nonarid climates. Since only about 10,000 gal/day of concentrate must be evaporated, the cost should be low.

Costs for a typical operation, based on this conceptual design, have been estimated and are presented in the section which follows.

The most valuable and most abundant cation in the leachate is potassium, which if suitably recovered might be profitably marketed. The agricultural grade of potassium sulfate has a market price of $77 per metric ton of potassium oxide. Recovery of potassium from cement dust was practiced during World War I to free the U.S. of a monopoly exercised by the German Industry. One cement plant reportedly recovered 17.5 kg of potassium sulfate for each metric ton (35 lb/ton) of cement produced.

In 1959, Patzias (37) made a study of a method for extraction of potassium sulfate from cement dust. By leaching at high temperatures in a pressurized vessel he achieved 84% recovery of alkalies from the dust.

After filtration the leachate was concentrated by evaporation, neutralized with sulfuric acid, and evaporated to dryness. For a plant treating 180 metric tons/day of dust containing 1.66% of potassium sulfate the calculated capitalized payout for the process was 0.44 year and the calculated net profit was $101,304. There would be no discharge from this process because all of the water from the leachate is evaporated.

While a process based on this concept appears technically sound, it apparently has not been exploited by the industry. The economic feasibility reevaluated in view of present costs indicates a recovery cost of about twice the present market price. A flow sheet illustrating this concept is shown in Figure 35.

FIGURE 35: FLOW SHEET FOR THE RECOVERY OF K_2SO_4 FROM CEMENT KILN DUST

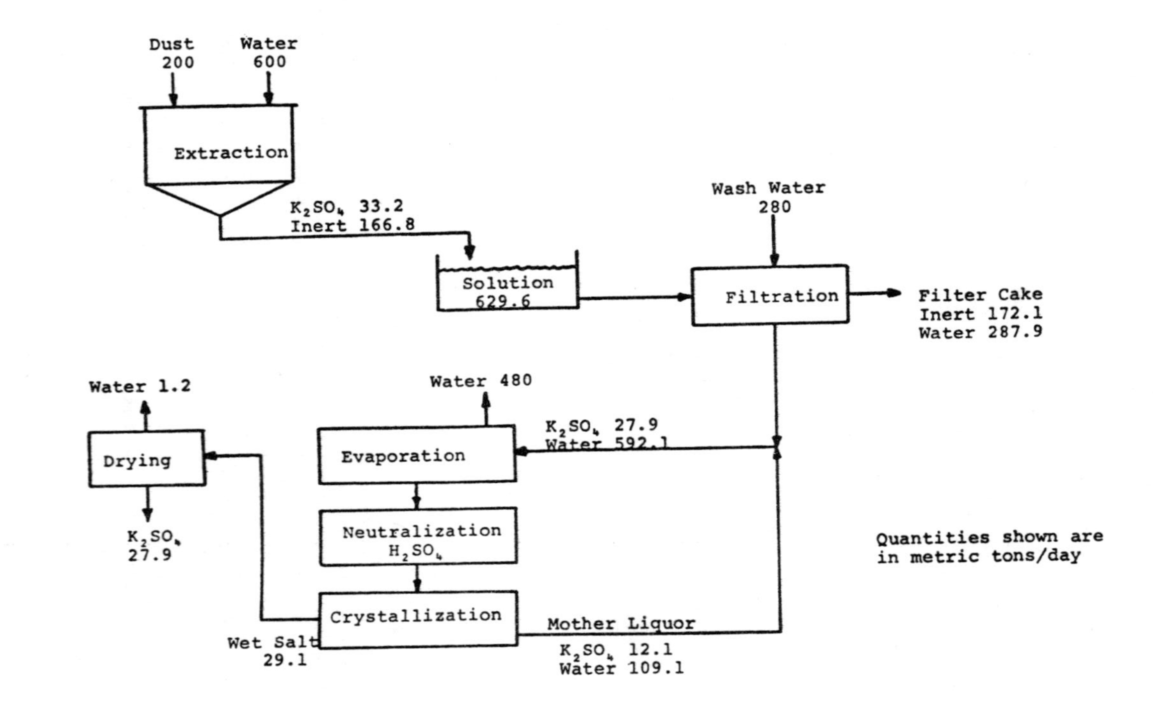

Source: EPA (22)

Cost and Reduction Benefits of Alternative Control and Treatment Technologies

A detailed analysis of the costs and pollution reduction benefits of alternative control and treatment technologies applicable to both subcategories of this industry is given in this section of the report. Table 39 summarizes the results of the analysis.

Nonleaching Plants: The present waste loadings from a typical nonleaching plant are shown in Table 39. These values represent the median of all values greater than 0.005 kg/kkg (0.01 lb/ton) of product reported by nonleaching plants.

Alternative A – Recycling and reuse of all water used in manufacturing, and containment or treatment of runoff from kiln-dust piles.

This alternative will result in essentially no discharge of pollutants. The investment cost of implementing this technology at a typical plant will be about $200,000 including a cooling tower ($94,000) or spray pond ($91,000), the necessary piping ($76,000), and containment dikes for coal piles and kiln-dust piles ($35,000). If an evaporative cooling pond is used, it would cost about $160,000 includiug piping, but not including land cost.

The operating costs of Alternative A will range from about $20,000 to $30,000 per year including maintenance, sludge removal, chemicals, labor, cost of power, and taxes and insurance. Power costs are limited to pumping and amount to $13,000 per year.

Alternative B – Limited reuse and in-plant controls

This alternative consists of isolation of cooling streams from possible contamination, reuse of cooling water in feed slurry (wet-process plants), retention and reuse or treatment of miscellaneous wastewater (e.g., truck washing) and containment or treatment of runoff from coal piles and kiln-dust piles and would also result in essentially no discharge of pollutants in manufacturing effluents.

Cost of implementing this alternative at individual plants may vary widely but on the average will be comparable to that for Alternative A. About 35 of 154 plants in the nonleaching subcategory (23%) are now achieving essentially no discharge of pollutants under either one of the alternatives described above.

Leaching Plants: The present wasteloading from a typical leaching plant is shown in Table 39. These typical loadings are substantially higher than those from the typical nonleaching plant and reflect the added presence of the leachate stream.

Alternative C – Segregation and Treatment of Leachate Stream

The nonleaching streams of leaching plants are treated like those of nonleaching plants under this alternative. Treatment of the leachate stream consists of neutralization of the leachate with stack gases to pH 9.0 followed by secondary sedimentation to remove both the residual suspended solids that were present in the leachate and the suspended solids (calcium carbonate) created by the neutralization with carbon dioxide.

This alternative will result in an acceptable pH of less than 9.0, and a suspended solids level of not more than 0.15 kg/kkg (0.30 lb/ton) of dust leached. Dissolved solids will remain at about their present level.

The cost of implementing Alternative C will be about $425,000 including $165,000 for the control of nonleaching streams and the cost of installing a stack-gas neutralization system and a clarifier ($260,000). Operating costs of Alternative C will range from about $35,000 to $45,000 per year.

One of the 12 plants in the leaching subcategory is presently equipped to implement this alternative with minor adjustments in operative procedures, this plant could meet the limitations of this alternative.

TABLE 39: WATER EFFLUENT TREATMENT COSTS AND POLLUTION REDUCTION BENEFITS

ALTERNATIVE		NON-LEACHING PLANTS A	NON-LEACHING PLANTS B
DESCRIPTION OF ALTERNATIVE	Present State No Added Controls	Installation of Cooling Tower or Spray Pond and Containment of Dust Pile Runoff	Isolation of Cooling Streams, limited reuse
INVESTMENT		$205,000	$205,000
ANNUAL COSTS			
Capital		$16,400	$16,400
Depreciation		$20,500	$20,500
Operation and Maintenance		$30,000	$20,000
Energy and Power		$13,000	$5,000
Total		$79,900	$61,900
EFFLUENT QUALITY in kg/kkg of cement except thermal and pH			
Alkalinity	0.12	No Discharge of Pollutants	No Discharge of Pollutants
Suspended Solids	0.075		
Dissolved Solids	0.19		
Sulfate	0.045		
Potassium	0.08		
Maximum pH	11		
Thermal (ΔT) in °C	2-11	3	3

(continued)

TABLE 39: (continued)

ALTERNATIVE		LEACHING PLANTS C	LEACHING PLANTS D
DESCRIPTION OF ALTERNATIVE	Present State No added Controls	Recycle and Reuse of Cooling and Miscellaneous Water, Neutralization and Settling of Leachate	Same as C plus Electrodialysis of Leachate to reduce TDS and Recycling of Leachate
INVESTMENT		$425,000	$645,000
ANNUAL COSTS			
Capital		$34,000	$51,000
Depreciation		$42,500	$64,500
Operation and Maintenance		$40,000	$68,000
Energy and Power		$13,000	$41,000
Total		$129,500	$224,500
EFFLUENT QUALITY in kg/kkg of cement except thermal and pH			
Alkalinity	1.38	1.38	No Discharge of Pollutants
Suspended Solids	0.905	0.15(a)	
Dissolved Solids	6.62	6.62	
Sulfate	3.66	3.66	
Potassium	3.3	3.3	
Maximum pH	12.5	9	
Thermal (ΔT) in °C	2-11	3	3

a. Based on quantity of leached dust.

(continued)

TABLE 39: (continued)

ALTERNATIVE	LEACHING PLANTS E
DESCRIPTION OF ALTERNATIVE	Abandonment of Dust Leaching
INVESTMENT	$205,000
ANNUAL COSTS	
Capital	$16,400
Depreciation	$20,500
Operation and Maintenance	$30,000
Energy and Power	$13,000
Dust Disposal	$165,000
Total	$244,900
EFFLUENT QUALITY in kg/kkg of cement except thermal and pH	
Alkalinity	No Discharge of Pollutants
Suspended Solids	
Dissolved Solids	
Sulfate	
Potassium	
Maximum pH	
Thermal (ΔT) in °C	3

Source: EPA (22).

Alternative D – Recycling of Leachate Water

This alternative consists of reducing the dissolved solids in the leachate stream by means of electrodialysis and recycling the partially demineralized leachate. The technology of Alternative C must be implemented to provide a stream acceptable for electrodialysis. The concentrated brine resulting form this treatment may be evaporated for the recovery of potassium salts or contained in a suitable pond. Implementation of Alternative D will result in essentially no discharge of pollutants. None of the plants in the leaching subcategory, however, is employing the technology described as Alternative D.

Alternative E – Abandonment of Existing Leaching Operations

Under this alternative, plants that presently leach kiln dust would abandon the practice and adopt either Alternative A or B which will result in no discharge of pollutants. A contractor would haul the dust for about $0.50 per ton. The value of the wasted dust would be about $2.00 per ton (46). Therefore, the annual cost of wasting 200 tons per day of dust that is presently leached would be $165,000.

The investment cost of $200,000 involved in implementing control and treatment technology at an existing nonleaching plant represents 0.5 to 1% of the estimated replacement cost of the plant ($20 to $40 million). In terms of plant size, these costs represent about $0.35 per metric ton of capacity. For plants in the leaching category, these figures may be approximately doubled.

The increased cost of manufacturing cement will range from about $0.10 per metric ton at nonleaching plants to about $0.21 at leaching plants.

One industry consultant has provided the typical production cost figures for 14 plants presented (5). The production cost ranges from $15.11 to $21.20 with an average of $17.52 per metric ton. The added cost of water pollution control will thus increase production cost by less than 1.5% at plants operating at full capacity. Since these costs are largely fixed costs and, thus, must be borne at any level of production, production at less than full capacity will reflect higher, added costs.

CERAMIC CLAY PRODUCTS MANUFACTURE

The manufacture of ceramic clay involves the conditioning of the basic ores by several methods. These include the separation and concentration of the minerals by screening, floating, wet and dry grinding, and blending of the desired ore varieties (see Figure 36). The basic raw materials in ceramic clay manufacture are kaolinite ($Al_2O_3 \cdot 2SiO_2 \cdot 2H_2O$) and montmorillonite [$(Mg,Ca)O \cdot Al_2O_3 \cdot 5SiO_2 \cdot nH_2O$] clays. These clays are refined by separation and bleaching, blended, kiln dried, and formed into such items as whiteware, heavy clay products (brick, etc.), various stoneware, and other products such as diatomaceous earth, which is used as a filter aid.

FIGURE 36: CERAMIC CLAY MANUFACTURING PROCESSES

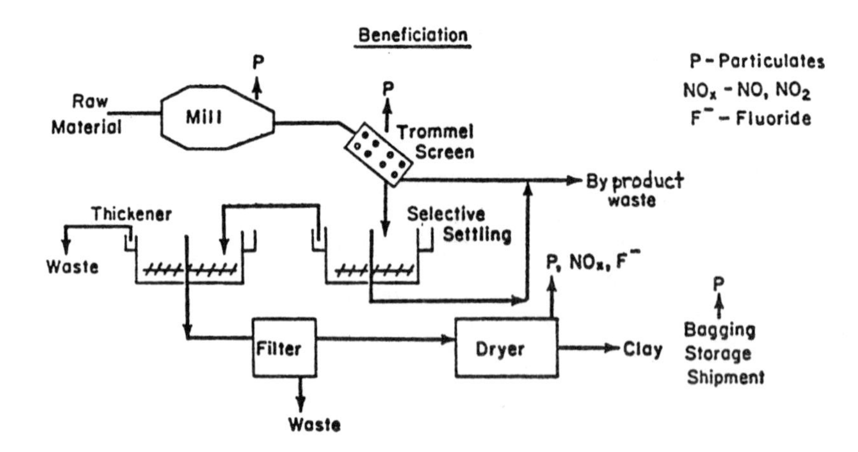

(continued)

FIGURE 36: (continued)

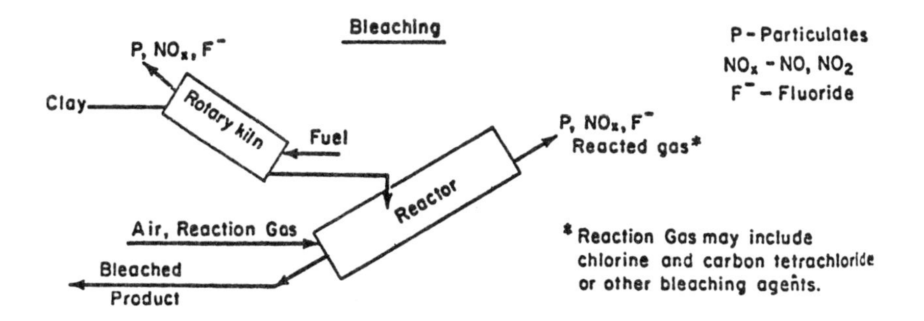

Source: A.E. Vandegrift, L.J. Shannon, E.W. Lawless, P.G. Gorman, E.E. Sallee and M. Reichel (20)

Halide bleaching for preparation of kaolinite utilizes the reactivity of the halide to remove the chemically active and unwanted constituents of the clay ore leaving behind a purified white product suitable for ceramics manufacture. Figure 36 includes a schematic diagram of this process.

The manufacture of filter and activated clays includes grinding and wet or acid treating, followed by drying and regrinding. The drying is accomplished in rotary kilns, which reduce moisture content from 15-20% to 10%. Ceramic clay is manufactured from a mixture of wet talc, whiting, silica clay, and other ceramic materials. This mixture is dried in a spray dryer.

AIR POLLUTION

Emissions from ceramic clay products manufacture consist primarily of particulates, but some fluorides and acid gases are also emitted in the drying process. The high temperatures of the firing kilns are also conducive to the fixation of atmospheric nitrogen and the subsequent release of NO, but no published information has been found for gaseous emissions. Particulates are also emitted from the grinding process and from storage of the ground product.

Factors affecting emissions include the amount of material processed, the type of grinding (wet or dry), the temperature of the drying kilns, the gas velocities and flow direction in the kilns, and the amount of fluorine in the ores.

Common control techniques include settling chambers, cyclones, wet scrubbers, electrostatic precipitators, and bag filters. The most effective control is provided by cyclones for the coarser material, followed by wet scrubbers, bag filters, or electrostatic precipitators for dry dust. Emission factors for ceramic clay manufacturing are presented in Table 40.

Additional data on exhaust gases from ceramic clay manufacture as regards flow rate, velocity and temperature have been presented by Engineering Science, Inc. (18).

Particulate emissions from heavy clay products manufacture occur during handling of raw materials, grinding, screening and blending, and during cutting and shaping operations. Fluorides, largely in a gaseous form, are also emitted from brick manufacturing operations.

TABLE 40: PARTICULATE EMISSION FACTORS FOR CERAMIC CLAY MANUFACTURING[a]

Type of process	Uncontrolled		Cyclone[b]		Multiple-unit cyclone and scrubber[c]	
	lb/ton	kg/MT	lb/ton	kg/MT	lb/ton	kg/MT
Drying	70	35	18	9	7	3.5
Grinding	76	38	19	9.5	–	–
Storage	34	17	8	4	–	–

[a]Emission factors expressed as units per unit weight of input to process.
[b]Approximate collection efficiency: 75 percent.
[c]Approximate collection efficiency: 90 percent.

Source: EPA (15)

The extent of raw material handling and processing greatly affects the dust emissions from this phase of the manufacturing process. Emissions when firing and/or curing the formed bricks are affected by the temperature in the ovens and the type and quantity of trace components in the brick. Thus, sulfur and/or fluoride compounds may be emitted when the bricks are subjected to high temperatures. The type of fuel used to heat the ovens also has a direct bearing on the combustion emissions.

Particulate emissions in kiln fired refractory plants occur from raw materials handling, crushing, calcining, drying, mixing, and burning operations. Emissions from the calcining and drying operations depend upon the type and quantity of material charged, kiln and dryer types, and final moisture content.

Particulate emissions from the manufacture of castable refractories are created by the drying, crushing, handling, and blending phases of this process, the actual melting process, and in the molding phase. Fluoride emissions, largely in the gaseous form, may also occur during the melting operations. Particulate emissions are affected by the amount of material handling and pretreatment required before melting, and by the components in the melt. Generally, increasing concentrations of silicon will cause increased particulate emissions.

Table 41 presents a summary of emission rates from the manufacture of clay products. Current emissions are estimated at 467,000 tons. Data available on emission rates, processing variations, and control equipment practices and utilization are meager, and emission figures in Table 41 are considered to be engineering estimates.

Common control techniques for the ceramic clay manufacturing processes include settling chambers, cyclones, wet scrubbers, electrostatic precipitators and bag filters. Cyclones for the coarser material followed by wet scrubbers, bag filters or electrostatic precipitators for dry dust are the most effective control techniques.

A variety of control systems may be used to reduce both particulate and gaseous emissions from heavy clay products manufacturing. Almost any type of particulate control system will reduce emissions from the materials handling process. However, good design and hooding are required to capture the emissions. Fluoride emission can be reduced to very low levels by using a water scrubber.

The general types of particulate controls may be used on the materials handling aspect of refractory manufacturing. However, emissions from the electric-arc furnace are largely condensed fume and consist of very fine particles, largely 2μ or smaller. Baghouses may be used to control particulate emissions from the furnace.

Multicyclones, baghouses, and electrostatic precipitators have been used on rotary and vertical kilns in kiln fired refractory plants.

TABLE 41: PARTICULATE EMISSIONS–CERAMIC CLAY PRODUCTS

Source	Quantity	Emission Factor	Efficiency of Control C_c	Application of Control C_t	Net Control $C_c \cdot C_t$	Emissions tons/year
Ceramic clay	7,870,000 tons					
Grinding	60% ceramics	76 lb/ton prod.	–	–	0.60	72,000
Drying	100% ceramics	70 lb/ton prod.	–	–	0.60	110,000

Source: A.E. Vandegrift, L.J. Shannon, E.W. Lawless, P.G. Gorman, E.E. Sallee and M. Reichel (20)

CONCRETE INDUSTRY

Concrete batching involves the proportioning of sand, gravel, and cement by means of weigh hoppers and conveyors into a mixing receiver such as a transit mix truck. The required amount of water is also discharged into the receiver along with the dry materials. In some cases, the concrete is prepared for on-site building construction work or for the manufacture of concrete products such as pipes and prefabricated construction parts.

Concrete is used in a vast number of construction applications such as highways, office buildings, water dams, bridges and foundations. Concrete plants are located throughout the United States; the Washington, D.C. area alone has over 200 concrete contractors.

The major pollutant is dust which is generated during the unloading of cement, gravel and sand, transport and filling operations. The recommended air pollution abatement equipment is the baghouse; however, most plants do not have any control devices. Typical operating times are 8 hours per day and 5 days per week. Figure 37 is a sketch of a plant.

One type is used to charge sand, aggregate, cement, and water to transit-mix trucks, which mix the batch en route to the site where the concrete is to be poured; this operation is known as "wet batching." Another type is used to charge the sand, aggregate, and cement to flat bed trucks, which transport the batch to paving machines where water is added and mixing takes place; this operation is known as "dry batching." A third type employs the use of a central mix plant, from which wet concrete is delivered to the pouring site in open dump trucks (14).

Dry concrete batching plants are used in road construction work. Because of advances in freeway construction in recent years, plants such as these are located in metropolitan areas, often in residential zones. The plants are portable, that is, they must be designed to be moved easily from one location to another. This is, of course, a factor in the design of the air pollution control equipment.

The central mix plant is being used more and more extensively by the concrete industry. In a central batch operation, concrete is mixed in a stationary mixer, discharged into a dump truck, and transported in a wet mixed condition to the pouring site.

The handling of aggregate and cement at these plants is similar to that at the other concrete batch plants. Sand, aggregate, cement, and water are all weighed or metered as in a wet concrete batching plant and discharged through an enclosed system into the mixer.

FIGURE 37: FLOW DIAGRAM OF A CONCRETE BATCHING PLANT

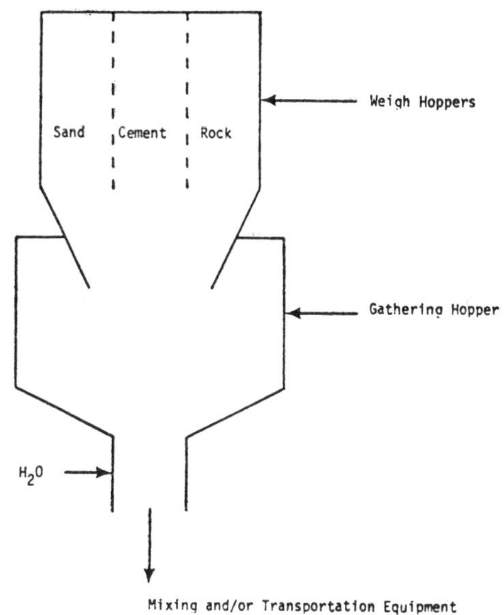

Source: Engineering Science, Inc. (18)

AIR POLLUTION

Particulate emissions consist primarily of cement dust, but some sand and aggregate gravel dust emissions do occur during batching operations. There is also a potential for dust emissions during the unloading and conveying of concrete and aggregates at these plants and during the loading of dry batched concrete mix. Another source of dust emissions is the traffic of heavy equipment over unpaved or dusty surfaces in and around the concrete batching plant.

Control techniques include the enclosure of dumping and loading areas, the enclosure of conveyors and elevators, filters on storage bin vents, and the use of water sprays. Table 42 presents emission factors for concrete batch plants.

Dust, the air contaminant from wet concrete batching, results from the material used. Sand and aggregates for concrete production come directly from a rock and gravel plant where they are washed to remove silt and clay-like minerals. They thus arrive at the batch plant in a moist condition and hence do not usually present a dust problem. However, when lightweight aggregates are used, they do pose a problem. These materials are formed by thermal expansion of certain materials. They leave the aggregate plant very dry and create considerable dust when handled. The simplest way to deal with this problem is to wet each load of aggregate thoroughly before it is dumped from the delivery truck. Attempts to spray the aggregate as it is being dumped have had very limited effectiveness.

Therefore, if wet or damp aggregate is used, practically all the dust generated from concrete batching operations originates from the cement. Particle size distribution and other characteristics of the dust vary according to the grade of cement. A range of 10 to 20%

by weight of particles of 5 micron size or less is typical for the various grades of cement. Bulk density ranges from 50 to 65 pounds per cubic foot of cement.

Cement dust can be emitted from several points. The receiving hopper, the elevator, and the silo are the points of possible emission from the cement receiving station. Other points of possible dust emissions are the cement weigh hopper, the gathering hopper, and the mixer.

TABLE 42: PARTICULATE EMISSION FACTORS FOR CONCRETE BATCHING*

	Emission	
Concrete Batching	Pound per Cubic Yard of Concrete	Kilogram per Cubic Meter of Concrete
Uncontrolled	0.2	0.12
Good control	0.02	0.012

*One cubic yard of concrete weighs 4,000 pounds (1 cubic meter = 2,400 kilograms). The cement content varies with the type of concrete mixed, but 735 pounds of cement per yard (436 kilograms per cubic meter) may be used as a typical value.

Source: EPA (15)

Dry batching poses a much more difficult dust control problem than wet batching does. Since most plants that do dry batching also do wet batching, the gathering hopper must be set high enough to accommodate transit mix trucks. Since the receiving hopper of most transit mix trucks is several feet higher than the top of the flat bed trucks used in dry batching, there is a long free fall of material when a dry batch is dropped. This produces a considerable amount of dust, sufficient to violate most codes that have an opacity limitation applicable to this type of operation.

From an air pollution standpoint, the dust to be collected has characteristics similar to those of the cement dust already discussed for wet concrete batching plants. In dry batching, however, volumes of dust created are considerably greater because the amount of concrete batched is large, no water is used, and the batches are dropped rapidly into the waiting trucks to conserve time.

From an air pollution control standpoint, the central mix type of operation is preferable to dry batching. The dust is more easily captured at the batch plant, and further, there is no generation of dust at the pouring site. The operation is also preferable to wet batching because designing control equipment for a stationary mixer is easier than it is for a transit mix truck loading area.

Control Equipment—Wet Batching

A typical cement receiving and storage system is shown in Figure 38. The receiving hopper is at or below ground level. If it is designed to fit the canvas discharge tube of the hopper truck, little or no dust is emitted at this point. After a brief initial puff of dust, the hopper fills completely and the cement flows from the truck without any free fall. Cement elevators are either the vertical screw type or the enclosed bucket type. Neither emits any dust if in good condition. The cement silo must be vented to allow the air displaced by the cement to escape. Unless this vent is filtered, a significant amount of dust escapes.

Figure 38 shows one type of filter. It consists of a cloth tube with a stack and weather-cap for protection. The pulley arrangement allows it to be shaken from the ground so that the accumulated layer of dust on the inside of the cloth tube can be periodically removed. The cloth's area should be sufficient to provide a filtering velocity of 3 feet per minute, based upon the displaced air rate.

FIGURE 38: CEMENT RECEIVING AND STORAGE SYSTEM

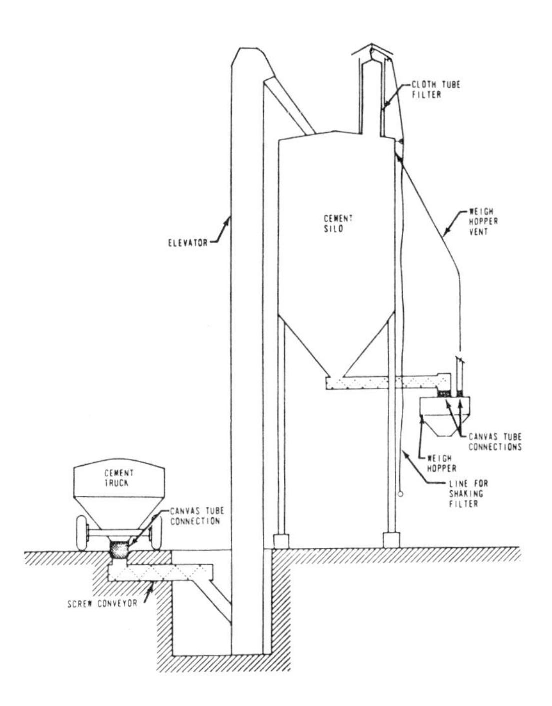

Source: J.A. Danielson (14)

Many concrete batch plants now receive cement pneumatically from trucks equipped with compressors and pneumatic delivery tubes. In these plants, a single filtered vent used for the gravity filling of cement has proved inadequate, and other methods of control are required. In this pneumatic delivery, the volume of conveying air is approximately 350 cubic feet per minute during most of the loading cycle and increases to 700 cubic feet per minute at the end of the cycle.

To control this volume of air, it is best to install a small conventional cotton sateen baghouse with a filtering area of 3 feet per minute (approximately 200 square feet of cloth area) to vent the cement silo. The baghouse should be equipped with a blower to relieve the pressure built up within the silo. A mechanical shaking mechanism also should be provided to prevent cement from blinding the filter cloth of the baghouse.

Another less expensive type of control device is to mount a bank of approximately four simple filtered vents atop the silo. The filtering area should not exceed 7 feet per minute, giving an area of approximately 100 square feet for the 700 cubic feet per minute of air encountered at the end of the cycle. The filter design must include a shaking mechanism to prevent blinding of the filter cloth.

The major disadvantage of using a bank of several simple filter vents as just described is the possibility of pressure build-up within the silo. If, for some reason, the filter should become blinded, there is danger of rupturing the silo. Therefore, proper maintenance and regular inspection of the filter are necessary.

Where baghouses are used to control other larger cement dust sources such as those existing in a dry concrete batching plant or in a central mix plant, then the cement silo can easily be vented to the same baghouse.

The cement weigh hopper may be a compartment in the aggregate weigh hopper or it may be a separate weigh hopper. Cement is usually delivered from the silo to the weigh hopper by an enclosed screw conveyor. To permit accurate weighing, a flexible connection between the screw conveyor and weigh hopper is necessary. A canvas shroud is usually used, and if properly installed and maintained, prevents dust emissions at this point. The weigh hopper is filled at a fairly rapid rate, and the displaced air entrains a significant amount of dust. This dust may be controlled by venting the displaced air back to the cement silo or by installing a filtered vent on the weigh hopper as described for cement silos.

The vent should be of adequate size to provide a filtering velocity of about 3 feet per minute, based upon the cement's volumetric filling rate. For example, if a weigh hopper is filled at the rate of 1,500 pounds in one-half minute, and the density of cement is 94 pounds per cubic foot, the displaced air rate equals 1,500/(94)(0.5), or 32 cubic feet per minute. The required cloth area would then be 32/3 or 10.7 square feet.

The dropping of a batch from the weigh hopper to the mixer can cause cement dust emissions from several points. In the loading of transit mix trucks, a gathering hopper is usually used to control the flow of the materials. Dust can be emitted from the gathering hopper, the truck's receiving hopper, and the mixer. The design and location of the gathering hopper can do much to minimize dust emissions. The hopper should make a good fit with the truck receiving hopper, and its vertical position should be adjustable. Figure 39 illustrates a design that has been used successfully in minimizing dust emissions.

FIGURE 39: AN ADJUSTABLE GATHERING HOPPER

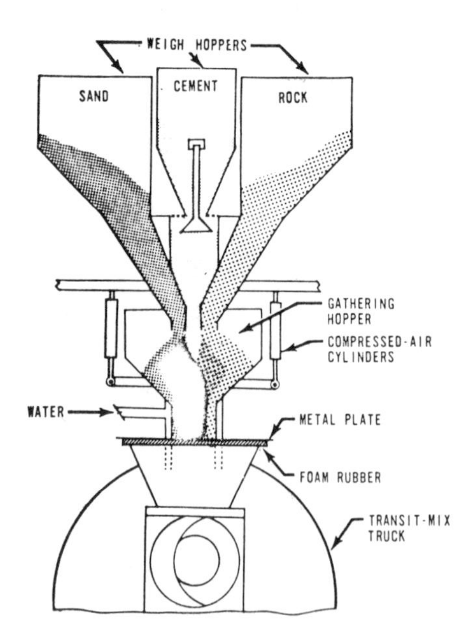

Source: J.A. Danielson (14)

Compressed air cylinders raise and lower the gathering hopper to accommodate trucks of varying heights. A steel plate with a foam rubber backing is attached to the bottom of the gathering hopper and is lowered until it rests on the top of the truck's receiving hopper. Water for the mix is introduced through a jacket around the discharge spout of the gathering hopper and forms a dust reducing curtain.

Discharge of the cement hopper into the center of the aggregate stream, and choke feed between the weigh hopper and the gathering hopper suppress dust emissions from the top of the gathering hopper.

Control Equipment–Dry Batching

A local exhaust system with an efficient dust collector is required to control a dry batching plant adequately. This is a difficult operation to hood without interfering with the truck's movement or the batch operator's view. The truck bed is usually divided into several compartments, a batch being dropped into each compartment. This necessitates repeated spotting of each truck under the direction of the batch operator; hence, he must be able to see the truck at the drop point.

A canopy type hood just large enough to cover one compartment at a time provides effective dust pickup and affords adequate visibility. The sides are made of sheets of heavy rubber to permit contact with the truck bed without damage. This hood is mounted on rails to permit it to be withdrawn to allow wet batching into transit mix trucks.

The exhaust volume required to collect the dust varies with the shape and position of the hoods. With reasonably good hooding, the required volume is approximately 6,000 to 7,000 cubic feet per minute.

A baghouse is the most suitable type of dust collector for this service. Scrubbers have been used, but they have been plagued with difficulties such as low collection efficiency, plugged spray nozzles, corrosion, and wastewater disposal problems. A baghouse for this service should have a filtering velocity of 3 feet per minute. It may be of the intermittent shaking type, since sufficient opportunities for stopping the exhauster for bag shaking are usually available. The drop area tunnel is enclosed on the sides and partially on the ends.

In many instances the greatest source of dust from the operation of a concrete batch plant is that created by the trucks entering and leaving the plant area. If possible, the yard and access roads should be paved or oiled; or if this is not feasible, they should be watered frequently to suppress the dust.

Control Equipment–Central Mix Plant

Effective control at the discharge end of the mixer in a central mix plant is a function of good hood design and adequate ventilation air. A hydraulically operated, swing-away, cone shaped hood is normally used with a 2-inch clearance between the hood and the mixer. This installation employs a mixer with a capacity of 8 cubic yards. The discharge opening of the mixer is 40 inches in diameter. Ventilation air was found to be 2,500 cubic feet per minute.

For a hood of this type, in-draft face velocities should be between 1,000 and 1,500 feet per minute. Velocities such as these are required for handling the air discharged from the mixer, which is displaced air and inspirated air from the aggregate and cement falling into the mixer.

A baghouse is required to collect the dust emissions. A filtering velocity of 3 feet per minute is adequate. Other baghouse features are similar to those previously discussed for dry concrete batching plants.

Proprietary Control Processes

A process developed by R.W. Strenlow (38) is one in which a concrete batch plant is provided with a subatmospheric air carried dust collection system which discharges under positive pressure into the body of the aggregates being processed through the plant. This aggregate body functions as a continuously renewed filter bed.

A process developed by A.A. Mills, Jr. et al (39) provides a dust control system for collecting dust as a transit mixer is charged from a batch plant. It includes a shroud which is moved into cooperation with the inlet of the mixer and a dust filtering system connected to the shroud. The filtering system includes a pair of filtering chambers, through which the dust is passed for collection during the loading of the mixer and in which the flow is reversed in alternate chambers during cleaning periods.

WATER POLLUTION

The ready mixed concrete industry takes processed sand, gravel and crushed stone material and adds cement, special additives and water. This mixture becomes active concrete which will begin to set within a few hours, so trucks and other plant equipment must be continually washed clean to prevent build-up of concrete and the breakdown of machinery. There are approximately 8,000 ready mixed concrete plants in the United States with a national production of 186,000,000 cubic yards per year. This mixed concrete is valued at $2,930,000,000.

Water used for making the product amounts to six billion gallons per year. Since this water is totally consumed in the product, it does add to the national water usage, but not to water pollution. The water used to wash the trucks and equipment does become a potential water pollutant.

The average ready mix truck load is 6.8 cubic yards and each truck averages 4.2 trips per day. Each truck and its portion of the plant equipment will use an average of 500 gallons of water per day (40).

The range of water use per truck per day varies widely because some operators wash out trucks once a day, while others wash out after each load. Availability of water also affects usage. When water is abundant and low in cost, operators use larger quantities. However, they wash off about the same amount of solids in any case, so the potential stream pollution load is not increased. Based on 500 gallons per day the annual water usage is as follows.

$$\text{Annual wash water} = \frac{186{,}000{,}000 \text{ yards per year} \cdot 500 \text{ gallons per day}}{6.8 \text{ yards per load} \cdot 4.2 \text{ loads per day}} = 4 \text{ billion gallons per year}$$

The aggregate and ready mixed concrete industries used 136 billion gallons of water in 1969 and the rate is increasing by 4% per year. This figure agrees closely with other surveys showing the stone and vitreous products using 2% of the total industrial water. The aggregate and ready mixed concrete industries amount to about one-half of the total stone and vitreous products.

This analysis gives a perspective of the volume of water used by the aggregate and ready mixed concrete industries and the comparison of the total water use. The analysis shows that although the aggregate and ready mixed concrete industries are heavy water users, they use only a small percentage of the total water used in the United States.

Aggregate and ready mix plants may raise the pH factor, increase turbidity and reduce the dissolved oxygen in wastewaters discharged. Each year, from the aggregate produced, the aggregate producing industry washes off many tons of fine sand, silts and clays. Wastewater from the ready mix industry contains fine sand, lime and cement. The uncontrolled and untreated discharge from both of these industries impairs the aesthetic quality of our streams and may be a hazard to marine life.

With the increase in production of these industries, an additional volume of water is required of approximately 4% per year. The requirements for cleaner products require more thorough washing and scrubbing which in turn increases the amounts of suspended particles in the wastewater. This increase in volume and concentration of suspended particles in the effluent water can only add to the existing problem. The aggregate and ready mixed concrete industries, as well as all other water users, must have clarification systems in order to improve the quality of water which they discharge into our streams and rivers.

Many of these plants do have effective systems for clarifying the water or recycling it. It is the purpose of this study to make these systems known to all the industry so proven systems can be put into operation quickly.

The ready mixed concrete plant uses water primarily in the batching of the concrete and for washing out and cleaning the central mixer and/or mixing trucks. The concrete batch consists of a mixture of specific proportions of sand, gravel, cement and water. These are weighed or metered directly into the mixer truck or into a central mixer which premixes the material before discharging into the trucks. Chemical additives may be included in this mixture to increase the workability of the concrete, to aid in the setting, to protect it from frost or to color the concrete.

The general practice is to order slightly more concrete than estimated to do the job to prevent running short. In some cases the overage is dumped on the job site. In most cases, however, it returns to the plant in the truck to be either incorporated in the following order or flushed out as waste material.

Disposition depends on the amount left in the truck, the length of time since batching, the consistency of the following order, and other factors. Most trucks are washed out and cleaned only at the close of the day's run or when discharging the waste material. The amount of waste varies, but has been estimated at 600 pounds.

The amount of water used in washing out and cleaning the trucks varies from 50 gallons per truck to several hundred gallons per truck, depending on the housekeeping practices of the individual companies. Water discharged from the washing operation contains sand and gravel and cement slurry and contains a pH rating of between 11.0 and 12.0 on the average.

Clarification of this wastewater consists primarily of lowering the pH factor and settling of the sand, gravel and cement particles. The pH factor for water being discharged into waterways is usually required to fall below the 11.0 to 12.0 average. Lime causing the high pH does actually act as a coagulant clearing the water.

Of those plants in the ready mixed concrete industry which answered questionnaires, only a very small percentage had installed clarification systems. As in the aggregate industry, most of the plants use self contained holding or filtration ponds on their own sites into which they discharge their wastewater and material. The build-up of this waste material is used either to fill the pond or pit or is periodically removed and hauled away as landfill material.

Water clarification generally consists of first removing the sand and gravel from the wash water and then removing the suspended fines and cement particles. The wash water also has a pH factor of around 10.0 to 12.0 and requires an acid treatment to neutralize it. Chemicals may also have to be used to aid in settling out the cement particles.

Those companies having clarification systems use several varying methods and arrangements. The general arrangement consists of a truck washing facility, a primary basin or pad to settle out the coarse sand and gravel and secondary basins or ponds to settle out the fines.

Many of them also recycle all of the wash water. Recycled water may be used to wash out the mixer barrel, but when used for washing the exterior may leave a white film which may be undesirable. A supply line of clean water may be installed in addition to the recycled line for the final rinse to eliminate the film.

The truck washing facility is dependent upon the housekeeping practice of the company, the amount of water used per truck, the number of trucks, the arrangement of the primary basin and many other factors. Trucks generally discharge their waste load and water in one of the following arrangements.

[1] The truck discharges onto a waste pile where the water runs off or drains into the primary pond leaving the bulk of the coarse material on the waste pile. The truck in this case receives its water for rinsing from a separate source on the site.

[2] The truck may back up to the primary basin, fill from a water source at that point and dump directly into the primary basin.

[3] The truck may back into one of several stalls located along a sloped paved apron which extends the entire length or width of the primary basin. After washing from water lines at each stall, the truck may dump directly into the primary basin or onto a sloping ramp which drains into the primary basin.

In some plants the trucks discharge their waste into a manufactured classifier or recovery unit with the overflow water running into the primary basin. Primary basins seem to be generally of three styles. One type has a bottom which starts at the level of the washing pad and slopes away from the pad at about a 1:12 slope to a depth of around four to five feet. The wall at the deep end is slightly lower than the side walls and serves as a weir. The sloping bottom aids in retaining the solids as they settle out and also enables clean-out with a front end loader.

One other type of primary basin is constructed with a flat bottom and a depth of five to ten feet or more. Some use a poured slab for the bottom while others use a gravel bottom. One side or part of a side is lower and serves as a weir. In some cases where the bottom is omitted, the basin also is used as a seepage pit.

The primary basin can also be in the form of a settling pond or pit with the water flowing over a weir or channel into the secondary basin or pond. The main purpose of the primary basin is to settle out the bulk of the waste material, which is the gravel and sand. By so doing, the clean-out is confined primarily to this basin and the treatment of the water in the secondary basin becomes more efficient when coagulants are to be used.

The secondary basin is used to settle out the super-fine and cement particles. Again, styles vary from a large settling pond to a single basin holding a one day's supply of wash water. The minimum basin seems to be one which will hold all the wash water for a maximum day's use. The water remains in the basin overnight and is pumped from the top of the surface downward in the morning until it starts to become cloudy. The water in this case is generally pumped to a storage tank or basin and recycled as wash water. The quality of this water in most cases will not meet the requirements necessary for discharging into a waterway.

Some plants use two secondary basins. Water from the primary basin enters one or the other of the two basins on alternate days. This arrangement allows the water to rest a full day and night in each tank before it is pumped out. Clarification of the water is greatly improved by this arrangement. However, due to the fineness of the cement particles and the high pH factor or alkaline content of the water, chemical treatment is generally required before the water will meet the standards required for discharging into a stream or waterway.

Because of the difficulty in bringing the quality of the discharged water within the required limits, most companies which have been discharging wastewater into a stream or waterway are now discharging the water from the secondary basin into a holding basin and recycling the water back as wash water. The secondary basins may be constructed below ground level in the form of a filtration basin to allow a percentage of the water to seep into the ground and to prevent an accumulation of wastewater. Fresh water is used to augment the water from the basin.

When coagulants are used, it appears that they work most economically and efficiently if they are introduced into the flow of water from the primary to the secondary basin. The chemicals are prepared in a mixing tank and are metered into the flow of the wastewater. Thorough mixing of the chemical additive with the wastewater is of greatest importance in order to bring the chemical in contact with the greatest number of suspended particles.

The designing of a clarification system will require the consideration and analysis of all of the many variables. The first consideration probably would be the feasibility of eliminating the discharge of wastewater from the plant site by installing a closed loop system. By recycling all the water the problem would be isolated or restricted to the plant area. Or perhaps by partial recycling, the decreased amount of discharged wastewater could be treated to meet the required standards.

Sizes and arrangement of basins are dependent upon the amount of water used, amount of ground available and the extent of clarification required. The use and amount of chemicals also is dependent on similar factors. Each plant needs an independent study and analysis of these factors to determine its own solution to the problem.

One of the simplest clarification systems is a filter bed. The filter most often would be rock (three-quarter to six inches). Figure 40 shows a rock filter that has worked very well. This water clarification system was designed for four wash stalls. The daily water usage is 20,000 gallons. The trucks dump wash water and waste materials into a ten foot wide basin. This basin is sloped so a front end loader can get in to remove the material. Enough water must be used to dilute the waste concrete that is dumped into this basin so the loader can pick it up.

This coarse material basin needs to be cleaned about every two weeks. The cleanout is done after a weekend so the water has had time to settle out. A weir must be so designed as to allow slow drainage of the surface water as the basin becomes full. This will require some effort on the part of the operator to fill the entire length of the basin uniformly. The loader can remove the aggregate without running in water, which would damage brakes or bearings.

The filter pond is 60 feet by 60 feet by 16 feet, excavated out of a natural gravel deposit. This gravel provides an effective filter. Because cement does seal the walls and bottom, this seal must be broken about twice a year by excavating out the sediment and sealed gravel. The pumping pond is 40 feet by 40 feet by 20 feet and is also in a gravel deposit. The water in the pond is clear and suitable for washing both the drum and the outside of the trucks. The volume of wash water becomes important in the design of this system. If too little water is used, the coarse material will still retain cement and set up. This will create a problem in cleanout. Using too much water will increase the flow rate in the filter pond. Cement particles will flow across the pond and will quickly seal the wall next to the pumping pond. Settlement rates should be checked in order to size the filter pond.

Much has been done to develop workable settling basins for ready mix plants. The National Ready-Mixed Concrete Association has surveyed some plants and printed the results. Through this study additional plants have been surveyed.

This research has pointed out that the basins are quite similar, but each system seems to have some good and some weak points. For this reason, the systems will not be described exactly as they were put in, but rather one example will be cited here (Figure 41).

FIGURE 40: PLANT LAYOUT OF ROCK FILTER WATER CLARIFICATION PLANT

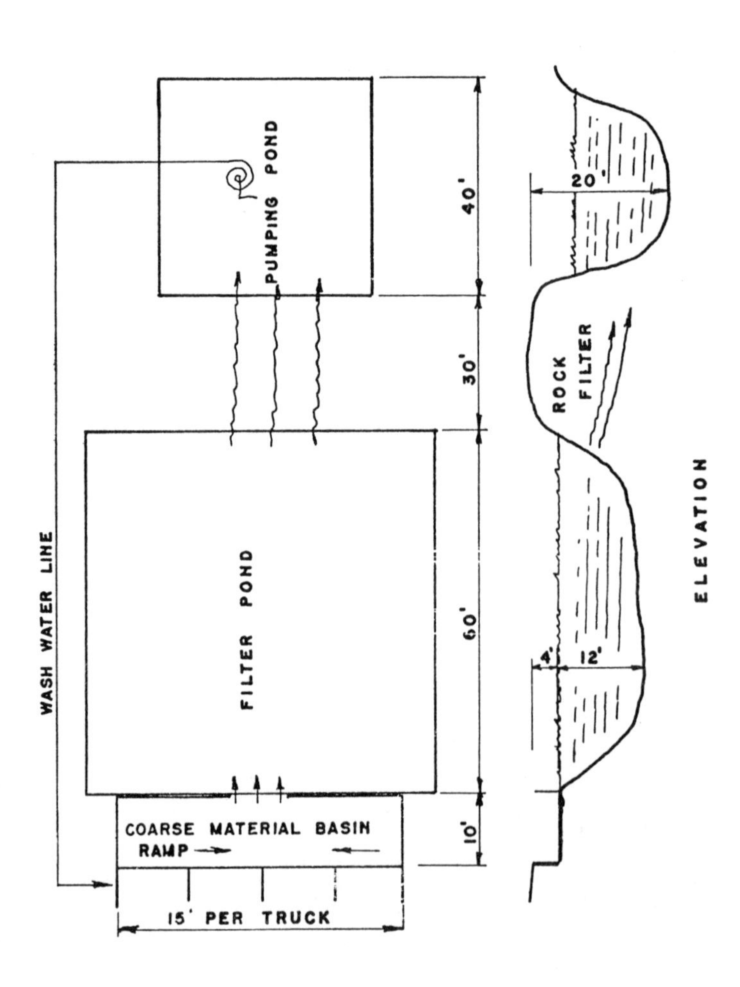

Source: R.G. Monroe (40)

FIGURE 41: PLANT LAYOUT OF SETTLING BASIN FOR WATER CLARIFICATION IN READY MIX PLANT

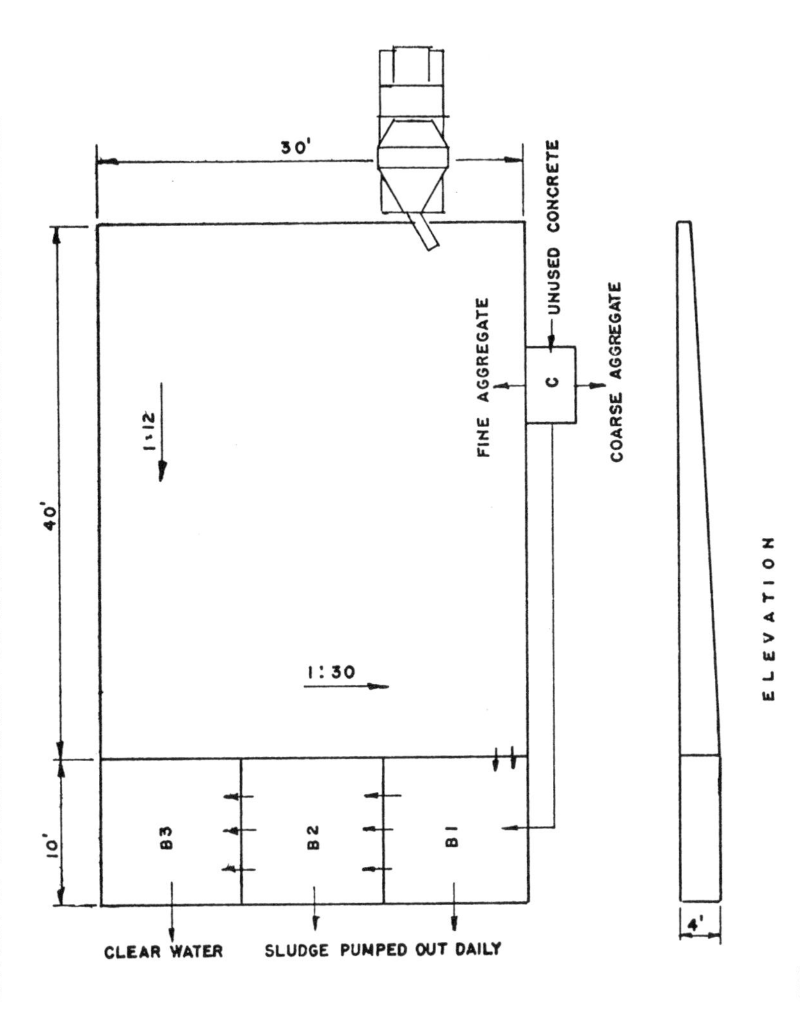

Source: R.G. Monroe (40)

The cost of operating the following described closed system is about $50.00 per day. This plant operates with a very small system and uses very little water—70 gallons per day are used for drum washing and 20 gallons per day are used for outside rinsing for each truck. A fleet of 35 trucks uses approximately 3,000 gallons of wash water per day.

Figure 41 shows arrangement of the water clarification system. An aggregate reclaiming unit and a settling basin are used to process mixer wash water and unused concrete. The coarse aggregate tends to remain on the upper portion of the slab while the cement, fine aggregate and water flow toward the lower end where the fine solids partially settle out. Cement laden water flows around the end of the wall at the lower end of the slab into settling basin **B1** which is ten feet by ten feet by four feet deep. Cement settles in this basin leaving partially clarified water to flow from the basin as more water enters. The clarified water flows to a second basin **B2** and is allowed additional settling time. Water then flows to **B3** and is stored for reuse as wash water to be pumped out as needed.

Unused concrete in amounts over one-half cubic yard is loaded into a reclaiming unit **C** which washes out and separates the coarse and fine aggregate. The coarse aggregate is collected in a dump truck and returned to stock for reuse. The fine aggregate from the reclaimer is dumped onto the inlcined slab to drain, and the cement and water flow to the settling basin. All the material collected on the inclined slab is picked up each morning with a front end loader and is either processed through the reclaimer or hauled to the dump.

Also each morning the clear water on top of the second basin is pumped to the third basin. The sludge from the first and second basins is pumped either into a tank truck and hauled immediately to landfill or discharged to allow drying just beyond the tank. Use of the tank truck eliminates the necessity of drying this material before hauling it away. In this case, one tank truck load per day is sufficient to clean the basin. The washing time is 4 hours per day, so the 3,000 gallons per day flow is at the rate of 12.5 gallons per minute or 1.67 cubic feet per minute.

$$\text{Horizontal velocity} = \frac{\text{volume of water}}{\text{effective cross section of } \mathbf{B1}} = \frac{\text{1.67 cubic feet per minute}}{\text{10 feet} \times \text{4 feet}} = 0.042$$

Experimental data show that 0.0001 inch material will settle at 0.04 feet per minute, so horizontal velocity is acceptable.

$$\text{Basin length} = \frac{\text{horizontal velocity} \times \text{depth}}{\text{vertical velocity}} = \frac{0.042 \times 4}{0.00148} = 11.4 \text{ feet}$$

Basins **B1** and **B2** act as settling ponds while **B3** is a holding basin. The combined length of basins **B1** and **B2** is 16 feet. Sixteen feet is greater than the required 11.4 feet, so the basins are adequate. Basin **B3** must be large enough to hold the surge which is a minimum of one day's water use. Most plants use much more water per truck so they would require much larger basins (**B1**, **B2** and **B3**).

Recently the manufacturers serving the ready mix concrete industry have developed wash out systems. Most of these units are quite new and many need certain improvements. The most popular type of clarifier is the drag chain type (see Figure 42). The truck drum is washed and dumped into the wash tank. The larger aggregates settle and are carried away by the first drag chain and stacked at the end of the washer. The finer aggregate and sand settle more slowly, thus carrying them to the second drag chain.

The water is then recycled into the next ready mix drum. This water is not clear enough to discharge into rivers, but will work well as a wash water for the insides of the mixer drums. The outside of the truck must be washed down with fresh water and some of this water can be collected and used for the make-up water for the washer. In most operations there will be an excess of water to clarify and discharge or use as mix water.

FIGURE 42: DRAG CHAIN WASHER

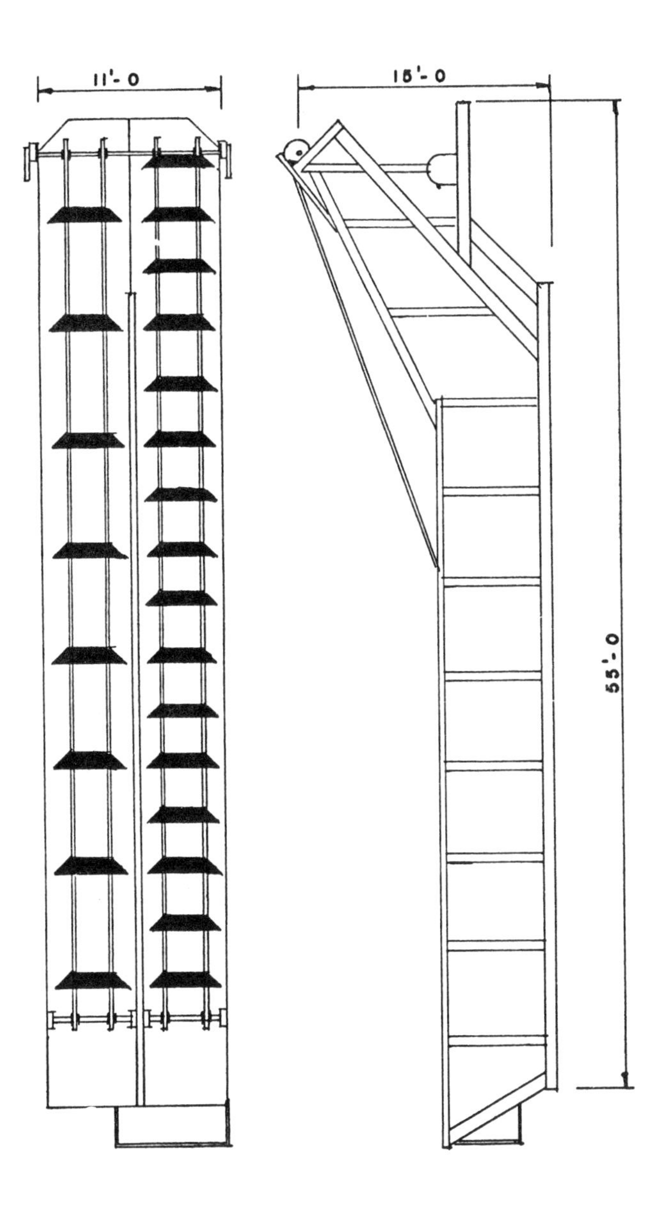

Source: R.G. Monroe (40)

Manufacturers can also furnish aggregate screws which will separate the coarse aggregate and most of the sand; see Figure 43. The fine sand and cement particles will remain in the water and can be recirculated into the ready mix truck drums or further clarified by weirs, settling basins, coagulants or filters. This system might be used by some operators because they may already have aggregate screws on hand.

A further refinement to the system is to discharge the material from the screw onto a screen; see Figure 44. This allows separation of waste aggregates so they can be returned to the proper stockpile for reuse.

If desired, the ready mix truck can discharge directly onto the screen and the aggregate is separated; see Figure 45. The sand and water feed into a sand screw where the sand and water are separated. The advantage of this arrangement is if the plant has a sand screw on hand, but lacks an aggregate screw, the existing equipment can be used. This system also provides a larger capacity because only the sand goes through the sand screw.

One disadvantage is that spray bars will be needed on the screen requiring additional water to be clarified. All these arrangements using screws will require a surge tank with capacity for as many trucks as will be washed at once. For example, if two trucks were washed at a time, each using a maximum of 500 gallons, the tank must hold at least 1,000 gallons. These washer units all salvage the aggregate for reuse. In the case of the screen, the material is separated by size for future use in the mix. The drag chain system, on the other hand, does not accurately grade the aggregate.

None of the units investigated clean the water sufficiently for discharge into the waterways. The content of suspended solids is too high as is the pH factor. Therefore, these units must be used as closed systems.

The washing of the outside of the trucks must be a separate system. The outsides of the trucks will need less washing as time goes on, due to increased demands for dust control at plant sites. With less dust in the air, trucks will require less frequent washing.

Although few plants are using wash water for mix water, this system should be considered because of possible economies and the fact that this system would eliminate all discharge into the waterways. The American Society for Testing and Materials C94-67 makes the following definition of acceptable mixing water.

> 3.1.3 Water – The mixing water shall be clear and apparently clean. If it contains quantities of substances which discolor it or make it smell or taste unusual or objectionable or cause suspicion, it shall not be used unless service records of concrete made with it or other information indicates that it is not injurious to the quality of the concrete.
>
> Note 3 – Information on the effects of questionable mixing water may be secured by testing mortar made with the water in question in comparison with mortar mixed with potable water of known acceptable quality in accordance with ASTM C87, Test for Effect of Organic Impurities in Fine Aggregate on Strength of Mortar.

Wash water from ready mix operations, when allowed to stand for short periods of time will become clear and apparently clean. There will be some cause for suspicion and if so, tests or service records could be used which should satisfy the user.

At least one plant has agitated the wash water to keep the cement suspended and has reported good concrete quality; however, clogged plumbing became a problem. Using slurries in such concentration to cause clogging would also raise a question as to the acceptance of the water as adequate mixing water. At least at the present, it appears more desirable to allow the aggregate, sand and cement to settle and be hauled away. The water would still be unacceptable for discharging into waterways, but should be very acceptable wash water. Extensive testing should be done so operators would be

FIGURE 43: SCREW WASHER

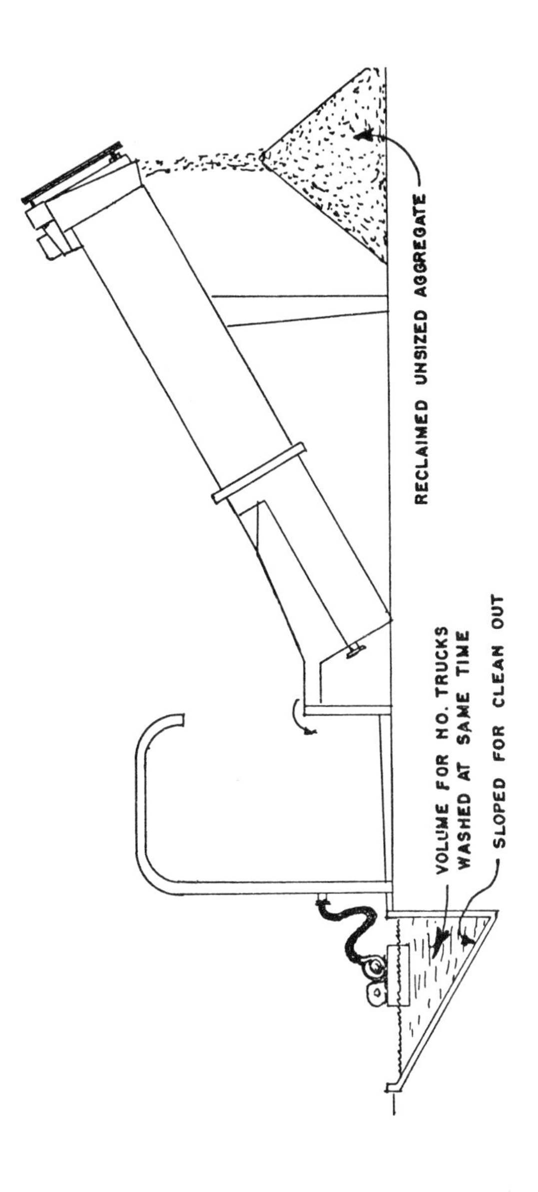

Source: R.G. Monroe (40)

FIGURE 44: SCREW WASHER AND SCREEN

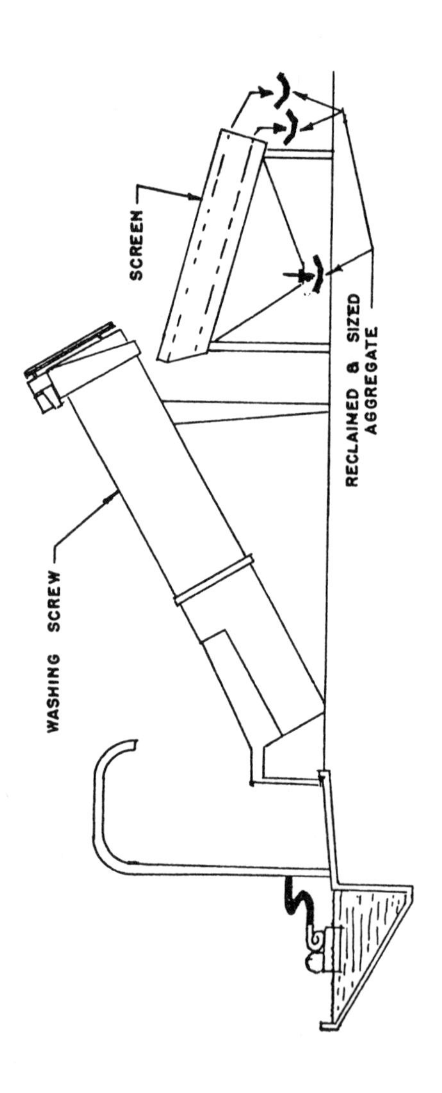

Source: R.G. Monroe (40)

FIGURE 45: WASHING SCREEN AND SAND SCREW

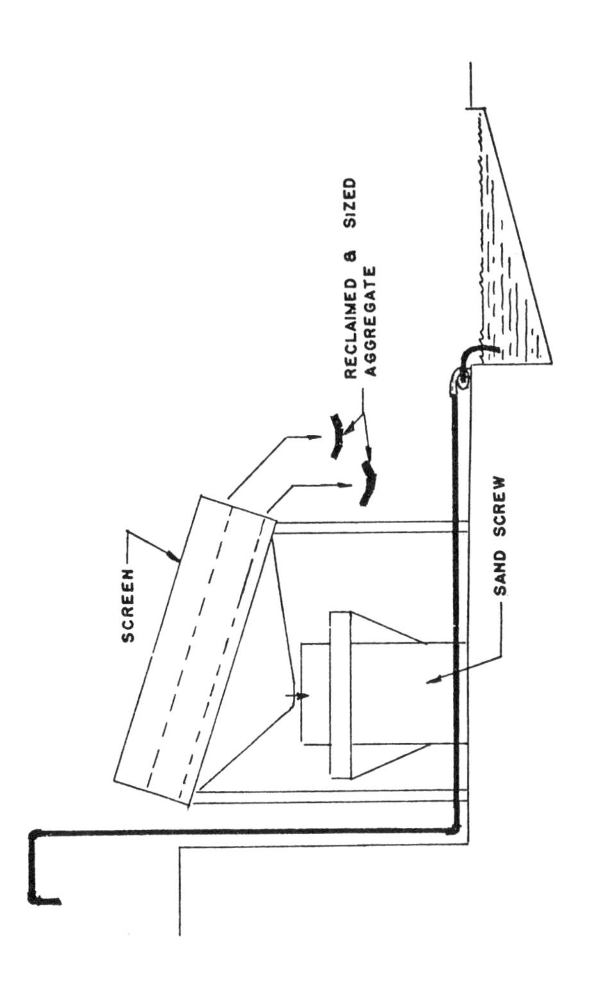

Source: R.G. Monroe (40)

able to furnish this evidence in gaining approval to use their wash water.

The cost of systems for clarifying water in the aggregate and ready mix industries varies widely. The plant owners seem unable to establish realistic prices on their systems. Some of the factors that must be considered in arriving at the cost are:

[1] value of land used;

[2] cost of land development;

[3] cost of equipment;

[4] cost of operation;

[5] cost of maintenance;

[6] effect on plant production;

[7] change in water and sewer charges; and

[8] cost of reporting to water control boards.

Many ready mix plants located on expensive property in or near cities have very limited space. To use any of this land for a clarification system will make the production facilities less efficient.

For this type of operation the only systems applicable and now developed are the manufactured units similar to the drag chain washer. A four-truck wash system of this kind, completely installed with paved area, water piping, electrical, concrete slab, pumps, etc., costs about $35,000.00.

Because most plants on high priced property are large operations, one unit would not be adequate. A thirty-truck fleet should have two units in order to have a smooth operation. The land required for equipment and truck wash area would be 110 feet by 200 feet with allowance for turning on one side, and a pond area for treating the excess water before discharging into the stream.

The cost of operation includes repairs, power consumption, removal of waste material by front end loader and trucks. The operating costs could vary from $10.00 to $100.00 per day, depending on the ability to sell some of the waste.

The ready mix batch plant and component costs also can vary widely from $50,000.00 to $250,000.00. The minimum property size would be about 60,000 square feet. Property value could vary from $30,000.00 per acre to $100,000.00 per acre.

There can be a wide variation in the cost per gallon of water to operate the plant inasmuch as the volume of water to be clarified ranges from 150 to 4,000 gallons of wash water per truck. It is apparent that the costs vary so widely that any averaging would only result in creating false impressions. The cost of clarification for some operators is a very small percent of the total cost of doing business, while for others the cost is very high.

FIBER GLASS INDUSTRY

Glass fiber products are manufactured by melting various raw materials to form glass (predominantly borosilicate), drawing the molten glass into fibers, and coating the fibers with an organic material. The two basic types of fiber glass products, textile and wool, are manufactured by different processes. Typical flow diagrams are shown in Figures 46 and 47.

In the manufacture of textiles, the glass is normally produced in the form of marbles after refining at about 2800°F (1540°C) in a regenerative, recuperative, or electric furnace. The marble-forming stage can be omitted with the molten glass passing directly to orifices to be formed or drawn into fiber filaments. The fiber filaments are collected on spools as continuous fibers and staple yarns, or in the form of a fiber glass mat on a flat, moving surface. An integral part of the textile process is treatment with organic binder materials followed by a curing step.

In the manufacture of wool products, which are generally used in the construction industry as insulation, ceiling panels, etc., the molten glass is most frequently fed directly into the forming line without going through a marble stage. Fiber formation is accomplished by air blowing, steam blowing, flame blowing, or centrifuge forming. The organic binder is sprayed onto the hot fibers as they fall from the forming device. The fibers are collected on a moving, flat surface and transported through a curing oven at a temperature of 400° to 600°F (200° to 315°C) where the binder sets. Depending upon the product, the wool may also be compressed as a part of this operation.

About 50 percent of the manufactured fiber glass is used for thermal and accoustical insulation and about 35 percent is used for textiles. The basic ingredients for continuous filament fiber include silicon dioxide, calcium oxide, aluminum oxide, and boric oxide. Increasing widespread uses continue to make the demand favorable. Glass marbles are melted then extruded through small holes (forming operation) making a filament. The filament travels through a curing oven before it is packed.

Plant size is determined by the number of lines or forming and curing machines. Since the raw materials for the process are only changed in form, one pound of ingredients will yield one pound of glass fiber material. Special properties like strength, elongation and durability are controlled by the addition of various basic ingredients. The major air pollutants from fiber glass manufacture are glass fiber particulates and phenol resins. Cyclones and fume incinerators are the common types of air pollution abatement equipment (18).

FIGURE 46: TYPICAL FLOW DIAGRAM OF TEXTILE-TYPE GLASS FIBER PRODUCTION PROCESS

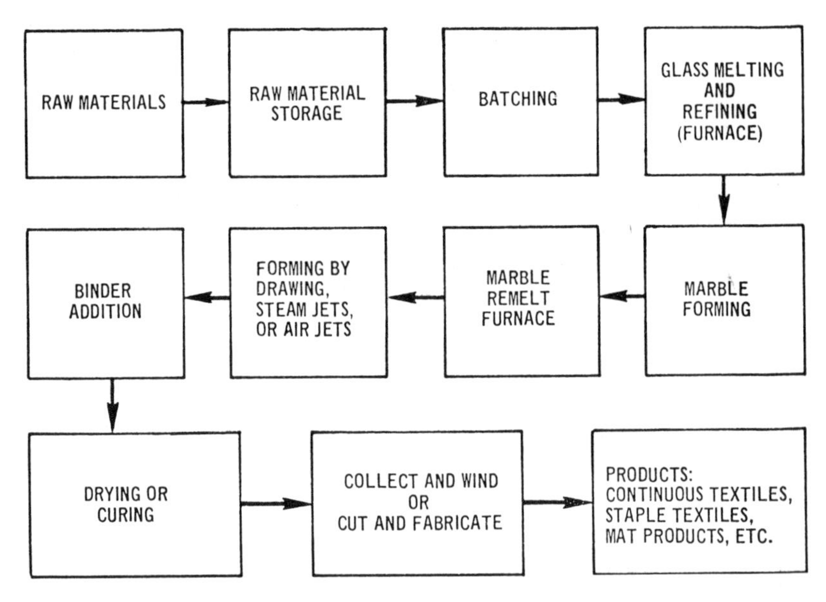

FIGURE 47: TYPICAL FLOW DIAGRAM OF WOOL-TYPE GLASS FIBER PRODUCTION PROCESS

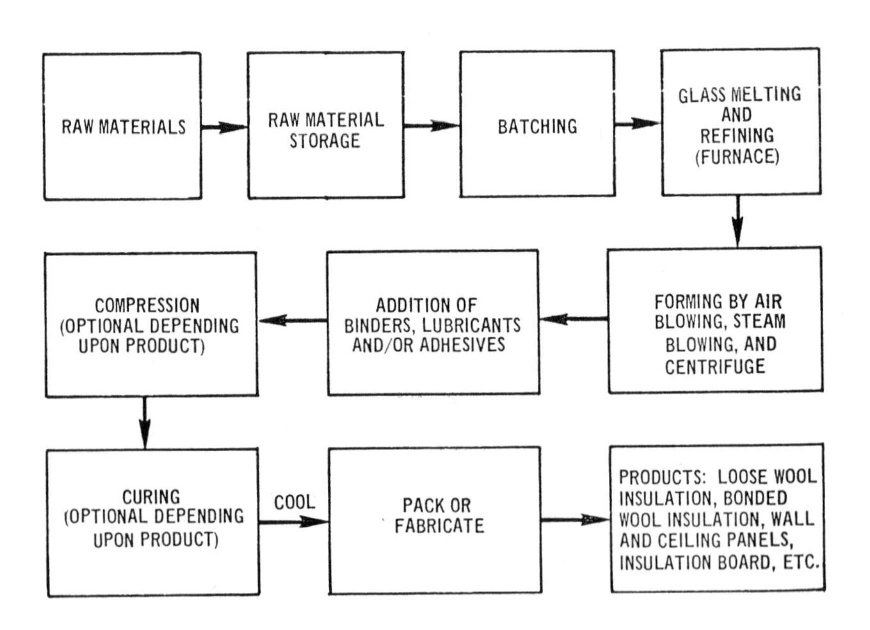

Source: EPA (15)

The industry covered in this section (41) is the insulation fiber glass manufacturing segment of the glass manufacturing source category. It encompasses a part of the Standard Industrial Classification 3296 in which molten glass is either directly or indirectly made, continuously fiberized and chemically bonded with phenolic resins into a wool-like insulating material. The scope of this subcategory also includes those products generally referred to as insulation fiber glass by the industry, that are produced by the same equipment and by the same techniques as thermal insulation. These include, but are not limited to, noise insulation products, air filters, and bulk wool products.

This category will be referred to as a primary process in contrast to a secondary operation in which waste textile fiber glass is processed into an insulation product. Such secondary operations are excluded because of their textile origin and the difference in processing techniques. Insulation fiber glass research and development laboratories are also excluded in this report, because the range of such research includes textiles, and a great diversity of experimentation not necessarily related to insulation products. The term insulation fiber glass is synonymous to the terms glass wool, fibrous glass, and construction fiber glass.

The modern fiber glass industry was born in 1935 when the Owens Illinois Glass Company and the Corning Glass Works combined their research organizations later forming Owens-Corning Fiberglas in 1938. The original method of producing glass fibers was to allow molten glass to fall through platinum bushings, forming continuous relatively thick threads of soft glass. The glass streams are then attenuated (drawn) into thin fibers by high velocity gas burners or steam. This process generally referred to as flame attenuation is pictured in Figure 48.

FIGURE 48: FLAME ATTENUATION PROCESS

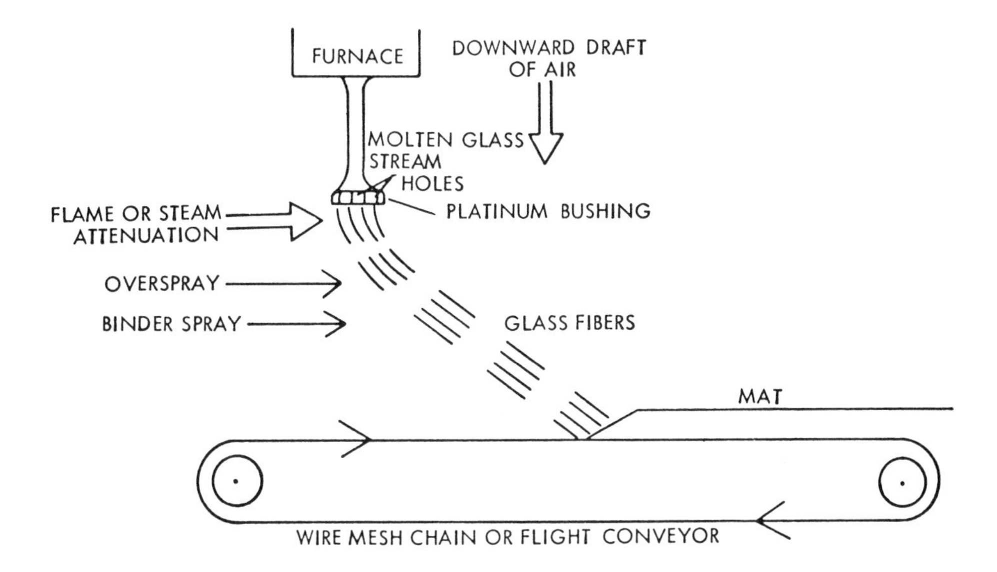

Source: EPA (41)

In the 1950s, Owens-Corning Fiberglas and the Cie de St. Gobain perfected the centrifugal or rotary process. A single steam of molten glass is fed into a rotating platinum basket which distributes the glass on an outer rotating cylindrical spinner. The spinner contains a large number of small holes arranged in rows in the wall. The molten glass is forced through the holes forming fibers which are then attenuated 90° from their forming direction by high velocity gas burners, air or steam as depicted in Figure 49. The output of a single spinner may range from 0.23 to 0.45 metric tons per hour (500 to 1,000 lb/hr) and up to 5 or 6 spinners are used to feed fiber to one line.

FIGURE 49: ROTARY SPINNING PROCESS

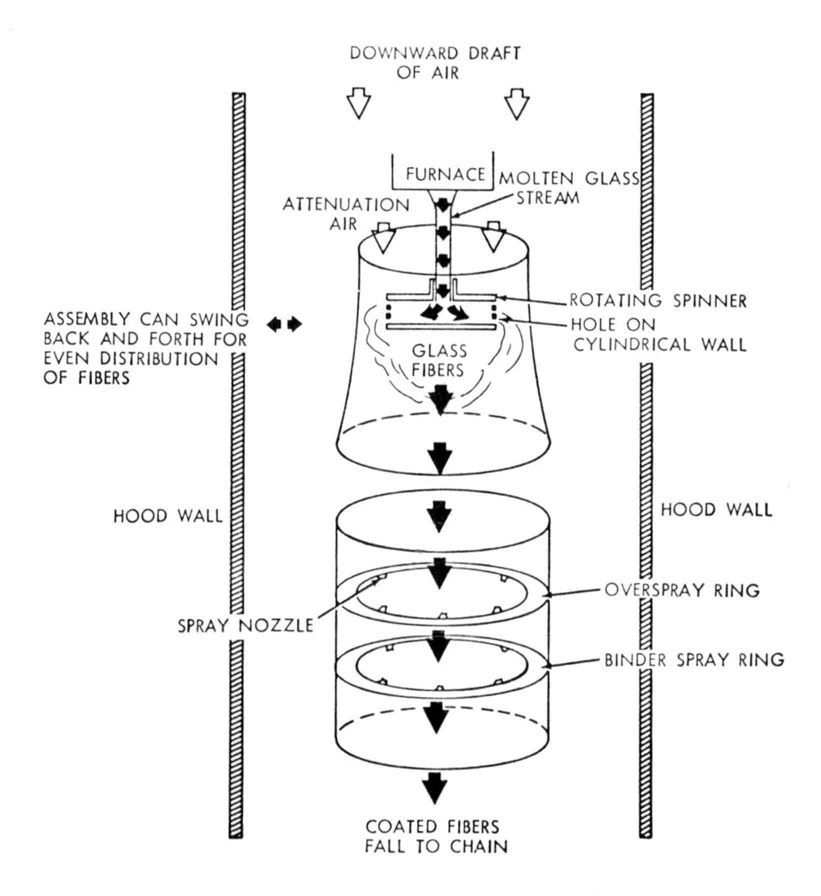

Source: EPA (41)

Figure 50 depicts the basic insulation fiber glass processes. They have their own individual merits. The flame attenuated product has greater longitudinal strength because the fibers are attenuated in the same direction (away from the gas or steam blower) and the lengths consequently align in one direction to give added tensile strength in that direction. This results in decreased damage to the product upon installation. Rotary spun fibers, on the other hand, are attenuated as they form on the circumference of a rotating disk. The fiber lengths thus assume random directions as they fall. Standard building insulation produced by the flame attenuated process generally uses less fiber (approximately 35%) to achieve the same thermal properties as rotary spun standard insulation. Since insulation is priced in

FIGURE 50: HOW INSULATION FIBER GLASS IS MADE

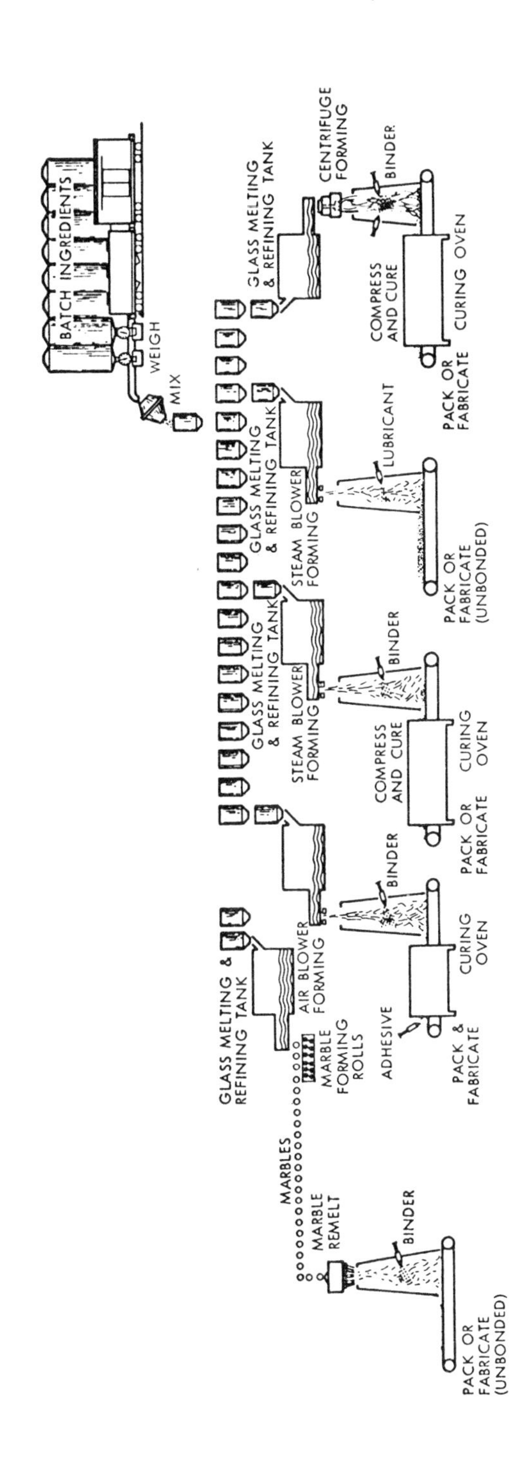

Source: EPA (41)

accordance to its thermal properties, annual production ratings and plant capacities measured in kilograms can be somewhat misrepresentative when comparing the economics of the two processes. All small plants utilize the flame attenuation process and are financially better off than an economic analysis based on overall industry, plant capacity would indicate. Rotary forming processes can produce more uniform fibers. They are also capable of producing huge tonnages of wool, and for these reasons the rotary process now dominates the industry.

Borosilicate glasses and low alkali silicate glasses are generally used in making glass fibers because of their chemical durability. The surface area to weight ratio of the glass fibers in glass wool products is so great that even atmospheric moisture could seriously weather common silicate glass fibers. Technological changes brought on by consumer demands has already made some of these products obsolete. The low thermal conductivity property of insulation fiber glass is not directly attributable to the glass, but rather to the ability of the glass fibers to establish stationary pockets of air. The fiber glass web in which these pockets are held minimizes heat transfer by air convection currents and limits it to conduction in air which is a much slower rate.

There are two methods of producing the molten glass (1260° to 1316°C) that feeds the fiberizing machine in the forming area. The older method involves first producing 2.5 cm (one inch) glass marbles and then feeding the marbles to a small remelt furnace which in turn feeds the fiberizer with molten glass. There can be several remelt pots to each production line. The marbles may either be produced at the plant site or made at a centrally located plant with a large furnace and shipped to other plants.

The original purpose of this seemingly redundant procedure is to insure glass uniformity before the fibers are made by visually inspecting the glass marbles. The mechanical problems caused by seeds and bubbles are more troublesome in fibers than massive glass because of the small glass diameters involved. The assurance of better quality control in the glass making stage, however, has led to the replacement of the intermediate glass marble process by direct feed furnaces. Currently only one company operates marble-feed processes for insulation products, since this company finds it less costly to ship marbles than to build and maintain glass making furnaces at every small plant.

Rotary processes are always fed by direct melt furnaces because rotary spinners have high volume production capabilities which can only be matched by direct melt furnaces. Furthermore the high cost of a glass furnace usually necessitates that it be large, which in turn requires a large plant capacity in order for the operation to be profitable. Both marble feed and direct melt processes feed flame attenuation forming processes.

When production changes occur in a direct melt process the molten glass flow is temporarily diverted from the fiberizers and quenched with water. The glass immediately solidifies and fractures into fragments resembling a mixture of sand and aggregate, which is termed cullet. A major portion of the cullet is collected at the machine in hoppers for reuse into the melting furnace. If the furnace is not bled by producing cullet, the lighter components in the molten glass will volatilize and the composition of the glass will be unpredictably altered. This is not a problem in the marble-feed process because of the very small volume of molten glass held in the remelt pots. This problem, along with other restrictions, requires that direct melt processes be operated 24 hours a day, all year round.

The quality of water needed for cullet cooling is not critical in that this water may be reused, with make-up water added to compensate for the water vaporized by contact with the hot glass. It is not important that the water be cooled, but sufficient suspended solids must be removed so as to prevent damage to the pumps. Colloidal silica suspensions are controlled by sufficient blowdown. After the molten glass is divided into fibers and attenuated, the fibers are sprayed in mid-air with a phenolic water-soluble binder (glue) and are forced by a downward air draft onto a conveyor chain. In many plants the newly formed fibers are oversprayed with water at the same time that the binder is applied. This overspray serves to cool the almost molten glass, minimizing both volatilization and earyl polymerization of the binder. The binder is a thermosetting resin composed of a dilute solution of phenols (resin) and other chemical additives which provide terminal crosslinking and stability of the finished product. The resin itself is a complex mixture of methylolphenols in both the monomer and polymer states. For some products lubricants are applied to the newly formed fibers singly or in addition to the binder. The lubricant, usually a mineral oil, is used to minimize skin irritation (fiber abrasion) of persons handling the insulation.

The binder is diluted with two to six times its volume in water before it is applied to the product. The quality of the dilution water is important in that it must not contain solids of such size as to plug the spray nozzles and in that it must not contain sufficient concentrations of chemicals that would interfere with the curing properties of the binder. For instance, magnesium and calcium found in hard water are incompatible with the binder.

The fibers fall to the chain where they collect in the desired mass and depth as required for the ultimate product. The density of the fiber mass (mat) on the conveyor is controlled by the fiber production rate and the speed of the conveyor chain. For the rotary forming process the chain speed will range from 127 to 508 linear cm/sec (50 to 200 ft/min). This mat then proceeds by conveyor through curing (200° to 260°C) and cooling ovens, it is compressed, and an appropriate backing (asbestos, paper, aluminum, etc.) may be applied as a vapor barrier. The product is then sized and/or rolled and packaged. The cured mat may instead be shredded to make blowing and pouring wool. This product is used where existing structures require insulating material that can be blown or poured into the walls. The thermal properties, however, are inferior to backed insulation.

The cured phenolic resin imparts a yellow color to the glass wool which may not be appealing to the customer. Consequently, various colored dyes are applied to the fiber glass in the binder spray and other than esthetics do not beneficiate the product.

Two types of chains are employed in the forming area. Flexible wire mesh conveyor belts were originally used, but many have since been replaced by flight conveyors. These are hinged steel plates that contain numerous holes or slits. The air stream which transports the glass fibers to the conveyor also contains droplets of resinous binder which have not adhered to the glass fibers. Many of these droplets deposit resin on the chain, and if not removed, the resin build-up will eventually restrict passage of the air stream. When the deposit becomes sufficiently great, insulation fiber glass formation is no longer possible, necessitating replacement of the conveyor.

At present only three companies produce fiber glass insulation in nineteen plants. Because a high volume production is necessary and the glass fiber operation is difficult to scale down, there are no very small plants when compared to other industries. The smallest plant produces 2,270 metric tons (5 million pounds) of specialty products per year.

AIR POLLUTION

The major emissions from the fiber glass manufacturing processes are particulates from the glass-melting furnace, the forming line, the curing oven, and the product cooling line. In addition, gaseous organic emissions occur from the forming line and curing oven. Particulate emissions from the glass-melting furnace are affected by basic furnace design, type of fuel (oil, gas, or electricity), raw material size and composition, and type and volume of the furnace heat-recovery system. Organic and particulate emissions from the forming line are most affected by the composition and quality of the binder and by the spraying techniques used to coat the fibers; very fine spray and volatile binders increase emissions. Emissions from the curing ovens are affected by oven temperature and binder composition, but direct-fired afterburners with heat exchangers may be used to control these emissions. Emission factors for fiber glass manufacturing are summarized in Table 43.

TABLE 43: EMISSION FACTORS FOR FIBER GLASS MANUFACTURING WITHOUT CONTROLS[a, b]

Type of process	Particulate		Sulfur oxides (SO_2)		Carbon monoxide		Nitrogen oxides (NO_2)		Fluorides	
	lb/ton	kg/MT	lb/ton	kg/MT	lb/ton	kg/MT	lb/ton	kg/MT	lb/ton	kg/MT
Textile products										
Glass furnace[c]										
Regenerative	16.4	8.2	29.6	14.8	1.1	0.6	9.2	4.6	3.8	1.9
Recuperative	27.8	13.9	2.7	1.4	0.9	0.5	29.2	14.6	12.5	6.3
Electric	ND[d]	–	–	–	–	–	–	–	–	–
Forming	1.6	0.8	–	–	–	–	–	–	–	–
Curing oven	1.2	0.6	–	–	1.5	0.8	2.6	1.3	–	–
Wool products[e]										
Glass furnace[c]										
Regenerative	21.5	10.8	10.0	5.0	0.25	0.13	5.0	2.5	0.12	0.06
Recuperative	28.3	14.2	9.5	4.8	0.25	0.13	1.70	0.9	0.11	0.06
Electric	0.6	0.3	0.04	0.02	0.05	0.03	0.27	0.14	0.02	0.01
Forming	57.6	28.8	–	–	–	–	–	–	–	–
Curing oven	3.5	1.8	ND	–	1.7	0.9	1.1	0.6	–	–
Cooling	1.3	0.7	–	–	0.2	0.1	0.2	0.1	–	–

[a]Emission factors expressed as units per unit weight of material processed.
[b]Owens-Corning Fiberglas Corporation communication to EPA.
[c]Only one process is generally used at any one plant.
[d]No data available.
[e]In addition, 0.09 lb/ton (0.05 kg/MT) phenol and 3.3 lb/ton (1.7 mg/MT) aldehyde are released from the wool curing and cooling operations.

Source: EPA (15)

Insulation fiber glass plants experience both air particulate and odor problems. Particulate emissions are found in the glass furnace, forming area, and curing and cooling ovens exhaust gases. The principal source of odors is volatilized phenols in the curing and cooling ovens exhaust gases. Several methods, involving both wet and dry processes, are being investigated in an effort to reduce the air emissions. At the present time the industry considers air pollution control to be a more serious problem than water pollution control.

Total process water recirculation systems have no adverse impact on air emissions. In only one case has this been the exception. In this case inadequately treated water is recycled as air scrubber water and may actually transfer contaminants to the air. However, this plant will soon be installing additional water treatment equipment which should correct the problem.

The type of system used to control air pollution will definitely affect the water treatment scheme. In plants where dry air pollution control equipment is adequate, high pressure, low volume chain sprays are feasible, easing the water treatment problem. Evaluation of the economics indicates that the cost differences of the water treatment systems as they apply to air emissions control systems are not a factor. At this time contaminated scrubber water is being accommodated by process water total recycle systems, and it is not necessary to subcategorize according to methods of air pollution control.

Proprietary Control Processes

A process developed by F.E. Warner et al (42) provides for the removal of phenolic air pollutants in the production of glass fiber products. The polluted air is passed sequentially through at least two low energy contacting zones in each of which it is contacted with descending scrubbing liquor, each succeeding contacting zone having at its bottom individual liquor collecting means. In passing from one contacting zone to the next zone the air stream undergoes acceleration in a transfer passage in which no interzone flow of liquor occurs, and then impinges on baffle means disposed in its path as it enters the next zone at a level above the pool of scrubbing liquor collected at the bottom of the next zone in the liquor collecting means. Figures 51 through 53 show the various steps in the present process in some detail.

The concentration and quantities of pollutants leaving a glass fiber product manufacturing line can vary greatly from day to day and in accordance with changes in the particular product being manufactured and the plant illustrated is intended to be capable of dealing with this situation. The production line includes the forming conveyor **11** and from this there issues the hot polluted air stream **12**. Upon leaving the forming conveyor the product passes through an oven (not shown) maintained at an elevated temperature suitable for curing the phenolic resin binder; fumes are evolved in this curing oven and it is to be understood that the polluted air stream for treatment, as indicated at **12**, will ordinarily include these oven fumes as well as the conveyor effluent. Provision also has to be made for washing the conveyor and the conveyor wash liquor can be recirculated from a collecting tank **13** by a pump **14**.

The polluted air stream **12** passes into an expansion box **15** where it is treated with wash liquor introduced through sprays **16**. This serves to remove glass fibers carried over in the air and also lowers the concentration of pollutants from the resin binder. The expansion box wash liquor is collected in a tank **17** and can be recirculated to the sprays **16** by a pump **18**. From the expansion box **15** the air stream passes to the foot of the gas scrubbing column **19**. This column comprises three contacting zones **20, 21, 22** up through which the air passes in sequence to leave at **23** at the top of the column. Each zone contains Glitsch Grid packing **24** supported on a baffle plate **24a** and has a liquor-collecting tray **25** at its base.

The air enters each zone by passing up through the tray **25** by way of chimney risers **26** that project above the tray so as to prevent liquor draining down. The zones **20, 21, 22** have respective sprays **27** for introducing scrubbing liquor at the top of each zone.

The liquor collecting in the trays **25** drains into respective tanks **28** where it can be recirculated by pumps **29** to the respective sprays **27**. The tank **28** associated with the top zone **22** overflows into the tank for the middle zone **21** which in turn overflows into the tank for the bottom zone **20**. To control the concentration of the liquor in the scrubbing zones fresh liquor is introduced to the pump feeding the spray of the top zone **22** via a line **30**. This causes a proportion of the liquor circulating through the top zone to be displaced into the scrubbing liquor circuit for the middle zone which in turn displaces liquor from the middle zone circuit into the bottom zone circuit. Liquid thus displaced from the bottom zone circuit leaves via a line **31** and a pump **32**.

The effluent leaving the pump **32** can flow either via a dilute effluent line **33** to a biological treatment unit **34** or via the expansion box and forming conveyor wash liquor circuits to a concentrated effluent line **35** where it passes through a chemical treatment section **36** for reducing its concentration before it also is delivered into the biological treatment unit **34**.

FIGURE 51: GAS SCRUBBING COLUMN FOR TREATMENT OF EXHAUST AIR FROM FIBER GLASS MANUFACTURE

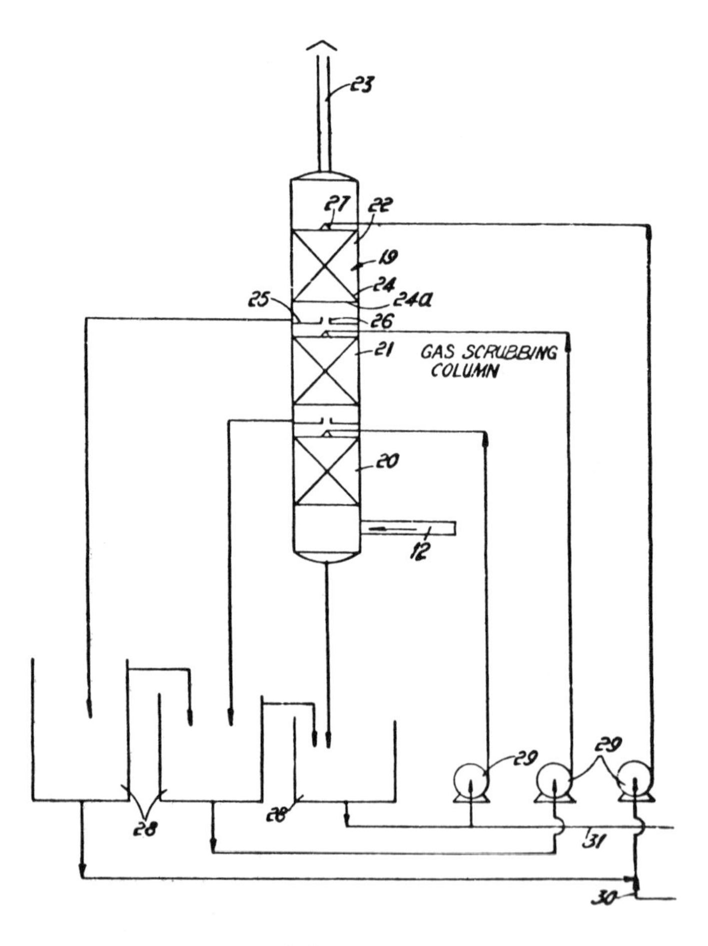

Source: F.E. Warner and A.P. Rice (42)

FIGURE 52: WASH LIQUOR CIRCUIT FOR FUME TREATMENT IN FIBER GLASS MANUFACTURE

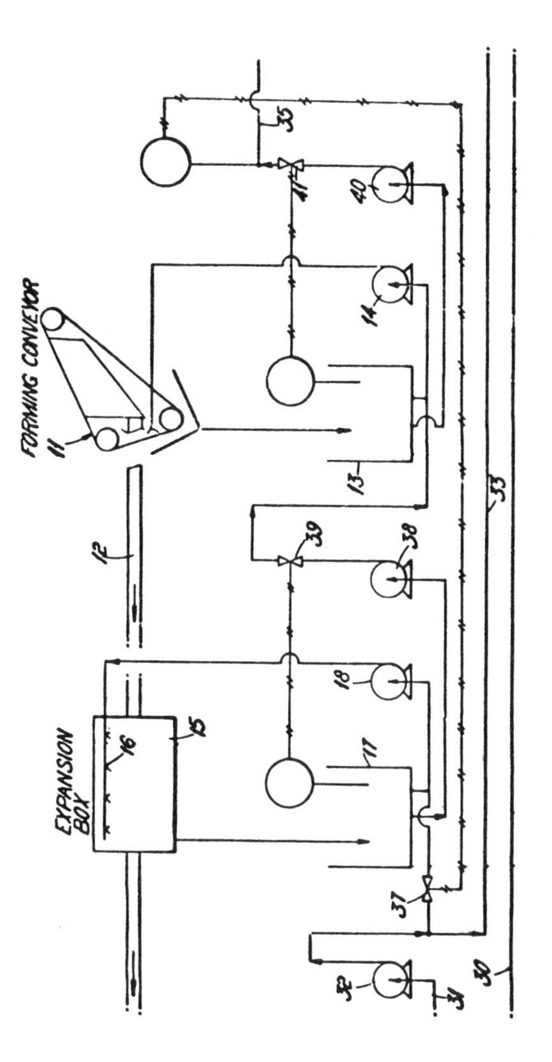

Source: F.E. Warner and A.P. Rice (42)

FIGURE 53: FILTRATION, BIOLOGICAL TREATMENT AND SLUDGE DISPOSAL SECTION OF FIBER GLASS WASH LIQUOR SYSTEM

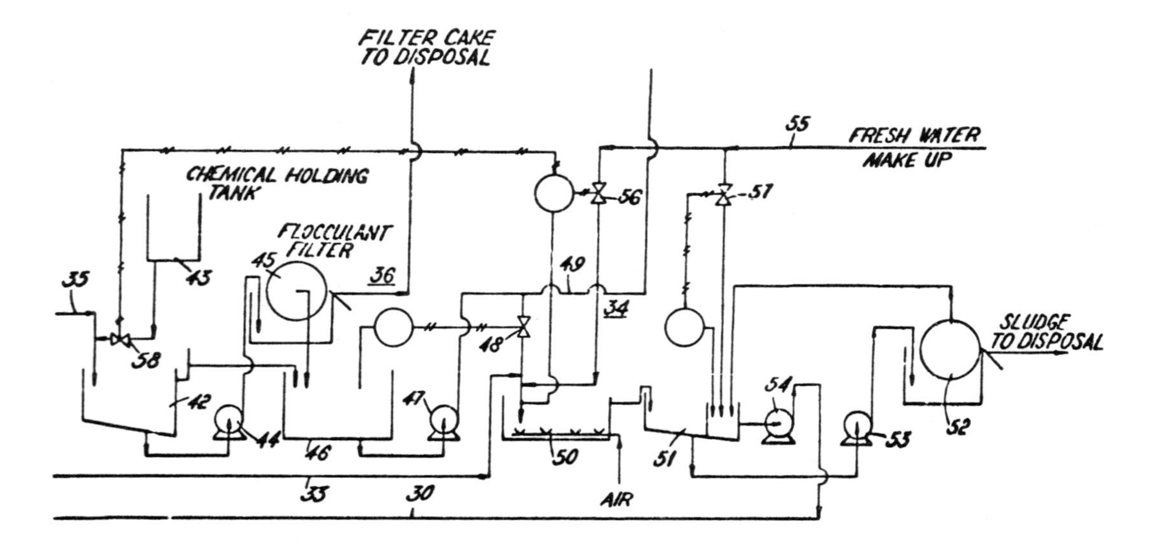

Source: F.E. Warner and A.P. Rice (42)

The pump **18** in the expansion box wash liquor circuit is able to draw its supply not only from the expansion box wash liquor tank **17** but also from the delivery of the scrubbing liquor effluent pump **32** via a process control valve **37**. A further pump **38** draws from the expansion box wash liquor tank **17**, and the pump **14** in the forming conveyor wash liquor circuit is able to draw its supply not only from the conveyor wash liquor tank **13** but also from the delivery of the pump **38** via a process control valve **39**. A proportion of the concentrated wash liquor in the conveyor wash liquor tank **13** is passed out by a pump **40** to the line **35** via a further process control valve **41**.

The proportion of the scrubbing liquor delivery from the pump **32** that passes through the expansion box conveyor wash circuits, as determined by the valve **37**, is regulated by a controller for the valve **37** which is responsive to the concentration of the liquor in outflow line **35**. The outflow from the expansion box and conveyor wash liquor circuits, via the respective valves **39** and **41**, is regulated by controllers for these valves responsive to the levels of the liquors in the expansion box wash liquor tank **17** and the conveyor wash liquor tank **13**, respectively.

The concentrated effluent line **35** delivers into a flocculation tank **42** where it is dosed with chemical from a supply tank **43** to cause flocculation and sedimentation of a proportion of the phenolic contents. The chemicals used can be ferric sulfate together with a polyelectrolyte flocculation to promote the rapid formation of large flocs, and lime to adjust the pH value to 7.5 to 8.0.

The liquor and sediment from the bottom of the tank **42** is pumped by a pump **44** to a rotary filter **45**. The filtrate liquor passes to a tank **46**; and any supernatant liquor overflowing from the tank **42** is passed direct into the tank **46**. A pump **47** withdraws liquid from the tank **46** and delivers a proportion into the biological treatment unit **34**, via a process control valve **48**, where it joins the dilute effluent from the line **33**. The remainder of the delivery of the pump **47** flows away in the line **49** for use in mixing the resin binder for the manufacturing process.

The biological unit **34** is of the activated sludge type employing aeration, the air being introduced at **50**, e.g., through submerged diffusers. The material from this unit is delivered into a sludge settling tank **51** where the sludge is pumped by a pump **53** to a rotary filter **52** the filtrate being returned to the settling tank. A pump **54** withdraws the clear supernatant liquor from the tank **51** and returns it to the scrubbing column **19** via the line **30**.

Since there is a net loss of water from the system by evaporation and by withdrawal for use in mixing of the resin binder, fresh make-up water is added from a line **55** into the biological treatment unit **34** via a control valve **56** and also into the settling tank **51** via a control valve **57**. This assists in bringing the liquor concentration in the biological unit into the best range for treatment. The fresh water flow into the biological unit **34** is regulated by a controller for the valve **56** which is responsive to the concentration of the liquor entering this unit. This controller also regulates the delivery of dosing chemical into the flocculation tank **42** by means of a control valve **58**. The fresh water flow into the settling tank **51** is determined by a controller for the valve **57** which responds to the level of the liquor in that tank.

The valve **48** determining the delivery from the tank **46** into the biological treatment unit is regulated in accordance with the level in the tank **46**. The main flow volume is determined by the fresh feed rate to the gas scrubbing column which reappears as effluent from the column with a phenolic concentration not exceeding about 1,100 ppm. This stream branches as described, the minor part passing to the expansion box and conveyor wash liquor circuits. The concentration of the liquor leaving these circuits is monitored and maintained at 3,000 ppm by blending in, as required, the more dilute scrubbing tower effluent. The addition of a volume of scrubbing effluent displaces an equivalent volume of expansion box and conveyor wash liquor at the controlled concentration of 3,000 ppm. This concentrated liquor is subject to chemical treatment to reduce its concentration to 1,500 ppm before it rejoins the main stream of the scrubbing column effluent at the inflow to the biological treatment plant. To keep the concentration of phenols entering the biological unit at a desirable low level it may be advantageous to make the hydraulic capacity of the plant greater than the fresh feed flow rate required by the scrubber. This enables a stream of treated effluent at very low concentration to bypass the scrubber and be available for dilution of the stream entering the biological unit. Alternatively, the feed rate to the scrubber can be increased with corresponding reduction in the phenolic concentration of the scrubber effluent.

The system described enables the operating conditions of the scrubber column to be held substantially steady by making other process adjustments. For instance, a higher load of phenols in the polluted air stream can be dealt with by changing the concentration of the expansion box liquor rather than by changing the flows of the column itself.

If the column were operated alone it would have a considerable water consumption and there would be the problem of disposing of the liquid effluent. But in the system as described the total water usage is reduced by condensation from the air stream in the column and there is no liquid effluent disposal problem at all. The only wastes for disposal are the cake from the chemical treatment plant filter **45** and the cake or sludge from the biological plant filter **52.**

A process developed by J.A. Borst (43) provides an improved method and apparatus for effecting environmental control in apparatus for producing fibers coated with a binder. Heat softened material such as glass is centrifuged into primary filaments which are subsequently attenuated into fibers by means of a high velocity gaseous blast. A vaporizable medium is applied to fibers by spraying for cooling the fibers, binder is subsequently applied to the cooled fibers and the mass of binder coated fibers is collected on a conveyor. An air stream is passed downwardly through the collected mass and the conveyor to cool the mass and to remove excessive binder.

The temperature of the air stream is sensed by a thermocouple to determine the temperature of the mass and the sensed temperature is used to control application of the vaporizable medium to maintain the mass at a substantially constant predetermined low temperature.

A process developed by R.E. Loeffler (44) is one in which a binder impregnated glass fiber blanket is cured and shaped by passing the blanket through a series of spaced-apart heated platen assemblies. The curing of the blanket produces vapors, fumes, odors and other pollutants which must be prevented from escaping to the surroundings of the platen assemblies. Consequently, these gases are purged from the blanket as the blanket passes between adjacent platen assemblies.

Hot air is introduced from a plenum chamber onto one side of the blanket. The hot air and pollutants from the binder are drawn through and from the blanket by a suction chamber on the opposite side of the blanket. The gases from the suction chamber are then passed through an air filler and discharged to the atmosphere.

WATER POLLUTION

Historically, the wire mesh chain has been cleaned while in service by routing the chain through a shallow pan containing a hot caustic water solution (refer to Figure 54). Fresh caustic makeup to the pans created caustic overflow containing phenolic resin and glass fiber.

Another method of chain cleaning uses either fixed position pressurized water sprays or rotating water sprays. Unlike the caustic soda bath processes, the wastewaters from this method are amenable to treatment and recirculation. Water spray chain cleaning has replaced caustic chain cleaning at all but one plant which uses a combination of the two methods. Although both methods have been used to clean wire mesh chains, it is impractical to caustic clean flight conveyors. Unlike the flexible wire mesh chains, the hinged

FIGURE 54: WIRE MESH CHAIN CLEANING

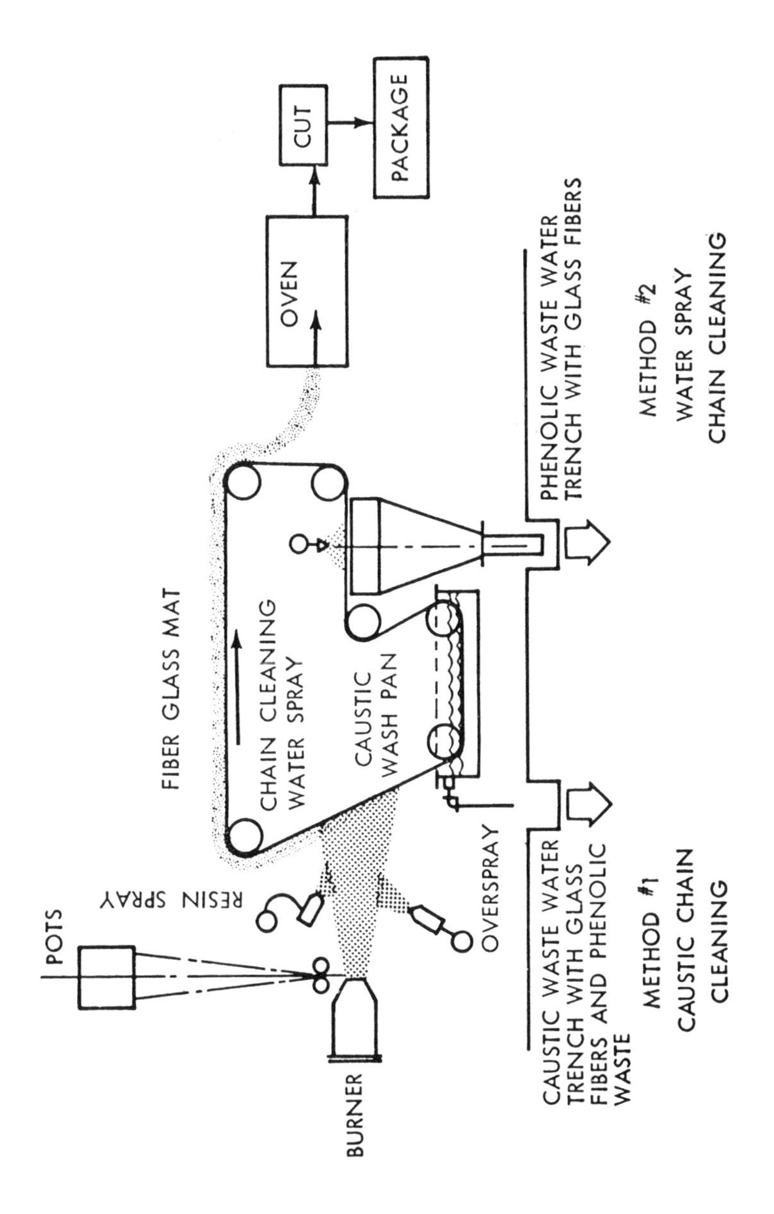

Source: EPA (41)

plates of the flight conveyor cannot be so easily routed through a pan. Furthermore, a flight conveyor is more expensive than a wire mesh chain, and corrosion caused by the caustic is of greater concern. Spray cleaning has the added advantage of cooling the forming chain, thereby decreasing both volatilization and polymerization of the phenolic resin.

Pipe insulation is made in various ways. One principal method involves wrapping uncured insulation about mandrels and curing them batchwise in ovens. The mandrel is a perforated pipe of the appropriate dimensions. Caustic is still used by the industry to batch clean mandrels. However, the volumes involved are much less than those required for chain washing and are consequently much less of a problem.

Another source of water pollution is hood wash water. The hood is either a stationary or rotating wall used to maintain the air draft in the forming area. It is necessary to wash the hood in order to keep any wool that has agglomerated there from falling onto the chain and causing nonuniformity of the product.

From evaluation of the available data it is concluded that the types of wastes generated in producing insulation fiber glass such as suspended solids, dissolved solids, phenols, and oxygen demanding substances are common to all such plants. The only exceptions are dyes and water treatment backwashes. The former parameter presents no problem in so far as quality of recycled water. The quality of water treatment backwashes varies considerably among the industry depending upon the intake water quality. The principal factor of concern to the industry is water hardness which will inhibit the bonding properties of the phenolic resins. The generally similar nature of the wastes generated in insulation fiber glass production indicates that the industry should be considered as a single subcategory.

From discussions with the industry and from plant inspections it was concluded that in a recycle system for an insulation fiber glass plant only three basic parameters in the process water affect its treatability, suspended solids, dissolved solids, and pH. The recycled waters can be adequately treated for reuse by coarse filtration, pH control (if necessary), and fine filtration or coagulation, settling. Sufficient blowdown as can be handled as overspray or binder dilution water is needed to check the buildup of dissolved solids.

Through proper design of the treatment system there should be no foreseeable reason other than plant expansion that these basic systems need to be altered in order to accommodate varying waste load characteristics. Therefore treatability of wastewater factors indicates that all insulation fiber glass plants fit into a single subcategory. As described above, there are two types of glass fiber forming processes, flame attenuation and rotary. Both processes are dry, and since the products are the same, water quality is not affected.

As described above also, there are also two basic methods for cleaning the forming chain of the glass fibers and phenolic resins. One method consists of dragging the wire mesh chain, on its return path to the forming area, through a hot caustic bath. The second method consists of spraying the wire mesh chain or flight conveyor with high velocity water.

The resultant wastes from caustic cleaning are extremely difficult to treat and unless considerable dilution is provided, the wastes are not suitable for recycling. The principle reason for this is that the only practical sink for the wastewaters in a completely closed system is for overspray and binder dilution, and that unless diluted, the caustics are incompatible with phenolic resins. The blowdown from spray washing is amenable to treatment and recycle.

Two subcategories therefore would seem appropriate. However, at the present time only one plant employs caustic chain washing. The remainder of the industry has switched to spray washing and has future plans to employ only spray washing equipment. The one existing plant that uses caustic baths does so in conjunction with spray washing equipment and it is not necessary in this case to blowdown from the caustic bath. The carryover caustic on the chain is so diluted by the wash water volumes that no problems are anticipated in the recycle system. For these reasons the industry cannot be meaningfully subcategorized according to chain cleaning techniques. It has been determined from industry data and from inspections that despite the wide range in plant capacities, plant size has no effect upon the quality of wastewaters. Plant size does affect the costs of installing total recycle systems because of the effect of plant size on the volume of water used. In the economic analysis section it is concluded that the cost of recycle per unit production will increase as much as threefold for plants producing less than 9,000 metric tons per year. However, plants of this size usually produce specialty products (e.g., pipe insulation) which command a higher price per unit weight than standard residential insulation. This factor will minimize the financial impact for the smaller plants. Therefore, subcategorization of according to plant size is not indicated.

Glass wool plants span an age of from 2 years to more than 25 years since plant start up. About 30 percent of the plants are 10 to 15 years old while 25 percent are less than 10 years old. All plants that are at least 5 years old have undergone considerable upgrading of the production processes and in many cases facilities have been expanded with installation of state of the art processes.

Wastewater characteristics are therefore similar for plants despite any difference in age. Except for old plants of large capacity, plant age should not significantly affect costs of installing the facilities. In large old plants space limitations and major pipe relocations will increase the capital costs. However, the capital cost of recycled water is lowest for large plants and this will help compensate for the increased installation costs. Hence, plant age is not an appropriate basis for subcategorization.

The raw materials required for wool glass are much the same as for standard massive glass, 55 to 73 percent silica and 27 to 45 percent fluxing oxides (e.g., limestone and borates). Once the glass is made either as fibers or cullet, it is for all practical purposes inert in water, and thus will not chemically affect wastewater quality.

The type of resin used, however, will exert some influence on both air and water quality. The industry is continually formulating new binder mixtures in an effort to minimize problems. However, the industry cannot be meaningfully subcategorized according to type of binder used for the following reasons. Different products can require different binder formulations, and these products can be made at different times on the same line. Composition changes in the binder can occur at any time, as the industry tries to improve the product and decrease raw material costs. No matter what formulation of resin is used, the general waste characteristics are the same and a chemical-physical treatment system will not be affected.

The type of product made will affect the chain wash water quality in that different products may require different resin formulations. However, for the same reasons given in the paragraph above, the industry cannot be meaningfully subcategorized on this topic. A general water flow diagram for an insulation fiber glass plant is pictured in Figure 55. Nonprocess waters identified in this diagram include boiler blowdown and water treatment backwashes.

Those parameters that are likely to be found in significant quantities in each of the waste streams are listed in Table 44. A more detailed analysis of each waste flow (i.e., concentration ranges) is not possible since the combined waste stream only has been of interest to the industry from whom most of the data was obtained. The principal process waste streams within the process are the chain cleaning water and water sprays used on the exiting forming air.

The principal uses for steam are for building heating and steam attenuation. In the latter case the industry has been converting to compressed air attenuation. The accompanying boiler blowdown in this case is replaced by noncontact cooling water for air compressors. Water usages vary significantly between plants. Factors such as design of furnace, method of chain cleaning and method of air emissions control will affect quantities of water. For example, plants at which marbles are remelted require very little furnace cooling water, since the remelt furnaces are small melting pots.

FIGURE 55: GENERAL WATER FLOW DIAGRAM FOR AN INSULATION FIBER GLASS PLANT

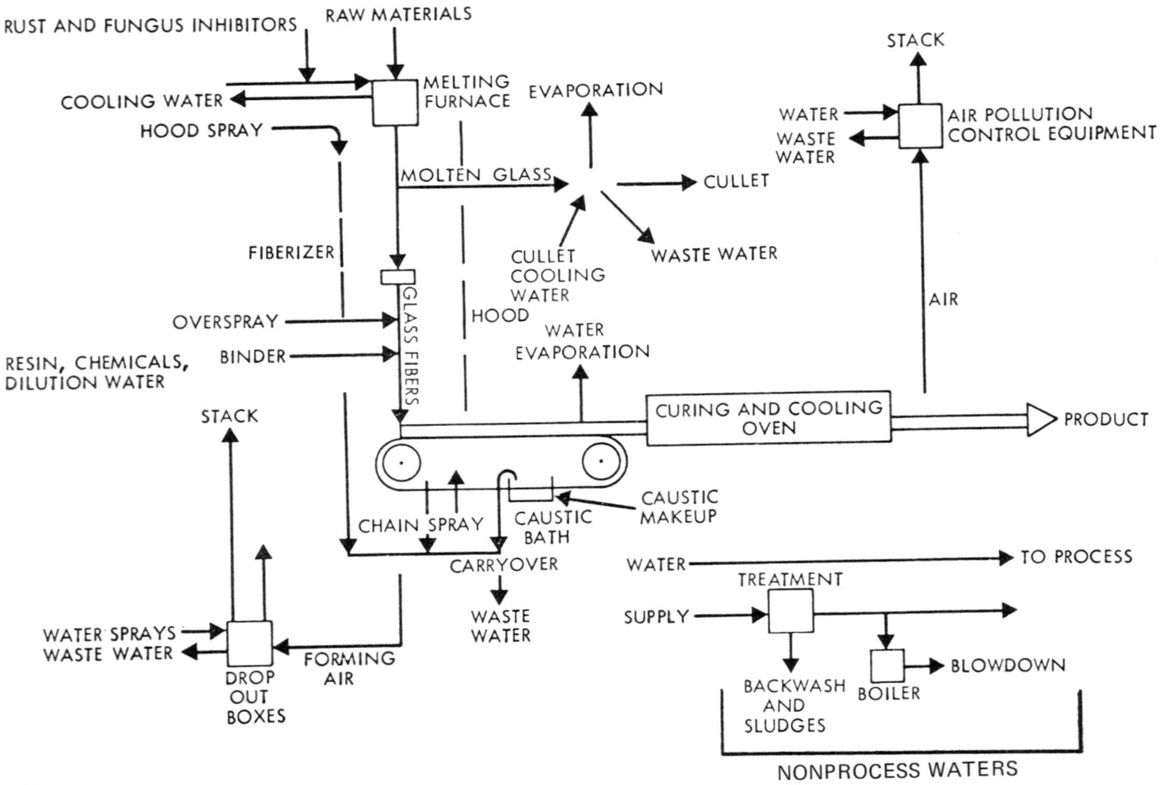

Source: EPA (41)

TABLE 44: CONSTITUENTS OF INSULATION FIBER GLASS WASTE STREAMS

Waste Stream	Phenols	BOD_5	COD	Dissolved Solids	Suspended Solids	Oil & Grease	Ammonia	pH	Color	Turbidity	Temperature	Specific Conductance
Air Scrubbing	x	x	x	x	x				x	x	x	x
Boiler Blowdown				x	x					x	x	x
Caustic Blowdown				x	x			x		x		
Chain Spray	x	x	x	x	x	x	x	x	x	x		x
Cullet Cooling				x	x						x	
Fresh Water Treatment				x	x			x		x	x	x
Hood Spray	x	x	x	x	x	x	x	x	x	x		x
Noncontact Cooling Water			x	x							x	x

Source: EPA (41)

Large continuous drawing furnaces, however, need large quantities of water to control oven temperatures and to protect the furnace bricks. Table 45 lists chain wash water flows for plants of various sizes. Again there is no correlation between plant size and water usage for chain washing, because each of the three insulation fiber glass producers uses chain wash water at different pressures and therefore at different flow rates.

TABLE 45: CHAIN WASH WATER USAGE

Plant	Plant Size*: Thousands of Metric Tons per Year	Plant Size*: Million Pounds per Year	Water Usage: liters/second	Water Usage: gpm
A	120	270	44	700
B	34	75	38	600
C	35	77	14	200
D	32	71	63	1,000
E	18	41	50	800
F	16	35	8	120
G	2	5	3	48

*All production figures are estimates

Source: EPA (41)

Annual raw waste loads for several plants in metric tons are computed in Table 46. The values are based upon an average of five parameters at four plants.

TABLE 46: ANNUAL RAW WASTE LOADS

Plant	Estimated Size (1000 metric tons per yr.)	Phenol	Suspended Solids	BOD_5	COD	Dissolved Solids
		Kilograms Pollutant Per Metric Ton Product				
A	120	0.36	1.29	1.67	11.0	
E	18	0.06	4.45	8.90	31.5	18.0
H	16	0.90	0.40	8.1		
I	131	0.33	5.60	6.65	24.2	14.1
Average		0.41	2.90	4.40	18.7	16.0
Annual Raw Waste Load[1] (Metric tons per yr.)		316	2240	3390	14,400	12,300

[1] Derived by multiplying kg/metric ton by 771,000 metric tons product per year by 1/1000 metric ton per kg.

Table 47 summarizes the raw waste concentrations for several plants. Although the numbers are not completely comparable because of treatment differences and different blowdown percentages, the table nevertheless shows a wide variance in wastewater composition. Other factors affecting the raw waste load include binder composition, chain temperature, and other thermal and time factors affecting the rate of resin polymerization.

TABLE 47: RAW WASTE LOADS FOR INSULATION FIBER GLASS PLANTS

Plant	Phenol mg/l	BOD5 mg/l	COD mg/l	TSS mg/l	TDS mg/l	TURBIDITY	pH	Percent Blowdown[3]
H	363	156	2500-4000	116-561				
F[1]	2564	7800	43,603	360	3000-5000			8.3
G	4.11			76	822		7.7-8.9	13.0
A	212	991	6532	769	10,000-20,000[2]			1.5
B[1]	240	6200	23,000	200	16,000	200	8.0	1.0
D[1]					40,000[2]			2.3
I	11-98	900	3,290	690	2,080		6.1-12.2	

1 - Sample taken from water recirculation system

2 - Given by company with no backup data

3 - Defined as percent total process water used as overspray or binder dilution

Source: EPA (41)

One particular waste stream addressed by this report is cullet cooling water. Suspended solids concentrations are extremely variable and depend upon how many fiberizers are being bypassed. Concentrations in the wastewater can range from a few hundred to tens of thousands mg/l even after settling. A size distribution study of the suspended solids resulting from cullet cooling appears in Table 48.

TABLE 48: SIEVE ANALYSIS ON WASTE CULLET WATER

U.S. Sieve Number	μ Equivalent	% By Weight Retained
50	297	98.30
100	149	1.20
140	105	0.30
200	74	0.05
325	44	0.01
400	37	0.05
Finer Passed		0.09
		100.00%

Source: EPA (41)

As seen from this table 99.50 percent of the cullet should be amenable to primary settling. However, especially at high cullet producing times, an appreciable amount of minus 100 mesh glass particles can remain suspended in the wastewater. Visual inspections at some plants noted cullet scattered about the river banks below discharges of cullet cooling water.

In summary, water usage and raw waste loads are not relatable in a practical manner to production levels or techniques. Of the nineteen existing plants, there may be as many different formulas for relating these factors. There are significant differences between plants even within the same company. A compensating factor, however, is the fact that all such wastes are amenable to the same general type of chemical and/or physical treatment.

In only one insulation fiber glass plant has secondary or more advanced treatment been applied to an effluent. Historically plants have discharged their waste streams to publicly owned treatment works. Use of biological end-of-pipe treatment for phenolic wastewaters was attempted at Plant A.

The treatment scheme (Figure 56) consisted of equalization, alum coagulation, nutrient addition, temperature control, extended aeration, post chlorination, aerobic sludge digestion, and vacuum filtration. It is noteworthy that the recirculation of chain wash waters was practiced thirteen years ago at this plant and that only blowdown from this recycled water received biological treatment.

Table 49 summarizes the performance of the system. Despite the percent removal efficiencies of the treatment system, objectionable concentrations of phenol and COD were still discharged. In addition the parameter of color received no treatment other than dilution.

FIGURE 56: BIOLOGICAL TREATMENT AT PLANT A

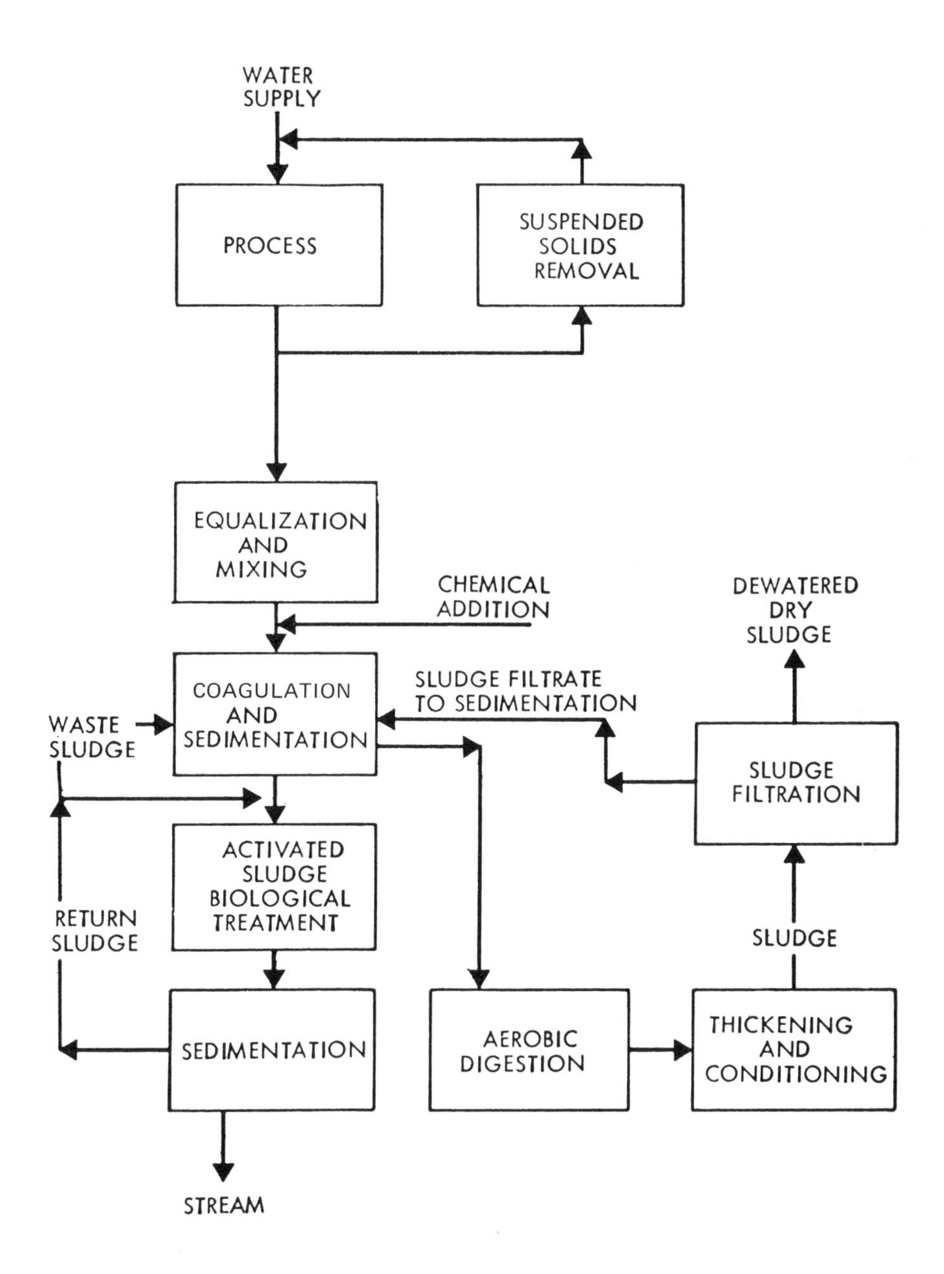

Source: EPA (41)

TABLE 49: BIOLOGICAL TREATMENT SYSTEM AT PLANT A

Parameter	Raw Waste mg/l	Final Effluent mg/l	Percent Removal
Phenol	212	1.48	99.3
Suspended Solids	769	27	92.0
COD	6532	298	94.4
BOD5	991	19.8	97.7

Note: Flow was 0.57 million liters per day

Source: EPA (41)

The company researched use of activated carbon absorption in an effort to remove the dye and the remaining phenol and COD in the effluent. However, this approach proved more costly than total recycle of process waters.

The only parameter that may interfere with a biological publicly owned treatment works is phenol. Only certain strains of microorganisms effectively remove phenols from wastewaters and their effectiveness is confined to low concentration ranges. Therefore, if sufficient dilution water is not present, wide variations of phenol in the raw waste load, due to process changes, may adversely affect the populations of these organisms.

The industry already has long realized that recirculation of chain wash water is feasible and that a blowdown is necessary to control the buildup of solids in the system. The industry also recognizes that suitable treatment of the blowdown for reuse as overspray or binder dilution water is less costly than performing advanced treatment to a final effluent. In the total recirculation scheme the contaminants in the blowdown essentially go onto the product as the binder and overspray waters evaporate from the hot fiber glass.

There has been no noticeable effect on product quality due to the small addition of these extra solids on the fiber glass. As an alternate method of blowdown disposal, some plants because of favorable climatic conditions and space availability have employed evaporation ponds.

The amount of water necessary to effectively clean the chain can be reduced by use of increased water pressures. However, sufficient concentrations of suspended and dissolved solids can in turn limit this pressure due to problems of increased pump maintenance and spray nozzle clogging. Since the dissolved solids concentration in the chain wash system is determined by the blowdown rate and degree of resin polymerization it is the more difficult of the two parameters to control.

The need to eliminate waste streams other than chain wash water by use as overspray or binder dilution will limit the blowdown rate of the recirculation chain wash system. This in turn will effect a steady state concentration of solids in the system which limits the wash water pressure.

The above methods constitute the current state of the art treatment technology employed by the industry. Table 50 lists the water pollution abatement status of all existing primary plants. In summary the table shows that three plants completely recycle all process waters. Another does the same except for cullet cooling water. Four plants recycle with three blowing down to evaporation ponds and the fourth to a spray field. Four plants recycle and discharge blowdown to publicly owned treatment works. Five discharge once-through waters to such works. Six plants have plans for complete recirculation of process or all waste streams.

TABLE 50: WATER POLLUTION ABATEMENT STATUS OF EXISTING PRIMARY INSULATION FIBER GLASS PLANTS

Plant	Status
A	Complete recirculation of process waters. Some indirect cooling water from an experimental air emissions control device discharged to stream
B	Complete recirculation
C	Discharge once-through waters to POTW.[1] Plans for recirculation
D	Complete recirculation except for discharge of cullet cooling water
E	Complete recirculation of phenolic wastes by 5-1-73. Other wastes to POTW
F	Complete recirculation
G	Completely recycle phenolic waters. Caustics and other waters to POTW
H	Recycle with blowdown to POTW, cooling waters to river. Plans for complete recirculation
I	Discharge once-through waters to POTW. Recycles cullet water. Plans for complete recirculation
J	Recycle on 1 line. Other lines discharge to river
K	Recycle with blowdown to evaporation pond
L	Evaporate wastes in pond
M	Discharge once-through water to POTW. Plans for recirculation
N	Wastes used for spray irrigation
O	Discharge to POTW
P	Recycle with blowdown to evaporation seepage ponds
Q	Discharge once-through waters to POTW. Plans for recirculation
R	Discharge once-through waters to POTW. Plans for recirculation
S	Recycle with blowdown to POTW

[1] POTW - Publicly Owned Treatment Works

Source: EPA (41)

All three insulation fiber glass producers operate plants in which process water is recirculated and in which blowdown is used as overspray or binder dilution. Thus the entire industry has the technology to apply the state of the art treatment technology.

Detailed descriptions of those plants that are currently practicing this technology follow. The plants described cover the entire range of types of plants: new and old; small, medium and large; flame attenuation and rotary spinning processes. The examples also illustrate how air pollution abatement methods can affect the water system.

It should be noted that technology transfer of specific items between plants is not always possible. This is especially true when comparing rotary and flame attenuation processes, which have widely different glass, binder, and air flow rates. This does not affect the conclusions that total process water recycle is practicable for all plants.

Cost Reduction Benefits of Alternate Treatment and Control Technologies

The three alternate treatment and control technologies considered are biological treatment, biological treatment and carbon adsorption, and complete recycle. All three treatment schemes consist of recycling chain wash water and treatment of only the blowdown. Consideration of treatment of once-through process water has long since been abandoned by the industry because of the large volumes involved and the amenability of chain wash water to treatment and recycle. Table 51 compares the costs and effluent qualities for the three alternate treatment schemes as they are estimated for Plant A.

TABLE 51: A COMPARISON BETWEEN THE ALTERNATE TREATMENT AND CONTROL TECHNOLOGIES[1]

	Raw Waste Load	Extended Aeration	Extended Aeration + Activated carbon	Total Recycle
Capital Costs ($1000)		1160	1320	785
Annual Operating Costs ($1000)[2]		540	556	508.5
Effluent Quality ($1000)				
BOD5 (mg/l)	991	20	10[3]	0[4]
COD (mg/l)	6532	298	50[3]	0[4]
Phenol (mg/l)	212	1.48	0.05[3]	0[4]
Suspended Solids (mg/l)	769	27	5[3]	0[4]
Color	yes	yes	no	no

1. All cost data based upon a 123,000 metric tons (270 million pounds) per year plant.
   Blowdown is 0.57 million liters per day.
2. Operating and maintenance costs and power costs for extended aeration and activated carbon are assumed to be the same for the total recycle system.
3. Estimated
4. No discharge hence no pollutants.

Source: EPA (41)

The table clearly indicates that total recycle is the best economic alternative of the three treatment schemes for best practicable control technology currently available, best available technology economically achievable, and best available demonstrated control technology. It is here assumed that the relationship between the costs of the three alternatives will hold for different plant sizes. Even if this were not true, it is quite significant that no discharge of pollutants can be achieved at costs comparable to end-of-pipe treatment technology.

Furthermore, best available technology economically achievable specifies application of technology which will result in reasonable further progress toward the national goal of eliminating the discharge of all pollutants. Total recycle of process waters both is economically achievable and meets the no discharge of pollutants goal. Total recycle of process waters is currently practiced by a significant portion of the industry.

Table 52 summarizes the water pollution abatement costs for a few insulation fiber glass plants. Investment costs have been interpolated to August 1971 dollars by using EPA tables of sewage treatment plant cost indexes (14). Two depreciation periods are used in calculating total annual cost. The first is the true depreciation period as determined by the company. For the second, a 10 year depreciation is used for the purpose of comparison. An economic study by one consultant (11) concluded that zero discharge is practical for the insulation fiber glass industry. The firm selected two basic forms of recycle systems.

TABLE 52: WATER POLLUTION ABATEMENT COSTS FOR TOTAL RECYCLE

	Plant									
	A	B	E	F	F[3]	G	I	L	O	Q
Capacity (Thousand Metric Tons/Yr.)	123	34	16.9	9	16	2.3	130	33	71	200
(Million Pounds/yr.)	270	75	35	20	35	20	287	73	157	444
Investment[1] ($1000)	785[4]	660[4]	483	325	340.5	245.4	1060[4]	316[4]	1220[4]	2700[4]
Investment/metric tons/yr.	6.4	19.4	28.6	36.2	21.3	100.6	8.2	9.6	17.2	13.5
Annual Costs										
Capital Costs ($1000)			2							
Depreciation ($1000)	78.5	66	24	23.7		17.5	106	31.6	122	270
Years Amortization	10	10	20	14		14	10	10	10	10
Operating and Maintenance ($1000)	382	100	55	36.5	44.5	13.8	200	50	137	438
Energy and Power Costs ($1000)	48	20	8	1.7	2.3	4.6	66	19	29	98
Total Annual Cost ($1000)	508.5	186	89	62		36	372	100.6	288	806
Adjusted Annual Cost[2] ($1000)	508.5	186	113	71	81	43	372	100.6	288	806
Adjusted annual cost/metric ton/yr.	4.1	5.5	6.7	7.9	5.1	18.7	2.9	3.1	4.1	4.0
Energy Consumption (100,000 kilowatt-hours/yr)	4.0	2.3	.551	.1658	.212	.512	6.9	2.3	4.6	12.7

1 Adjusted to August 1971 dollars using sewage treatment plant cost index (14).
2 Total Annual Cost using a 10 year amortization period.
3 After 1972 expansion to 4 lines, includes original oversized treatment system.
4 Estimated by company, not necessarily adjusted to August 1971 dollars.

Source: EPA (41)

Treatment A, coarse filtration, fine filtration and water recycle is practiced at Plant F. Treatment B, coarse filtration, flocculation, settling and water recycle, is practiced at Plant B. Table 53 lists the resultant fixed capital investment and annual operating costs for the two treatment schemes scaled to the four plant sizes considered by the consultant.

TABLE 53: ESTIMATED COST OF WASTEWATER TREATMENT FOR INSULATION FIBER GLASS MANUFACTURE

Plant Capacity			Type Treatment System	
Thousand metric tons/yr.	Million lb/yr		(A) Coarse Filtration Fine Filtration Water Recycle	(B) Coarse Filtration Flocculation Settling Water Recycle
200	440	Fixed Cap. Investment ($1000)	2000	1050
		Fixed Cap. Investment/Metric tons/yr.	10.0	5.2
		Annual Operating Cost $1000)	610	680
		Annual Operating Cost/Metric tons/yr	3.0	3.4
41	90	Fixed Cap. Investment ($1000)	800	400[1,2]
		Fixed Cap. Investment/Metric tons/yr.	19.5	9.8
		Annual Operating Cost ($1000)	200	200[3]
		Annual Operating Cost/Metric tons/yr	4.9	4.9
9	20	Fixed Cap. Investment ($1000)	325[1]	
		Fixed Cap. Investment/Metric tons/yr.	36.1	17.8
		Annual Operating Cost ($1000)	80	71
		Annual Operating Cost/Metric tons/yr.	8.9	7.9
2	5	Fixed Cap. Investment ($1000)	150	70
		Fixed Cap. Investment/Metric tons/yr.	65.2	30.4
		Annual Operating Cost ($100)	46	37
		Annual Operating Cost/Metric tons/yr.	20.0	16.1

1. Based on Costs reported by the Industry

2. Actual investment was closer to $600,000 but the existing system has more capacity than required.

3. Reported cost was closer to 0.3¢/lb., but reported treatment chemical cost seems high.

Source: EPA (14)

As a conservative estimate 80 percent production was used to calculate incremental capital and operating costs as shown in Table 54. Assumed selling prices and estimated current fixed capital investments were used. Figures 57 and 58 both clearly show that the investment cost of total process water recycle per unit production and the annual operating cost per unit production for the treatment systems are not linealy related to plant size. Therefore, the smaller plants will spend more per unit of product in order to maintain a closed water system than larger plants.

TABLE 54: SUMMARY OF CAPITAL AND OPERATING COST EFFECTS OF WOOL GLASS FIBER

Plant Capacity M metric ton	Plant Capacity (MM lb)	Type of Treatment Process	Plant* Output (MM lb)	Net Revenues ($MM)	Current Fixed Capital Investment ($MM)	Water Pollution Control Costs: Incremental Investment ($MM)	Incremental Investment as % of Current Investment	Incremental Operating Cost (¢/lb)	Incremental Operating Cost as % of Selling Price
200	440	(A) Coarse and Fine Filtration	352	98.5**	80	2.0	2.5	0.18	0.64
		(B) Flocculation and Settling	352	98.5	80	1.0	1.25	0.19	0.68
41	90	(A)	72	18.7***	26	0.8	3.8	0.27	1.04
		(B)	72	18.7	26	0.4	1.9	0.29	1.11
9	20	(A)	16	4.4**	10	0.325	3.25	0.50	1.78
		(B)	16	4.4	10	0.16	1.6	0.44	1.57
2.3	5	(A)	4	1.2****	4	0.15	3.75	1.15	3.83
		(B)	4	1.2	4	0.07	1.75	0.93	3.10

* @ 80% Yield
** @ 28¢/lb
*** @ 26¢/lb
**** @ 30¢/lb

Source: EPA (41)

TABLE 57: INVESTMENT COST OF TOTAL RECYCLE PER UNIT PRODUCTION

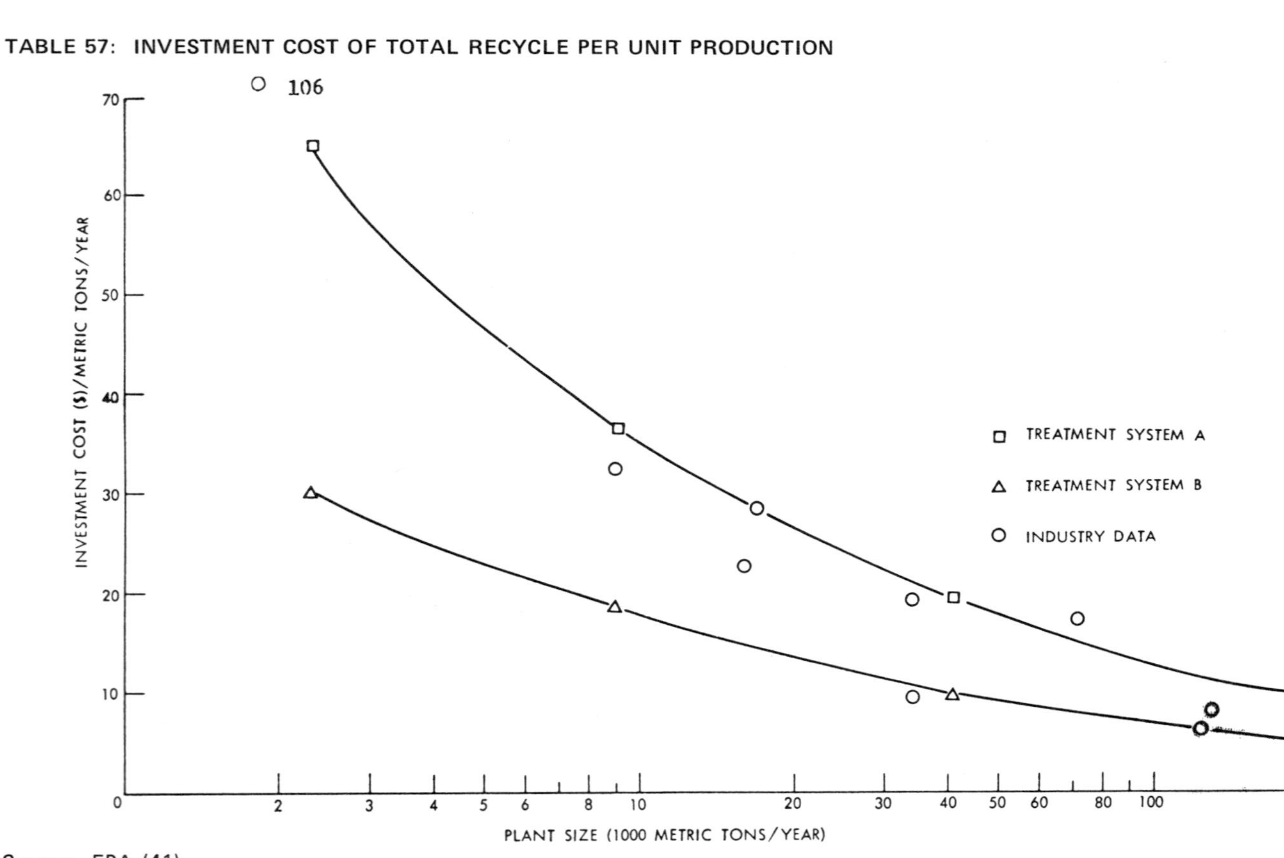

Source: EPA (41)

FIGURE 58: ANNUAL OPERATING COSTS OF TOTAL RECYCLE PER UNIT PRODUCTION

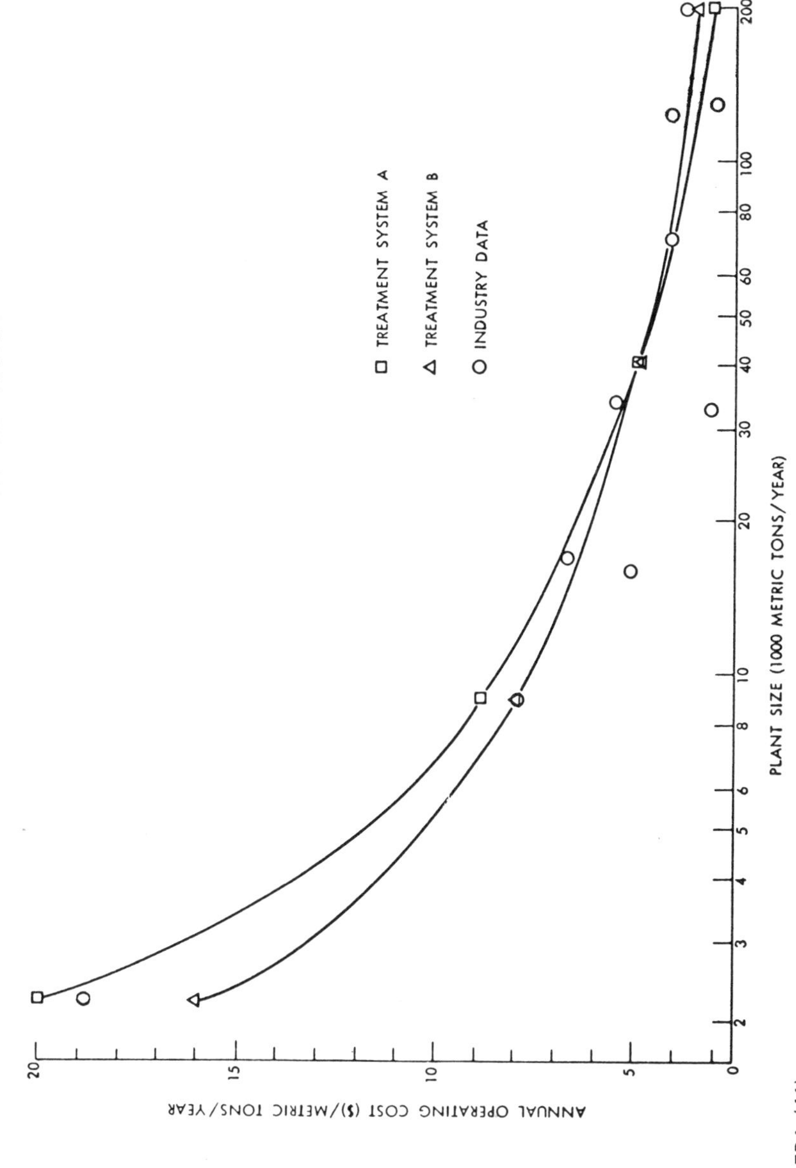

Source: EPA (41)

Assuming no price increases the relative effects on company and plant pretax earnings as a result of the incremental operating costs, will be equal to the proportion of selling price represented by these costs. If incremental costs are passed on, the current rate of profitability will be maintained. As current returns on investment are unknown for individual plants, the relative effects on returns on investment can only be obtained by assuming a certain level of profits on sales before taxes, and measuring sensitivity at various levels of returns on investment.

For this analysis, average pretax earnings are assumed to be 12 percent on sales for wool glass fibers. The current returns on investments tested are 5, 10, and 15 percent in Table 55. Thus for wool glass, a one percent increase in operating costs will reduce returns on investments by 8.3 percent of the current rate.

TABLE 55: EFFECTS ON RETURNS ON INVESTMENT—WOOL GLASS FIBER

Plant Size Capacity (M metric tons/yr)	Wastewater Treatment Type	Operating Cost as % of Selling Price	Predicted Effect on Return on Investment if Currently at		
			5%	10%	15%
200	A	0.64	4.7	9.5	14.2
	B	0.68	4.7	9.5	14.2
41	A	1.04	4.6	9.2	13.7
	B	1.11	4.5	9.1	13.6
9	A	1.79	4.3	8.5	12.8
	B	1.57	4.4	8.7	13.0
2	A	3.83	3.4	6.8	10.2
	B	3.10	3.7	7.4	11.1

Source: EPA (41)

Plants of any size that currently have a return on investment no better than 5 percent will become marginal and could possibly cease production. However, no such facilities currently exist. Plants operating at over 5 percent return on investment will continue to enjoy reasonable returns.

The capital that is needed for the industry to achieve no discharge, assuming that there are presently no treatment facilities, will range from 6.0 to 13.5 million dollars depending upon the recycle alternative, 10 million dollars being the estimated mean. Operating costs of pollution control equipment are estimated to be 3.7 million dollars per year for the industry. The consultant concluded that the insulation fiber glass industry has the financial capabilities to install total recycle facilities, and that this will have minimal effect on the selling price of its products.

The economic analysis of the consultant report was based upon treatment systems employed at only two plants of different companies. Figures 57 and 58 compare the costs of water treatment for different sizes of plants as determined from actual calculations and the estimations by the consultant previously mentioned. As seen, actual costs lie within or below the limits estimated by the consultant report, and it can be assumed that the conclusions of the consultant study generally hold true for the entire insulation fiber glass industry.

Further analysis reveals that annual operating and investment of recycle per annual production rating roughly double from the largest plant, 200,000 metric tons per year, to plants producing 9,000 metric tons per year. Eighty-five percent of the insulation fiber glass plants operate within this range and the relatively small cost variance should not give the large plants a particular advantage.

In fact, the largest plants which seemingly have the greatest cost advantage, are old plants which require considerable plant modifications not accounted for in the economic analysis. The costs of recycle systems increase at a much faster rate for plants smaller than 9,000 metric tons per year. However, plants in this size range produce specialty products that sell for a higher price than the standard building insulation that is most economically produced by medium and large size plants. The average price of industrial insulation which includes pipe insulation is 40 percent more than for building insulation. This means that the percentage cost increase relative to market price should vary less over the entire range of plant sizes than Figures 57 and 58 indicate. In fact, the smallest primary insulation plant has successfully recycled chain wash waters for 3½ years.

Subsurface disposal of process waters by seepage ponds, has caused ground water contamination at one insulation fiber glass plant. Evaporation ponds should therefore be lined or sealed. Insufficient information, regarding spray irrigation with process wastewaters, exists to judge this disposal method.

This type of treatment system does affect land requirements. The treatment systems employed at Plants A and B and proposed at Plant D require considerable space for flocculating and settling tanks, since low pressure, high volume wash systems are used. Emergency holding ponds are desirable but not practicable at many existing urban plants.

Proprietary Control Process

A process developed by J.E. Etzel et al (45) is one in which the wastewater is treated both mechanically and chemically to remove solid materials contained in the wastewater such as fiber glass, dust and other materials. The waters are also chemically treated with high molecular weight cationic polymers to reduce substantially the phenolic resin content of the waters. Recirculation of various water streams for utilization in the fiber glass manufacturing process including utilization of treated waters in the manufacture of binder solutions is also described. The process is such that wastewaters used in the manufacturing process can be totally reused thus requiring no disposal of waters to the environment.

SOLID WASTE DISPOSAL

In the progression from no treatment to recycle systems the industry has had to contend with increasing amounts of sludges consisting of cullet, glass fiber-resin masses, particulates removed from stack gases, and wasted product. Since these solids are in an unusable form, they are hauled to sanitary landfills. Restrictions at some sites prohibits burial of phenolic wastes because of the fear of ground water contamination. One company proposes to autoclave its sludges to insure complete polymerization of the phenols. It should be emphasized that the amount of solid wastes generated by the total recirculation system is no greater than if the industry were to employ alternate end of pipe wastewater treatment technologies.

FRIT MANUFACTURING

Ceramic coatings are generally divided into two classes, depending upon whether they are applied to metal or to glass and pottery. In the case of metal, the coating is widely referred to in this country as porcelain enamel. The use of the term vitreous enamel seems to be preferred in Europe. Glass enamel is sometimes used interchangeably with both terms. On the other hand, the coating applied to glass or pottery is known as ceramic glaze. Ceramic coatings are essentially water suspensions of ground frit and clay.

Frit is thus a ceramic base product which is mainly used (when combined with a large portion of fluxes) as porcelain enamel coating for bathtubs, sinks, latrines, and household cooking pots as well as in glazing porcelain and pottery. The preparation of frit is similar to the first steps of the manufacture of ordinary glass. In a typical plant, the raw materials consist of a combination of materials such as borax, feldspar, sodium fluoride or fluorspar, soda ash, zinc oxide, litharge, silica, boric acid, and zircon.

The raw materials are mixed in proper proportions and charged to a melting furnace. Frit smelting is a batch operation and has varying melt times, however 3 hours is typical (18). The molten materials are then poured into a quenching tank of cold water, shattering the melt into friable pieces. The frit is further ground with a ball mill, mixed with a variety of fluxes and sprayed on as a surface coating of specific products.

When suspended in a solution of water and clay, the resulting mixture is known as a ceramic slip. Enamel slip is applied to metals and fired at high temperatures in a furnace. Glaze slip is applied to pottery or glass and fired in a kiln. Frit manufacturing plants operate continuously, 24 hr/day, 7 days/week. A typical integrated enamel-frit-smelter, sheet-steel-enameling plant is shown in Figure 59.

AIR POLLUTION

Significant dust and fume emissions are created by the frit-smelting operation as shown in Figure 59. These emissions consist primarily of condensed metallic oxide fumes that have volatilized from the molten charge. They also contain mineral dust carryover and sometimes contain noxious gases such as hydrogen fluoride. In addition, products of combustion and glass fibers are released. The quantity of these air contaminants can be reduced by following good smelter-operating procedures. This can be accomplished by not rotating the smelter too rapidly, to prevent excessive dust carryover, and by not heating the batch too rapidly or too long, to prevent volatilizing the more fusible elements before they react with the more refractory materials.

FIGURE 59: ENAMEL FRITTING–UNCONTROLLED PROCESS MODEL

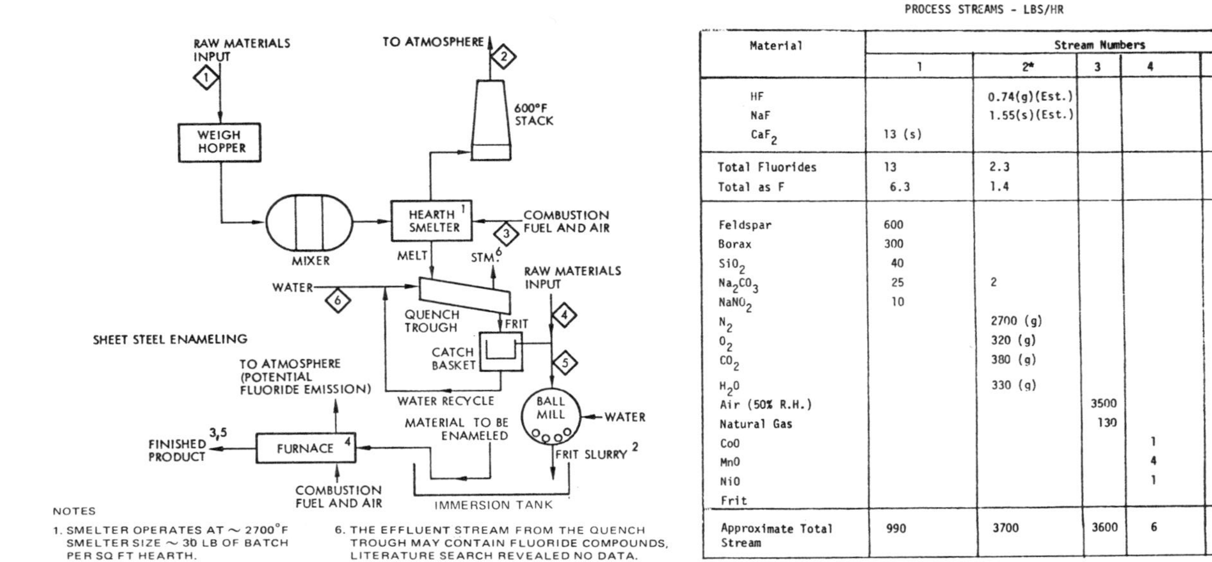

NOTES

1. SMELTER OPERATES AT ~ 2700°F SMELTER SIZE ~ 30 LB OF BATCH PER SQ FT HEARTH.
2. PARTICLE SIZE OF SOLIDS IN THE FRIT SLURRY ARE:
   GROUND COAT (1ST COAT) = 95% THRU 200 MESH
   COVER COAT (2ND COAT) = 98% THRU 325 MESH
3. SHEET STEEL COATINGS USUALLY 5 TO 8 MILS THICK.
4. FIRING TEMPERATURE = 1500°F FOR 5 MIN
5. MATERIALS BEING ENAMELED MUST BE RECYCLED THROUGH A SECOND COVER COAT AFTER THE INITIAL GROUND COATING.
6. THE EFFLUENT STREAM FROM THE QUENCH TROUGH MAY CONTAIN FLUORIDE COMPOUNDS, LITERATURE SEARCH REVEALED NO DATA.

BASIS - 1000 LBS/HR FEED (STREAMS 1 & 4), SHEET STEEL ENAMELING
PROCESS STREAMS - LBS/HR

Material	Stream Numbers					
	1	2*	3	4	5	6
HF		0.74(g)(Est.)				
NaF		1.55(s)(Est.)				
CaF_2	13 (s)				10 (s)	
Total Fluorides	13	2.3			10	
Total as F	6.3	1.4			4.9	
Feldspar	600					
Borax	300					
SiO_2	40					
Na_2CO_3	25	2				
$NaNO_2$	10					
N_2		2700 (g)				
O_2		320 (g)				
CO_2		380 (g)				
H_2O		330 (g)			100	100
Air (50% R.H.)			3500			
Natural Gas			130			
CoO				1		
MnO				4		
NiO				1		
Frit					890	
Approximate Total Stream	990	3700	3600	6	1000	100(A)

*Gaseous effluent stream

(A) Plus make-up H_2O for steam loss

Soluble fluoride evolution factor = 3.15 lb F/ton dry frit (215 lb F/ton CaF_2 fed)

Source: J.M. Robinson, G.I. Gruber, W.D. Lusk, and M.J. Santy (19)

A typical rotary smelter, for example, discharges 10 to 15 pounds of dust and fumes to the atmosphere per hour per ton of material charged. In some cases, where ingredients require high melting temperatures (1500°F or higher), emissions as great as 50 pounds per hour per ton of material have been observed. Depending upon the composition of the batch, a significant visible plume may or may not be present. Tables 56 through 59 indicate the extent of emissions from uncontrolled, rotary frit smelters for various-sized batches and compositions.

TABLE 56: DUST AND FUME DISCHARGE FROM A 1,000-POUND, ROTARY FRIT SMELTER

Test data	Test No.		
	1	2	3
Process wt, lb/hr	174[a]	174[a]	174[a]
Stack vol, scfm	1,390	1,540	1,630
Stack gas temp, °F	450	750	900
Concentration, gr/scf	0.118	0.387	0.381
Stack emissions, lb/hr	1.41	5.11	5.32
CO, vol % (stack condition)	0.002	0.001	0.002
N_2, vol % (stack condition)	76.9	75.10	73.50
	4	5	6
Process wt, lb/hr	292[b]	292[b]	292[b]
Stack vol, scfm	1,310	1,400	1,480
Stack gas temp, °F	960	950	930
Concentration, gr/scf	0.111	0.141	0.124
Stack emissions, lb/hr	1.25	1.79	1.57
CO, vol % (stack condition)	0	0	0
N_2, vol % (stack condition)	73	72.60	73.30

[a]These three tests represent approximately the 1st, 2d, and 3d hours of a 248-minute smelting cycle. The total charge amounted to 717 pounds of material consisting of borax, feldspar, sodium fluoride, soda ash, and zinc oxide.

[b]These three tests represent approximately the 1st, 2d, and 3d hours of a 195-minute smelting cycle. The total charge amounted to 949 pounds of material consisting of litharge, silica, boric acid, feldspar, fluorspar, borax, and zircon.

Source: J.A. Danielson (14)

TABLE 57: DUST AND FUME DISCHARGE FROM A 3,000-POUND, ROTARY FRIT SMELTER

Test data	Test No.		
	7	8	9
Process wt, lb/hr	472[a]	472[a]	472[a]
Stack vol, scfm	2,240	2,270	2,260
Stack gas temp, °F	630	800	840
Concentration, gr/scf	0.143	0.114	0.172
Stack emissions, lb/hr	2.70	2.20	3.30
CO, vol % (stack condition)	0.02	0.02	0.02
N_2, vol % (stack condition)	75.30	75.60	76.30

[a]These three tests represent approximately the 1st, 2d, and 3d hours of a 248-minute smelting cycle. The total charge amounted to 1,951 pounds of material consisting of litharge, silica, boric acid, feldspar, whiting, borax, and zircon.

Source: J.A. Danielson (14)

TABLE 58: FLUORIDE DISCHARGE FROM A ROTARY FRIT SMELTER

Test data	Test No.			
	10	11	12	13
Process wt, lb/hr	174[a]	174[a]	162[b]	162[b]
Stack vol, scfm	1,400	1,600	1,000	1,000
Stack gas temp, °F	530	840	480	480
Concentration, gr/scf	0.061	0.035	0.196	0.058
Stack emissions, lb/hr	0.73	0.48	1.68	0.50

[a] These two tests were of 90 minutes' duration each and represented approximately the first half and the second half of a 248-minute smelting cycle. The total charge amounted to 717 pounds of material consisting of borax, feldspar, sodium fluoride, soda ash, and zinc oxide.

[b] These two 60-minute tests represented approximately the 1st and the 4th hours of a 450-minute smelting cycle. The total charge amounted to 1,213 pounds of material consisting of sodium carbonate, calcium carbonate, pyrobar, and silica. The test was specifically conducted for a batch containing maximum carbonates (19%) and no litharge.

TABLE 59: DUST AND FUME DISCHARGE FROM A 2,000-POUND ROTARY FRIT SMELTER

Test data	Test No.			
	14	15	16	17
Process wt, lb/hr	857[a]	857[a]	890[b]	890[b]
Stack vol, scfm	2,430	2,430	4,347	4,347
Stack gas temp, °F	600	600	340	340
Concentration, gr/scf	0.130	0.112	0.111	0.103
Stack emissions, lb/hr	2.710	2.340	4.150	3.820

[a] These two 60-minute tests represent the 1st and 2d hours of a 140-minute smelting cycle. The total charge amounted to 2,000 pounds of material containing silica, litharge, and whiting.

[b] These two 60-minute tests represent the 1st hour and 37 minutes of a 135-minute smelting cycle. The total charge amounted to 2,000 pounds of material containing silica, litharge, and whiting.

Source: J.A. Danielson (14)

Soluble fluoride emissions from the enamel frit industry were estimated at 700 tons in 1968, and are forecast at 1,060 tons in 2000, assuming continuation of the use of venturi and wet scrubber control at the current level (19). It is estimated that only 20% of the operational facilities currently utilize wet scrubbers. If wet scrubber devices are applied throughout the industry and provide abatement at the 99% efficiency level, the soluble fluoride tonnage emitted by the frit industry would drop to about 13 tons in 2000.

Emission of soluble fluorides from the hearth smelter used for frit production, on the basis of the proprietary thermochemical analyses program, follows a like mechanism to that involved in evolution and emission of soluble fluorides from opal glass furnaces. At the high temperatures present in the hearth, the volatilized fluorides react with water vapor so that at equilibrium, roughly equal molal concentrations of gaseous HF and NaF are formed. When the smelter gas passes into the stack and cools off, the NaF forms a particulate dispersoid fume.

Rotary smelters require a detached canopy-type hood suspended from the lower end of a vertical stack. It is suspended far enough above the floor to trap the discharge gases from the smelter when in the horizontal position. Refractory-lined, it is of sufficient size to prevent gases from escaping into the room, its size varying with the size of the smelter. The typical hood opening area ranges from 3 to 5 square feet.

The stack should be of sufficient height to obtain good draft, about 20 feet, if it is not vented to air pollution control equipment. If it is vented to control equipment, ventilation requirements are approximately 3,000 scfm for a 2,000-pound batch smelter as an example. Hood indraft velocity should be about 500 fpm.

Crucible and hearth smelters do not require hoods but do require a 20- or 25-foot stack to conform with good chimney design practice if not vented to air pollution control equipment. Some crucible smelters are vented directly into the room. If vented to air pollution control equipment a canopy hood must be used on the crucible smelter. Hood indraft velocities should be approximately 200 fpm. The requirement for a hearth (box) type smelter is approximately 4,000 scfm for a 3,000-pound batch smelter. As a general rule, about 70 scfm is required for each square foot of hearth area.

The two most feasible control devices for frit smelters are baghouses and venturi water scrubbers. Of these devices, baghouses are more effective. Glass bags cannot be used, however, owing to the occasional presence of fluorides in the effluent. The discharge gases must be cooled by heat exchangers, quench chambers, cooling columns, or by some other device to a temperature compatible with the fabric material selected. Filtering velocities should not exceed 2.5 fpm.

A venturi-type water scrubber is satisfactory if at least 20 to 25 inches of pressure drop is maintained across the venturi throat. The throat velocity should be between 15,000 and 20,000 fpm. The water requirement at the throat is about 6 gallons per minute for each 1,000 cubic feet of gas treated. Power consumption is high owing to the high pressure drop.

A baghouse installation can handle vent gases from four rotary, gas-fired frit smelters. The production capacity of one of the smelters is 3,000 pounds while that of the other three is 1,000 pounds each. Maximum gas temperatures encountered in the discharge stack at a point 20 feet downstream of the smelters are approximately 950°F while the average temperature is 780°F.

The baghouse is a cloth-tubular, pullthrough type, containing 4,400 square feet of cloth area. It is equipped with an exhaust fan that delivers 9,300 cfm at approximately 170°F. The filtering velocity is 2.2 fpm.

Radiation cooling columns are used to reduce the effluent gas temperature from 585°F at the inlet to the cooling columns to 185°F at the baghouse inlet. Approximately 1,300 lineal feet of 30-inch-diameter, heavy-gauge steel duct with a surface area of 10,000 square feet is used. The average overall heat transfer coefficient is 1.35 Btu per hour per square foot per °F, as calculated from actual test data.

The cooling columns are not one continuous run, but consist of single, double, and triple runs. Thus, the gas mass velocity varies considerably throughout the unit, with resulting changes in heat transfer coefficients. Additional cooling is accomplished with dilution air at the detached hoods, which are suspended about 1 foot away from the discharge end of each smelter. The baghouse inlet temperature of 185°F is satisfactory for the Dacron cloth material used, and excellent bag life can be expected.

Smelting of the frit volatilizes gaseous and particulate fluorides and other very fine particulate matter. Emphasis has been on removal of the particulate material, and the most frequently used devices are baghouses and venturi scrubbers. The venturi scrubber approach will remove gaseous fluroides. Although central processes are currently applied to the smelter effluent, it may become necessary to control the quench trough and baking furnace effluents also. Mass balances and process flow diagrams are presented in Figure 60 for two currently employed control processes.

The economics of a characteristic frit production plant are summarized in Table 60. While there is a moderate volume of merchant frit production for sale, a considerable portion of the frit produced is for captive consumption. Return on investment without fluoride control processes, is estimated at 18.8% (Table 60). Tables 61 and 62 present economic analyses of control processes currently employed for abatement of fumes from frit manufacture. ΔROI because of emissions control is estimated as 12%, equivalent to a reduction in ROI to 16.5%, due to the added costs of pollution control.

FIGURE 60: ENAMEL FRITTING–CONTROLLED PROCESS MODEL

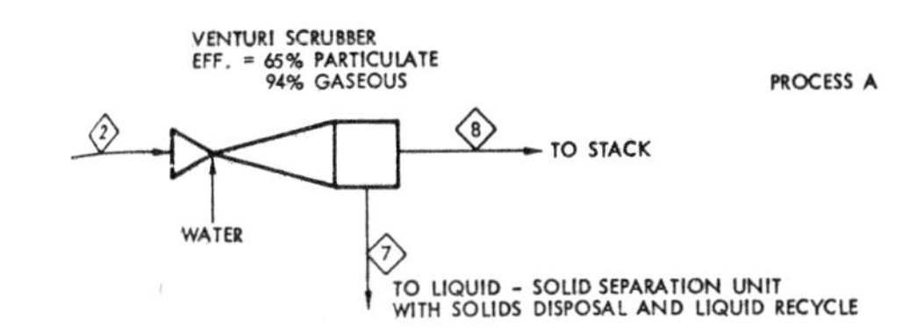

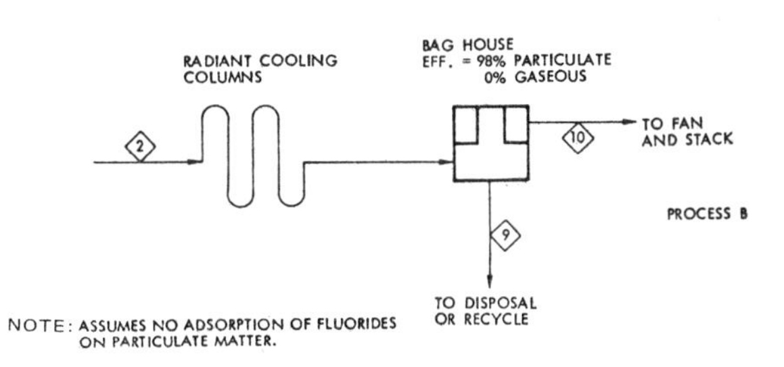

BASIS - 1000 LBS/HR FEED, SHEET STEEL ENAMELING PROCESS STREAMS - LBS/HR

Material	Stream Number				
	2	7	8*	9	10*
HF	0.74(g)	0.70 (l)	0.04(g)		0.74(g)
NaF	1.55(s)	1.01 (s)	0.54(s)	1.52(s)	0.03(s)
Total Fluorides	2.3	1.71	0.58	1.52	0.77
Total as F	1.4	1.12	0.28	0.69	0.71
Na_2CO_3	2(s)	1.3(s)	0.7(s)	1.96(s)	0.04(s)
N_2	2700(g)		2700(g)	2	2700(g)
O_2	320(g)		329(g)		320(g)
CO_2	380(g)		380(g)		380(g)
H_2O	330(g)	230(l)	100(g)(Est.)		330(g)
Approx. Total Stream	3700	230(A)	3500	2	3700

*Gaseous Effluent Stream

(A) Plus scrubbing water and recycled soluble fluorides.

Control Processes estimated to be utilized by 20% of the industry; remaining 80% uncontrolled.

Source	Soluble Fluoride Emission Factor - lb F/ton dry Frit	
	Process A	Process B
Smelter Emission	0.63	1.60
Assumed Fugitive Emission	0	0
Total Emission	0.63	1.60

Overall soluble fluoride emission factor = 2.64 lb F/ton dry frit (180 lb F/ton CaF_2 fed)

Source: J.M. Robinson, G.I. Gruber, W.D. Lusk, and M.J. Santy (19)

TABLE 60: ESTIMATED ECONOMICS OF ENAMEL FRIT PRODUCTION (POLLUTION CONTROL COST EXCLUDED)

Basis - 890 lb/hr Frit Production

Total capital investment	1.9	$MM
Production costs		
Direct costs		
Fluorspar (0.015 tons/net ton at $65.00/ton)	0.98	$/net ton
Feldspar (0.67 tons/net ton at $20.00/ton)	13.40	$/net ton
Borax (0.34 tons/net ton at $50.25/ton)	17.09	$/net ton
Silica (0.045 tons/net ton at $12.75/ton)	0.57	$/net ton
Soda ash (0.028 tons/net ton at $34.00/ton)	0.95	$/net ton
Soda Niter (0.011 tons/net ton at $213.00/ton)	2.34	$/net ton
Cobalt oxide (0.0011 tons/net ton at $4,400.00/ton)	4.84	$/net ton
Manganese oxide (0.0045 tons/net ton at $440.00/ton)	1.98	$/net ton
Nickel oxide (0.0011 tons/net ton at $2,700.00/ton)	2.97	$/net ton
Natural gas (5,300 scf/ton at $0.40/1,000 scf)	2.12	$/net ton
Water (190 gal/ton at $0.20/1,000 gal)	0.04	$/net ton
Labor (four positions at $4.00/hr)	35.96	$/net ton
Smelter repairs (1.2%)	6.47	$/net ton
Fringe benefits and supervision	35.96	$/net ton
General maintenance and supplies (1.25%)	6.74	$/net ton
Total direct costs	132.41	$/net ton
Indirect costs		
Depreciation (at 7.1%)	38.28	$/net ton
Interest (at 7%, 20% debt)	3.36	$/net ton
Local taxes and insurance (at 3%)	16.17	$/net ton
Plant and labor overhead	43.15	$/net ton
Total indirect costs	100.96	$/net ton
Total average costs	233.37	$/net ton
General and sales expenses	4.67	$/net ton
F.O.B. cost	238.04	$/net ton
Average product revenue	400.00	$/net ton
Average profit after taxes (at 50%)	80.98	$/net ton
Cash flow	0.42	$MM/yr
Return on investment	18.8	%

TABLE 61: ENAMEL FRITTING—ESTIMATED ECONOMICS OF CONTROL PROCESS A

(Basis - 10 Tons of Frit Produced Per Day)

		- - - - Capital Cost Estimates ($1,000) - - - - -		
Item Number	Description	Equipment F.O.B. Cost	Installation Factor	Equipment Installation Cost
1	Venturi Scrubber, t_{in} = 600°F, 0.04 lb solid/min loading, monel clad, 2,300 cfm, 31.5 in W.G. pressure drop	22	1.63	36
			Capital Subtotal	36
			Indirects (@ 15%)	5
			Contingency (@ 20%)	7
			Total Capital (as of January 1971)	48

(continued)

TABLE 61: (continued)

Item Number	Operating Cost ($/hr) Power Cost	Maintenance Cost	Equipment Operating Cost
1	0.9	0.13	0.22

Subtotal	0.22
Water	0.01
Disposal	-
Total Operating Cost	0.23

Total Operating Cost ($/hr)	0.23
Taxes and Insurance (2%, 330 days)	0.12
Capital (7.1%, 330 working days)	0.34
Pollution Control Cost ($/hr)	0.69
Pollution Control Cost ($/ton)	1.66

TABLE 62: ENAMEL FRITTING–ESTIMATED ECONOMICS OF CONTROL PROCESS B (Basis - 10 Tons of Frit Produced Per Day)

Item Number	Description	Capital Cost Estimates ($1,000) Equipment F.O.B. Cost	Installation Factor	Equipment Installation Cost
1	Radiant Cooling Columns, 1,300 ft^2, 2,300 cfm, 2 in W.G. pressure drop, carbon steel t_{in} = 300°F, t_{out} = 150°F	10	1.80	18
2	Baghouse, fabric filter, 2,300 cfm, 2.5 in W.G. pressure drop, 1.2 horsepower	20	4.13	83

Capital Subtotal	101
Indirects (@ 15%)	15
Contingency (@ 20%)	20
Total Capital (as of January 1971)	136

Item Number	Operating Cost ($/hr) Power Cost	Maintenance Cost	Equipment Operating Cost
1	-	0.10	0.10
2	0.01	1.26	1.27

Subtotal	1.37
Water	-
Disposal	-
Total Operating Cost	1.37

Total Operating Cost ($/hr)	1.37
Taxes and Insurance (2%, 330 days)	0.34
Capital (7.1%, 330 working days)	1.22
Pollution Control Cost ($/hr)	2.93
Pollution Control Cost ($/ton)	7.03

Source: J.M. Robinson, G.I. Gruber, W.D. Lusk and M.J. Santy (19)

GLASS INDUSTRY

Glass is thought to have been made in Persia 7,000 years ago and is known to have been produced 2,000 years ago in Egypt. It was first used for gems and was later made into hollow vessels such as jars and vases. A circular piece was used for a window in a bathhouse in Pompeii sometime between 600 BC, when the city was founded, and 79 AD, when it was destroyed by the eruption of Mt. Vesuvius. The glass was made by casting and then drawn with pincers.

The glass blowpipe was invented at the beginning of the Christian era and led to two important methods for manufacturing flat glass; the crown process and the cylinder process. In the crown process (which was thought to have been invented by the Syrians), a sphere was blown, an iron rod was attached to the sphere opposite the blowpipe, and the blowpipe was cracked off. The iron rod was then used to spin the reheated sphere until it opened to a flat circular sheet. Glassmaking was introduced in America by the English, and the first glass factory in America was erected by the beginning of the seventeenth century at Jamestown, Virginia.

During the nineteenth century, the crown process of making flat glass was replaced by the cylinder process in which a cylinder was blown, the ends were cracked off, the cylinder was split along the side, and then reheated so that it could be opened into a flat sheet. The cylinder process did away with the thick center and thick edge that were characteristic of the crown process. In addition, larger sheets could be formed. Various improvements were made in mechanizing the process, including using compressed air for blowing.

In 1904, a patent was granted to Emile Fourcault in Belgium for a process in which a flat sheet of glass could be drawn directly from a bath of molten glass. Two other methods were developed for making sheet glass at about the same time in America. These were the Colburn (or Libbey-Owens) process and the Pittsburgh process. All three processes are still in use and many improvements have been made since the original development.

Although sheet glass has a high surface finish, the surfaces are inherently wavy and are unsuitable for mirrors or large windows in which undistorted vision is desired. This fault can be overcome by grinding and polishing the sheet glass, although ground and polished glass is produced today by the plate process and, to a lesser extent, by the rolled process.

The earliest plate glass was produced by a casting process invented in France in the middle of the seventeenth century. The glass was melted in pots, poured onto a casting table and then leveled to the required thickness with a roller. The glass was allowed to cool and was then ground with sand and water using finer grades of sand as the grinding progressed.

The glass was then polished with felt-covered wheels fed with a fine abrasive slurry of iron oxide. These basic grinding and polishing steps are in use today, although continuous processes are now employed in manufacturing ground and polished plate glass. The latest method for producing high optical-quality glass is the float process, introduced by Pilkington Brothers Limited in 1959. The method gets its name from that part of the process in which the glass is drawn across a bath of molten tin. Heat is applied and, together with the effect of gravity, a distortion-free sheet of glass is produced which has the high surface quality of sheet glass. Float glass is rapidly replacing ground and polished plate glass.

There are a total of 36 plants owned by 11 companies which manufacture flat glass and fabricate automobile window glass in the United States with a combined daily processing capacity of 10,700 metric tons (11,800 short tons) of primary flat glass products and 173,000 square meters (1.86 million square feet) of automotive glass products (46). The daily capacity of an average plant engaged in primary flat glass manufacturing is 413 metric tons (455 short tons). These plants range in size from 54 metric tons (60 short tons) per day to 1,090 metric tons (1,200 short tons) per day.

The daily capacity of an average plant engaged in automotive glass fabrication is 10,800 square meters. These plants range in size from 2,000 square meters (22,000 square feet) per day to 24,700 square meters (266,000 square feet) per day.

Total employment in the industry is 24,000 with an average of 670 employees per plant. Plant employment ranges from about 100 to 2,900. It should be noted that employment figures are based on plant totals. Many of the plants carry on production processes (such as architectural glass fabricating) which are not covered by the study.

The glass pressing machine was introduced in America in 1827. In this process the molten glass is pressed into a mold manually with a plunger. Several other glass manufacturing innovations occurred in the 19th century. The first successful bottle-blowing machine was invented by Ashley of England in 1888. Several other bottle-making machines were developed during the next years. In 1889 Michael J. Owens conceived the first fully automatic bottle machine, which in less than 20 years revolutionized glass container manufacturing.

Another major manufacturing breakthrough was the introduction of the Corning ribbon machine in the early 1900's. The ribbon machine can manufacture as many as 2,200 bulbs per minute. The Hartford IS (Individual Section) machine, developed in 1925, remains the most popular method for manufacturing glass containers. Techniques for forming glass containers and machine-pressed products have essentially remained the same since the 1920's. Most recent developments in the glass industry are in the application of glass into new areas such as conductive coatings, electrical components, and photosensitive glasses.

There are four manufacturing steps that are common to the entire pressed and blown glass industry. The four steps include weighing and mixing of raw materials, melting of raw materials, forming of molten glass, and annealing of formed glass products. Forming methods vary substantially, depending on the product and subcategory, and range from hand blowing to centrifugal casting of picture tube funnels.

Following forming and annealing, the glass may be prepared for shipment or may be further processed in what is referred to as finishing. There is little or no finishing involving wastewater in the glass container and machine-pressed and blown subcategories, while extensive finishing is required in television picture tube envelope, incandescent lamp envelope, and hand-pressed and blown glassware manufacturing.

There are approximately 30 firms with a total of 140 plants presently manufacturing glass containers in the United States (47). The eight largest firms in the industry produce about 78% of the glass container shipments and operate two-thirds of the individual plants. Plants are located throughout the United States to service regional customers, but a large number are concentrated in the northeastern United States. The industry originally located in the Northeast because convenient sources of raw materials and fuel were available.

The glass container industry employs over 70,000 persons and has a daily processing capacity of 50,500 metric tons (55,500 tons) of glass pulled. The average glass container plant capacity is 388 metric tons (427 tons). Plants range in size from 122 metric tons (134 tons) per day to 1,320 metric tons (1,450 tons) per day.

There are about 50 machine-pressed and blown glass manufacturing plants in the United States and the average capacity is 91 metric tons (100 tons) pulled per day. Machine-pressed and blown ware plant capacities range from 40 metric tons (44 tons) to 349 metric tons (384 tons).

About 30 plants manufacture glass tubing in the United States. Production, expressed as furnace pull per day, ranges from 40 metric tons (44 tons) per day to 164 metric tons (180 tons) per day and averages 100 metric tons (110 tons).

Approximately 10 television picture tube envelope factories are located in the United States. The average amount of glass pulled per day is 208 metric tons (229 tons). Plant production varies from 142 metric tons (156 tons) pulled per day to 255 metric tons (280 tons) pulled per day.

Incandescent lamp envelopes are manufactured at 18 plants in the United States. Plant production in terms of furnace pull ranges from 141 metric tons (155 tons) per day to 245 metric tons (270 tons) per day. The average plant production is 193 metric tons (212 tons) per day.

Hand-pressed and blown glass manufacturing plants are small and primarily located in West Virginia, western Pennsylvania, and Ohio. Approximately fifty handmade glassware plants are located in the United States. A number of hand-pressed and blown ware plants also have facilities to manufacture machine-made glassware. The average amount of finished product produced per day at a hand-pressed and blown ware plant is 3.6 metric tons (4.0 tons). The range of production varies from 0.7 metric tons (0.8 tons) per day to 6.5 metric tons (7.2 tons) per day.

Nearly all glass produced commercially is one of five basic types: soda-lime, lead, fused silica, borosilicate, and 96% silica. Soda-lime glass represents by far the largest tonnage of glass made today (about 90%) and serves for the manufacture of containers of all kinds, flat glass, auto glass, tumblers, and tableware. Soda-lime glass is produced on a mass scale in large direct-fired continuous melting furnaces. Plant capacities range from a few pounds to several tons per day. The basic raw materials for soda glass include: silica sand, cullet, soda ash, limestone, niter, salt cake, arsenic, and a decolorizer.

The raw materials are brought to the plant on railroad cars and unloaded into storage bins. The materials are withdrawn from the storage bins, batch weighed and blended in a mixer. The mixed batches are fed continuously to the furnace and brought to a molten state 2700°F. The glass is drawn from the furnace and then passed through a forming machine, which either presses, blows, rolls, draws or casts the molten glass to its final shape. Glass manufacturing plants operate continuously, 24 hours per day (18).

Material-handling systems for batch mixing and conveying materials for making soda-lime glass normally use commercial equipment of standard design. This equipment is usually housed in a structure separate from the glass-melting furnace and is commonly referred to as a batch plant. A flow diagram of a typical batch plant is shown in Figure 61. In most batch plants, the storage bins are located on top, and the weigh hoppers and mixers are below them to make use of the gravity flow. Major raw materials and cullet (broken scrap glass) are conveyed from railroad hopper cars or hopper trucks by a combination of screw conveyors, belt conveyors, and bucket elevators, or by pneumatic conveyors (not shown in Figure 61) to the elevated storage bins. Minor ingredients are usually delivered to the plant in paper bags or cardboard drums and transferred by hand to small bins.

Ingredients comprising a batch of glass are dropped by gravity from the storage bins into weigh hoppers and then released to fall into the mixer. Cullet is ground and then mixed with the dry ingredients in the mixer. Ground cullet may also bypass the mixer and be mixed instead with the other blended materials in the bottom of a bucket elevator. A typical batch charge for making soda-lime flint glass (14) in a mixer with a capacity of 55 cubic feet consists of:

	Weight in lb
Silica sand	2,300
Cullet	650
Soda ash	690
Limestone	570
Niter	7
Salt cake	12
Arsenic	2
Decolorizer	1
	4,232

Raw materials are blended in the mixer for periods of 3 to 5 minutes and then conveyed to a charge bin located alongside the melting furnace. At the bottom of the charge bin, rotary valves feed the blended materials into reciprocating or screw-type furnace feeders.

FIGURE 61: PROCESS FLOW DIAGRAM OF A GLASS MANUFACTURE BATCH PLANT

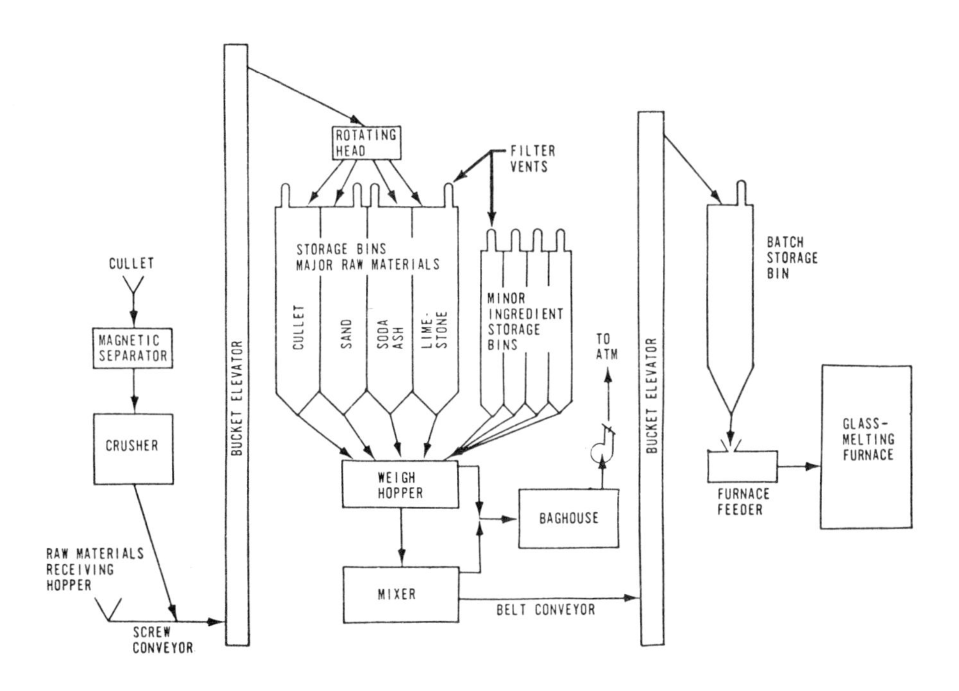

Source: J.A. Danielson (14)

In a slightly different arrangement of equipment to permit closer control of batch composition, blended materials are discharged from the mixer into batch cans that have a capacity of one mixer load each. Loaded cans are then conveyed by monorail to the furnace feeders. Trends in batch plant design are toward single reinforced-concrete structures in which outer walls and partitions constitute the storage bins. Complete automation is provided so that the batch plant is under direct and instant control of the furnace foreman.

While limited quantities of special glasses such as lead or borosilicate are melted in electrically heated pots or in small-batch, regenerative furnaces with capacities up to 10 tons per day, the bulk of production, soda-lime glass, is melted in direct-fired, continuous, regenerative furnaces. Many of these furnaces have added electric induction systems called boosters to increase capacity. Continuous, regenerative furnaces usually range in capacity from 50 to 300 tons of glass per day; 100 tons is the most common capacity found in the United States.

Continuous, regenerative, tank furnaces differ in design according to the type of glass products manufactured. All have two compartments. In the first compartment, called the melter, the dry ingredients are mixed in correct proportions and are continuously fed onto a molten mass of glass having a temperature near 2700°F. The dry materials melt after floating a third to one-half of the way across the compartment and disappearing into the surface of a clear, viscous liquid glass.

Glass flows from the melter into the second compartment, commonly referred to as the refiner, where it is mixed for homogeneity and heat conditioned to eliminate bubbles and stones. The temperature is gradually lowered to about 2200°F. The amount of glass circulating within the melter and refiner is about 10 times the amount withdrawn for production.

Regenerative furnaces for container and tableware manufacture have a submerged opening or throat separating the refiner from the melter. The throat prevents undissolved materials and scum on the surface from entering the refiner. Glass flows from the semicircular refining compartment into long, refractory-lined chambers called forehearths. Oil or gas burners and ventilating dampers accurately control the temperature and viscosity of the glass that is fed from the end of the forehearth to glass-forming machines.

Continuous furnaces for manufacturing rod, tube, and sheet glass differ from furnaces for container and tableware manufacture in that they have no throat between the melter and refiner. The compartments are separated from each other by floating refractory beams riding in a drop arch across the entire width of the furnace. Glass flows from the rectangular-shaped refiner directly into the forming machines.

Regenerative firing systems for continuous glass furnaces were first devised by Siemens in 1852, and since then, nearly all continuous glass furnaces in the United States have used them. In Europe, continuous glass furnaces employ both recuperative and regenerative systems.

Regenerative firing systems consist of dual chambers filled with brick checkerwork. While the products of combustion from the melter pass through and heat one chamber, combustion air is preheated in the opposite chamber. The functions of each chamber are interchanged during the reverse flow of air and combustion products. Reversals occur every 15 to 20 minutes as required for maximum conservation of heat.

Two basic configurations are used in designing continuous, regenerative furnaces—end port, and side port. In the side port furnace, combusion products and flames pass in one direction across the melter during one-half of the cycle. The flow is reversed during the other half cycle. The side port design is commonly used in large furnaces with melter areas in excess of 300 square feet. In the end port configuration, combustion products and flames travel in a horizontal U-shaped path across the surface of the glass within the melter. Fuel and air mix and ignite at one port and discharge through a second port adjacent to the first on the same end wall of the furnace. While the end port design has been used extensively in smaller furnaces with melter areas from 50 to 300 square feet, it has also been used in furnaces with melter areas up to 800 square feet.

Continuous furnaces are usually operated slightly above atmospheric pressure within the melter to prevent air induction at the feeders and an overall loss in combustion efficiency. Furnace draft can be produced by several methods: induced-draft fans, natural-draft stacks, and ejectors.

AIR POLLUTION

The economics and techniques connected with mass production of glass articles have led to the construction of glass-manufacturing plants near or within highly populated areas. Unfortunately, airborne contaminants generated by these glass plants can contribute substantially to the air pollution problem of the surrounding community. Control of dust and fumes has, therefore, been, and must continue to be, inherent to the progress of this expanding industry.

Air pollution control is necessary, not only to eliminate nuisances, but also to bring substantial savings by extending the service life of the equipment and by reducing operating expenses and downtime for repair. Reduction in plant source emissions can be accomplished by several methods, including control of raw materials, batch formulation, efficient combustion of fuel, proper design of glass-melting furnaces, and the installation of control equipment.

Silica sand, dry powders, granular oxides, carbonates, cullet (broken glass), and other raw materials are transferred from railroad hopper cars and trucks to storage bins. These materials are withdrawn from the storage bins, batch weighed, and blended in a mixer. The mixed batch is then conveyed to the feeders attached to the side of the furnace. Although dust emissions are created during these operations, control can be accomplished by totally enclosing the equipment and installing filter vents, exhaust systems, and baghouses.

Screw- or reciprocating-type feeders continuously supply batch-blended materials to the direct-fired, regenerative furnace. These dry materials float upon the molten glass within the furnace until they melt. Carbonates decompose releasing carbon dioxide in the form of bubbles. Volatilized particulates, composed mostly of alkali oxides and sulfates, are captured by the flame and hot gases passing across the molten surface. The particulates are either deposited in the checkers and refractory-lined passages or expelled to the atmosphere.

The mixture of materials is held around 2700°F in a molten state until it acquires the homogeneous character of glass. Then it is gradually cooled to about 2200°F to make it viscous enough to form. In a matter of seconds, while at a yellow-orange hot temperature, the glass is drawn from the furnace and worked on forming machines by a variety of methods including pressing, blowing in molds, drawing, rolling, and casting.

One source of air pollution is hydrocarbon greases and oils used to lubricate the hot delivery systems and molds of glass-forming machines. The smoke from these greases and oils creates a significant amount of air pollution separate from furnace emissions.

Immediately after being shaped in the machines, the glass articles are conveyed to continuous annealing ovens, where they are heat-treated to remove strains that have developed during the molding or shaping operations and then subjected to slow, controlled cooling. Gas-fired or electrically heated annealing ovens are not emitters of air contaminants in any significant quantity. After leaving the annealing ovens, the glass articles are inspected and packed or subjected to further finishing operations. Glass-forming machines for mass production of other articles such as rod, tube, and sheet usually do not emit contaminants in significant amounts.

The furnace is usually natural-gas-fired and is not a major source of air pollution. Grease and oils are used to lubricate the glass-forming machine which is the source of significant smoking. Often the molten glass is sprayed after forming with annealing reagents to provide specific strength properties. The pollutant from this annealing process is usually of a Cl^- structure like HCl gas or $SnCl_4$. The annealing furnace is either electric-heated or gas-fired and presents no air pollution problem. Very few plants will have any control devices for any of the operations associated with the glass in the molten state. Some plants use cyclones and baghouses to handle the fine dusts emanating from the raw material transport operations (18).

Emissions from the glass-melting operation consist primarily of particulates and fluorides, if fluoride-containing fluxes are used in the process. Because the dust emissions contain particles that are only a few microns in diameter, cyclones and centrifugal scrubbers are not as effective as baghouses or filters in collecting particulate matter. Table 63 summarizes the emission factors for glass melting.

TABLE 63: EMISSION FACTORS FOR GLASS MELTING

Type of Glass	- - - Particulates* - - -		- - - Fluorides** - - -	
	lb/ton	kg/MT	lb/ton	kg/MT
Soda-lime	2	1	4F***	2F***

*Emission factors expressed as units per unit weight of glass produced.
***J. Air Pollution Control Assoc.* 7 (2), 92–108 (August 1957).
***F equals weight percent of fluoride in input to furnace; e.g., if fluoride content is 5%, the emission factor would be 4F or 20 (2F or 10).

Source: EPA (15)

Various states have regulations for glass industry emissions which are pertinent here. Thus since March 27, 1972, the New Jersey regulation (N.J.A.C. 7:27-6.1 et seq) has allowed emissions in terms of either:

(a) a collection efficiency of 99% of the potential emission rate or
(b) emissions not exceeding 0.02 grains per standard cubic foot, whichever is greater.

Until April 27, 1974, the Pennsylvania emission standard applicable to the glass industry was 0.04 grains per standard cubic foot. On that date, the standard was changed to the higher of:

(a) 0.02 grains per standard cubic foot or
(b) the rate in pounds per hour calculated from the following formula: $(0.76)(50W)^{0.42}$, where W is the furnace process rate in tons per hour.

The exact form of the Pennsylvania standard is given by the formula:

$A = (0.76)E^{0.42}$, where
A = allowable emissions (lb/hr)
E = emission index = F x W (lb/hr)
F = process factor (pounds/unit of production), and
W = production or charging rate (units of production/hour)

For the glass industry, F is 50 pounds/ton of fill, and the unit of production for F and W is tons of fill. In addition, Pennsylvania restricts the emission of visible air contaminants from glass furnaces to an opacity of no greater than 20%, except that for no more than three minutes in any one hour the opacity may be greater than 20% but not in excess of 60%.

Emission and Control in Batch Plants

The major raw materials for making soda-lime glass—sand, soda-ash, and limestone—usually contain particles averaging about 300 microns in size. Particles less than 50 microns constitute only a small portion of the materials, but are present in sufficient quantities to cause dust emissions during conveying, mixing, and storage operations. Moreover, minor raw materials such as salt cake and sulfur can create dust emissions during handling. Dust is the only air contaminant from batch plants, and the control of dust emissions poses problems similar to those in industrial plants handling similar dusty powder or granular materials.

Dust control equipment can be installed on conveying systems that use open conveyor belts. A considerable reduction in the size of the dust control equipment can be realized by totally enclosing all conveying equipment and sealing all covers and access openings with gaskets of polyurethane foam. In fact, by totally enclosing all conveying equipment, exhaust systems become unnecessary, and relatively small filter vents or dust cabinets can be attached directly to the conveying equipment and storage bins.

On the other hand, exhaust systems are required for ventilating the weigh hoppers and mixers. For example, a 60-cubic-foot-capacity mixer and a 4,500-pound-capacity mixer each requires about 600 cfm ventilation air. Seals of polyvinyl chloride should be installed between the rotating body of the mixer and its frame to reduce ventilation to a minimum.

Railroad hopper cars and hopper bottom trucks must be connected to sealed receiving hoppers by fabric sleeves so that dust generated in the hoppers during the loading operation is either filtered through the sleeves or exhausted through a baghouse.

Local exhaust systems for dust pickup are designed by using the recommended practice of the Committee on Industrial Ventilation (1960). For example, the ventilation rate at the transfer point between two open belt conveyors is 350 cfm per foot of belt width, with 200 fpm minimum velocity through the hood openings.

Because dust emissions contain particles only a few microns in diameter, cyclones and centrifugal scrubbers are not as effective as baghouses or filters are in collecting these small particles; consequently, simple cloth filters and baghouses are used almost exclusively in controlling dust emissions from batch plants.

Filter socks or simple baghouses with intermittent shaking mechanisms are usually designed for a filter velocity of 3 fpm, but baghouses with continuous cleaning devices such as pulse jets or reverse air systems can be designed for filter velocities as high as 10 fpm. Filtration cloths are usually cotton, though nylon, Orlon, and Dacron are sometimes used.

Dusts collected are generally noncorrosive. Filters or baghouses for storage bins are designed to accommodate not only displaced air from the filling operation but also air induced by falling materials. Filtration of air exhaust from pneumatic conveyors used in filling the bins must also be provided. Filters with at least a 1-square-foot area should be mounted on the hand-filled minor-ingredient bins.

Transfer chutes of special design are used for hand filling the minor-ingredient bins. They are first attached securely with gaskets to the top of the bins. The bags are dropped into a chute containing knives across the bottom. The knives split the bag, and as the materials fall into the bin, the broken bag seals off the escape of dust from the top of the chute.

Emissions and Control in Glass-Melting Furnaces

Particulates expelled from the melter are the result of complex physical and chemical reactions that occur during the melting process. Glass has properties akin to those of crystalline solids, including rigidity, cold flow, and hardness. At the same time, it behaves like a supercooled liquid. It has nondirectional properties, fracture characteristics of an amorphous solid, and no freezing or melting point. To account for the wide range of properties, glass is considered to be a configuration of atoms rather than an aggregate of molecules. The theory has been proposed that glass consists of an extended, continuous, three-dimensional network of ions with a certain amount of short-distance-ordered arrangement similar to that of a polyhedral crystal. These dissimilar properties explain in part why predictions of particulate losses from the melter based solely upon known temperatures and vapor pressures of pure compounds have been inaccurate. Other phenomena affect the generation of particulates. During the melting process, carbon dioxide bubbles and propels particulates from the melting batch.

Particulates are entrained by the fast-moving stream of flames and combustion gases. As consumption of fuel and refractory temperatures of the furnace increase with glass tonnage, particulates also increase in quantity. Particulates, swept from the melter, are either collected in the checkerwork and gas passages or exhausted to the atmosphere.

In a recent study, many source tests of glass furnaces in Los Angeles County were used for determining the major variables influencing stack emissions. Particulate emissions, opacities, process variables, and furnace design factors were noted. These particulate samples were obtained from the catch of a pilot baghouse venting part of the effluent from a large soda-lime container furnace.

Chemical composition of the particulates was determined by microquantitative methods or by spectrographic analysis. Five separate samples, four from a pilot baghouse, and one from the stack of a soda-lime regenerative furnace were analyzed. They were found to be composed mostly of alkali sulfates although alkalies are reported as oxides. The chemical composition was also checked by x-ray crystallography. In this analysis, the only crystalline material present in identifiable amounts was two polymorphic forms of sodium sulfate.

From the source test data available, particulate emissions did not correlate with the opacity of the stack emissions. Some generalizations on opacity can, however, be made. Opacities usually increase as particulate emissions increase. More often than not, furnaces burning U.S. Grade 5 fuel oil have plumes exceeding 40% white opacity while operating at a maximum pull rate, which is the glass industry's common term for production rate.

Plumes from these same furnaces were only 15 to 30% white opacity while burning natural gas of U.S. Grade 3 (PS 200) fuel oil. Somewhat lower opacities may be expected from furnaces with ejector draft systems as compared with furnaces with natural-draft stacks or induced-draft fans.

In order to determine the correct size of air pollution control equipment, the volume of dirty exhaust gas from a furnace must be known. Some of the more important factors affecting exhaust volumes include furnace size, pull rate, combustion efficiency, checker volume, and furnace condition.

As the furnace campaign progresses, dust carry-over speeds destruction of the checkers. Upper courses of the firebrick checker glaze when subjected to high temperatures. Dust and condensate collect on the brick surface and form slag that drips downward into the lower courses where it solidifies at the lower temperature and plugs the checkers. Slag may also act somewhat like flypaper, tenaciously clinging to the upper courses and eventually sealing off upper gas passages.

Hot spots develop around clogged checkers and intensify the destructive forces, which are reflected by a drop in regenerator efficiency and a rise in fuel consumption and horsepower required to overcome additional gas flow resistance through the checkers. Checker damage can finally reach a point where operation is no longer economical or is physically impossible because of collapse.

Thus, successful operation of modern regenerative furnaces requires keeping dust carry-over from the melter to an absolute minimum, which also coincides with air pollution control objectives by preventing air contaminants from entering the atmosphere. Aside from reducing air contaminants, benefits derived from reducing dust carry-over are many and include longer furnace campaigns, lower maintenance cost, and savings on fuel.

Design and operation of soda-lime, continuous, regenerative furnaces to alleviate dust carry-over and minimize particulate emissions are discussed in succeeding paragraphs. Advantages of all-electric, continuous furnaces for melting glass are also cited.

Although glassmakers have traditionally sought fine-particle materials for easier melting, these materials have intensified dust carry-over in regenerative furnaces. A compromise must be reached. Major raw materials should be in the form of small particles, many of them passing U.S. 30-mesh screen, but not more than 0.3 weight percent passing U.S. 325 mesh screen.

Because crystals of soda ash, limestone, and other materials may be friable and crush in the mixer, producing excessive amounts of fines, screen analyses of individual raw materials should not be combined for estimating the screen analyses of the batch charge. Crystalline shape and density of raw materials should be thoroughly investigated before raw material suppliers are selected.

Since particulate emissions from soda-lime regenerative furnaces increase with an increase in equivalent sulfate (SO_3) present in the batch charge, sulfate content should be reduced to an absolute minimum consistent with good glass-making. Preferably, it should be below 0.3 weight percent. Equivalent sulfate (SO_3) content of the batch includes all sulfur compounds and elemental sulfur. Compounds of fluorine, boron, lead, and arsenic are also known to promote dust carry-over, but the magnitude of their effect upon emissions is still unknown. In soda-lime glass manufacture, these materials should be eliminated or should be present in only trace amounts.

From the standpoint of suppressing stack emissions, cullet content of the batch charge should be kept as high as possible. Plant economics may, nevertheless, dictate reduction in cullet where fuel or cullet is high in cost or where cullet is in short supply. Some manufacturing plants are able to supply all their cullet requirements from scrap and reject glassware.

There are a number of ways to condition a batch charge and reduce dust carryover. Some soda-lime glass manufacturers add moisture to the dry batch, but the relative merits of this process are debatable. Moisture is sprayed into the dry batch charge at the mixer as a solution containing 1 gallon of surface-active wetting agent to 750 gallons of water. Surface tension of the water is reduced by the wetting agent so that the water wets the finest particles and is evenly distributed throughout the batch.

Fluxing materials such as salt cake appear more effective, since the unmelted batch does not usually travel so far in the melter tank before it melts. Moisture content of the batch is normally increased to about 2% by weight. If the moisture content exceeds 3%, batch ingredients adhere to materials-handling equipment and may cake in storage bins or batch cans.

Other batch preparation methods have been employed on a limited-production or experimental basis to reduce dust carry-over from soda-ash glass manufacture. One method involves presintering the batch to form cullet and then charging only this cullet to the furnace. Advantages claimed are faster melting, better batch control, less seed formation, reduced clogging in the checkers, and lower stack losses. A Dutch oven doghouse cover also reduces dust carryover by sintering the top of the floating dry batch before it enters the melter. This method is probably not as efficient as is complete presintering in reducing dust carryover.

Other methods include: (1) Charging briquets, which are made from regular batch ingredients by adding up to 10% by weight of water; (2) charging wet batches containing 6% moisture, which are made by first dissolving soda-ash to form a saturated solution and mixing this solution with sand and the other dry materials; (3) charging the dry batch (submerged) in the melter; (4) enclosing batch feeders (Fabrianio, 1961); and (5) installing batch feeders on opposite sides of end port, regenerative furnaces and charging alternately on the side under fire.

The design concept of modern regenerative furnaces, with its emphasis on maximum use of fuel, is also indirectly committed to reducing dust carry-over. All things being equal, less fuel burned per ton of glass means less dust entrainment by hot combustion gases and flames flowing across the surface of the melting glass. Although container furnaces constructed over 15 years ago required over 7,000 cubic feet of natural gas per ton of glass at maximum pull rates, container furnaces built today can melt a ton of glass with less than 5,000 cubic feet of natural gas.

While several design changes are responsible for this improvment, one of the most important is the increase in checker volume. The ratio of checker volume (cubic feet) to meter area (square feet) has been rising during the years from about 5 in earlier furnaces to about 9 today. Enlarged checkers not only reduce fuel consumption and particulate formation but also present a more effective trap for dust particles that are expelled from the melter. Source tests conducted by a large glass-manufacturing company indicated that over 50% of the dust carry-over from the melter is collected by the checkers and gas passages instead of entering the atmosphere.

Of course, the economics connected with regenerative furnace operation dictates the checker volume. The law of diminishing returns operates where capital outlay for an added volume of checkers will no longer be paid within a specified period by an incremental reduction in fuel costs. Checkers have been designed in double-pass arrangements to recover as much as 55% of the heat from the waste gases.

Although dust collects within checkers by mechanisms of impingement and settling, the relationship among various factors influencing dust collection is unknown. These factors include: Gas velocity, brick size, flue spacing, brick setting, and brick composition. Checkers designed for maximum fuel economy may not necessarily have the highest collection efficiency. Further testing will be necessary in order to evaluate checker designs.

Checkers designed for maximum heat exchange contain maximum heat transfer surface per unit volume, a condition met only by smaller refractories with tighter spacing. Since gas velocities are also highest for maximum heat transfer, less dust collects by simple settling than by impingement. Dust collection is further complicated in that smaller brick increases the potential for clogging.

To prevent clogging in the checkers and ensure a reasonable level of heat transfer, checkers should be cleaned once per month or more often; an adequate number of access doors should be provided for this purpose. Compressed air, water or steam may be used to flush fine particles from the checkers. Virtually nothing can be done to remove slag after it has formed. Checkers can be arranged in a double vertical pass to reduce overall furnace height and make cleaning easier. Access doors should also be provided for removing dust deposits from the flues.

Further reductions in fuel consumption to reduce dust emissions may be realized by installing rotary, regenerative air preheaters in series with the checkers. Additional benefits include less checker plugging, reduced maintenance, and increased checker life. Rotating elements of the preheater are constructed of mild steel, low-alloy steel, or ceramic materials. Preheaters raise the temperature of the air to over 1000°F, and the increased velocity of this preheated air aids in purging dust deposits that block gas passages of the checkers.

Exhaust gases passing through the opposite side of the preheater are cooled below 800°F before being exhausted to the atmosphere. A heat balance study of a plate glass, regenerative furnace shows a 9% increase in heat use by the installation of a rotary, regenerative air preheater. To maintain heat transfer and prevent reentrainment, dust deposits on the preheater elements must be removed by periodic cleaning. Ductwork and valves should be installed for by-passing rotary air preheaters during the cleaning stage.

Slagging of the upper courses of checkerwork can be alleviated in most cases by installing basic (high alumina content) brick in place of superduty firebrick. Basic brick courses extend from the top downward to positions where checker temperatures are below 1500°F. At this temperature, firebrick no longer wets and forms slag with dust particles. Dust usually collects in the lower courses of firebrick in the form of fine particles that are easily removed by cleaning.

Although basic brick costs 3 or 4 times as much as superduty firebrick, some glass manufacturers are constructing entire checkerworks of basic brick where slagging and clogging are most severe. In some instances, basic refractories are replacing fireclay rider tiles and rider arches in checker supports. A word of caution, basic brick is no panacea for all ills of checkers. Chemical composition of the dust should be known, to determine compatibility with the checkers.

Regenerative furnaces can be designed to consume less fuel and emit less dust by proper selection and application of insulating refractories. A heat balance study of a side port, regenerative furnace shows that, in the melting process, glass receives 10% of heat transfer from convection and 90% from radiation. Of the radiation portion of heat transferred, the crown accounts for 33%. Since heat losses through the uninsulated crown can run as high as 10% of the total heat input, there is need for insulation at this spot.

Most crowns are constructed of silica brick with a maximum furnace capacity restricted to an operating temperature of 2850°F. Insulation usually consists of insulating silica brick backed with high-duty plastic refractory. Furnaces are first operated without insulation, so that cracks can be observed. Then the cracks are sealed with silica cement, and the insulation is applied.

Insulation is needed on the melter sidewall and at the port necks to prevent glassy buildup caused by condensation of vapors. Condensate buildup flows across port sills into the melter and can become a major source of stones.

While insulation of sidewalls shows negligible fuel reduction for flint glass manufacture, it does show substantial fuel reduction for colored glasses. The problem in manufacturing colored glass is to maintain a high enough temperature below the surface to speed the solution of stones and prevent stagnation. Insulation on sidewalls raises the mean temperature to a point where stones dissolve and glass circulates freely.

Six inches or more of electrofusion cast block laid over a clay bottom in a bed of mortar not only saves fuel but is also less subject to erosion than is fireclay block. Insulation is seldom needed on the refining end of the furnace since refiners have become cooling chambers at today's high pull rates. Nose crowns, however, are insulated to minimize condensation and drip. Checkers are sometimes encased in steel to prevent air infiltration through cracks and holes that develop in the refractory regenerator walls during the campaign.

Furnace size also has an effect upon use of fuel, with a corresponding effect on the emissions of dust. Large furnaces are more economical than are small furnaces because the radiating surface or heat loss per unit volume of glass is greater for small furnaces.

Slightly greater fuel economy may be expected from end port furnaces as compared with side port furnaces of equal capacity. Here again, the end port furnace has a heat loss advantage over the side port furnace because it has less exposed exterior surface area for radiating heat. Side port furnaces can, however, be operated at greater percentages in excess of capacity since mixing of fuel with air is more efficient through several smaller inlet ports than it is through only one large inlet port.

In fact, end port furnaces are limited in design to the amount of fuel that can be efficiently mixed with air and burned through this one inlet port. As far as dust losses are concerned, there are only negligible differences between end port and side port furnaces of equal size. Reduced fuel consumption to reduce dust carry-over can also be realized by increasing the depth of the melter to the maximum consistent with good-quality glass. Maximum depths for container furnaces are 42 inches for flint glass and about 36 inches for amber glass and emerald green glass.

Dust emissions as well as fuel consumption can also be reduced by firing practice. Rapid changes in pull rates are wasteful of fuel and increase stack emissions. Hence, charge rates and glass pull rates for continuous furnaces should remain as constant as possible by balancing loads between the glass-forming machines.

If possible, furnaces should be fired on natural gas or U.S. Grade 3 or lighter fuel oil. Particulate emissions increase an average of about 1 pound per hour when U.S. Grade 5 fuel oil is used instead of natural gas or U.S. Grade 3 fuel oil, and opacities may exceed 40% white.

Combustion air should be thoroughly mixed with fuel with only enough excess air present to ensure complete combustion without smoke. Excess air robs the furnace of process heat by dilution, and this heat loss must be overcome by burning additional fuel. Volume of the melter should be designed for a maximum fuel heat release of about 13,000 Btu per hour per cubic foot.

Furnace reversals should be performed by an automatic control system to ensure optimum combustion. Only automatic systems can provide the exact timing required for opening and closing the dampers and valves and for coordinating fuel and combustion airflow.

For instance, fuel flow and ignition must be delayed until combustion air travels through the checkers after reversal to mix with fuel at the inlet port to the melter. Furnace reversals are usually performed in fixed periods of 15 to 20 minutes, but an improvement in regenerator efficiency can be realized by programming reversal periods to checker temperatures measured optically. Reversals can then occur when checker temperatures reach preset values consistent with maximum heat transfer.

An excellent system for controlling air-to-fuel ratios incorporates continuous flue gas analyzers for oxygen and combustible hydrocarbons. With this system, the most efficient combustion and best flame shape and coverage occur at optimum oxygen with a trace of combustible hydrocarbons present in the flue gas. Sample gas is cleaned for the analyzers through water-cooled probes containing sprays. The system automatically adjusts to compensate for changes in ambient air density. Fuel savings of 6 to 8% can be accomplished on furnaces with analyzers over furnaces not so equipped.

Combustion of natural gas in new furnaces occurs efficiently when the oxygen content of the flue gases in the exhaust ports is less than 2% by volume. As the campaign progresses, air infiltration through cracks and pores in the brickwork, air leakage through valves and dampers, increased pressure drop through the regenerators, and other effects combine to make combustion less efficient.

To maintain maximum combustion throughout the campaign, pressure checks with draft gauges should be run periodically at specified locations. Fuel savings can also be expedited by placing furnace operators on an incentive plan to keep combustion air to a minimum.

Although melting glass by electricity is a more costly process than melting glass by natural gas or fuel oil, melting electrically is a more thermally efficient process since heat can be applied directly to the body of the glass.

Electric induction systems installed on regenerative furnaces are designed to increase maximum pull rates by as much as 50%. These systems are called boosters and consist of several water-cooled graphite or molybdenum electrodes equally spaced along the sides of the melter 18 or 32 inches below the surface of the glass. Source test results indicate that pull rates can be increased without any appreciable increase in dust carry-over or particulate emissions. Furnace temperatures may also be reduced by boosters, preventing refractory damage at peak operations.

Furnace capacity increase is nearly proportional to the amount of electrical energy expended. A 56-ton-per-day regenerative furnace requires 480 kilowatt-hours in the booster to melt an additional ton of glass, which is close to the theoretical amount of heat needed to melt a ton of glass. Electric induction can also be used exclusively for melting glass on a large scale. Design of this type of furnace is simplified since regenerative checkerworks and large ductwork are no longer required. One recently constructed 10-ton-per-day, all-electric furnace consists of a simple tank with molybdenum electrodes.

A small vent leads directly to the atmosphere, and dust emissions through this vent are very small. The furnace operates with a crown temperature below 600°F and with a thermal efficiency of over 60%. Glass quality is excellent, with homogeneity nearly that of optical glass.

After the first 11 months of operation, there was no apparent wear on the refractories. First costs and maintenance expenses are substantially lower than for a comparable-size regenerative furnace. An electric furnace may prove competitive with regenerative furnaces in areas with low-cost electrical power.

Air pollution control equipment can be installed on regenerative furnaces where particulate emissions or opacities cannot be reduced to required amounts through changes in furnace design, control of raw materials, and operating procedures. Regenerative furnaces may be vented by two types of common industrial control devices–wet centrifugal scrubbers and baghouses.

One typical control device is a low-pressure, wet, centrifugal scrubber containing two separate contacting sections within a single casing. Separate 50-horsepower, circulating fans force dirty gas through each section containing two to three impingement elements similar to fixed blades of a turbine. Although the collection efficiency of this device is considered about the highest for its type, source tests show an overall efficiency of only 52%. This low efficiency demonstrates the inherent inability of the low-pressure, wet, centrifugal scrubbers to collect particulates of submicron size.

On the other hand, baghouses show collection efficiencies of over 99%. Although baghouses have not as yet been installed on large continuous, regenerative furnaces, they have been installed on small regenerative furnaces. One baghouse alternately vents a 1,800-lb and a 5,000-lb batch regenerative furnace used for melting optical and special glasses used in scientific instruments. Bags are made of silicone-treated glass fiber. Off-gases are tempered by ambient air to reduce the temperature to 400°F, a safe operating temperature for this fabric.

Another baghouse, although no longer in operation, vented a 10-ton-per-day regenerative furnace for melting soda-lime flint glass. Stack gases were cooled to 250°F by radiation and convection from an uninsulated steel duct before entering the baghouse containing Orlon bags.

To determine the feasibility of using a cloth filtering device on large continuous, regenerative furnaces, a pilot baghouse was used with bags made of various commercial fabrics. An air-to-gas heat exchanger containing 38 tubes, each 1½ inches in outer diameter by 120 inches in length, cooled furnace exhaust gases before the gases entered the pilot baghouse. The baghouse contained 36 bags, each 6 inches in diameter by 111 inches in length, with a 432-net-square-foot filter area. A 3-horsepower exhaust fan was mounted on the discharge duct of the baghouse.

When subjected to exhaust gases from amber glass manufacture, bags made of cotton, Orlon, Dynel, and Dacron showed rapid deterioration and stiffening. Only Orlon and Dacron bags appeared in satisfactory condition when controlling dirty gas from flint glass manufacture and when the dirty gas was held well above its dew point. This difference in corrosion between amber and flint glass was found to be caused by the difference in concentrations of sulfur trioxide (SO_3) present in the flue gas. To reduce the concentration of SO_3 from amber glass manufacture, iron pyrites were substituted for elemental sulfur in the batch, but this change met with no marked success. Stoichiometric amounts of ammonia gas were also injected to remove SO_3 as ammonium sulfate. Ammonia injection not only failed to lessen bag deterioration but also caused the heat exchanger tubes to foul more rapidly. In all cases, the baghouse temperature had to be kept above the dew point of the furnace effluent to prevent condensation from blinding the bags and promoting rapid chemical attack. At times, the baghouse had to be operated with an inlet temperature as high as 280°F to stay above the elevated dew point caused by the presence of SO_3.

Additional pilot baghouse studies are needed to evaluate Orlon and Dacron properly for flint glass manufacture. Experiments are also required for evaluating silicone-treated glass fiber bags in controlling exhaust gases from regenerative furnaces melting all types of glass.

Information now available indicates that glass fiber bags can perform at temperatures as high as 500°F, well above the elevated dew points. They are virtually unaffected by relatively large concentrations of SO_2 and SO_3, and there is less danger from condensation. One advantage of glass fiber is that less precooling of exhaust gases is required because of the higher allowable operating temperatures. Reverse air collapse is generally conceded to be the best method of cleaning glass fiber bags, since this material is fragile and easily breaks when regular shakers are installed.

Furnace effluent can be cooled by several methods: Air dilution, radiation cooling columns, air-gas heat exchangers, and water spray chambers. Regardless of the cooling method selected, automatic controls should be installed to ensure proper temperatures during the complete firing cycle. Each cooling method has its advantages and disadvantages. Dilution of off-gases with air is the simplest and most trouble-free way to reduce temperature but requires the largest baghouse.

Air-to-gas heat exchangers and radiation and convection ductwork are subject to rapid fouling from dust in the effluent. Automatic surface-cleaning devices should be provided, or access openings installed for frequent manual cleaning to maintain clean surfaces for adequate heat transfer. If spray chambers are used, severe problems in condensation and temperature control are anticipated.

Emissions and Control in Glass Forming

Dense smoke is generated by flash vaporization of hydrocarbon greases and oils from contact lubrication of hot gob shears and gob delivery systems. This smoke emission can exceed 40% white opacity. Molds are lubricated with mixtures of greases and oils and graphite applied to the hot internal surfaces once during 10- to 20-minute periods. This smoke is usually 100% white in opacity and exists for 1 or 2 seconds. It rapidly loses its opacity and is completely dissipated within several seconds.

During the past decade, grease and oil lubricants for gob shears and gob delivery systems have been replaced by silicone emulsions and water-soluble oils at ratios of 90 to 150 parts of water to 1 part oil or silicone. The effect has been the virtual elimination of smoke. The emulsions and solutions are applied by intermittent sprays to the delivery system and shears only when the shears are in an opened position.

Cost of Alternative Control Technologies

Table 64 presents the estimated economics for producing opal glass at two plant sizes characteristic of current production practice. Labor rates and material costs are typical of Gulf Coast data. Return on investment prior to use of fluoride control processes is estimated to be 14.5% and 46.7% for the 20-ton and 60-ton-per-day plants.

Tables 65 through 67 show the estimated capital outlay and operating costs for three current process approaches to dust and fume abatement in the opal glass industry. Because of emissions control, ΔROI is estimated between -2% and -3%.

TABLE 64: ESTIMATED ECONOMICS OF OPAL GLASS PRODUCTION (POLLUTION CONTROL COST EXCLUDED)

	PLANT CAPACITY	
	20 Tons/Day	60 Tons/Day
TOTAL CAPITAL INVESTMENT[1]	3.2 $MM	5.8 $MM
PRODUCTION COSTS		
DIRECT COSTS		
GLASS SAND (.8 Tons/Net Ton @ 12.75 $/Ton)	10.20 $/Net Ton	10.20 $/Net Ton
SODA ASH (.15 Tons/Net Ton @ 34.00 $/Ton)	5.10	5.10
FLUORSPAR (.16 Tons/Net Ton @ 65.00 $/Ton)	10.40	10.40
BORAX (.01 Tons/Net Ton @ 50.25 $/Ton)	.50	.50
FELDSPAR (.25 Tons/Net Ton @ 20.00 $/Ton)	5.00	5.00
LIMESTONE (.12 Tons/Net Ton @ 4.00 $/Ton)	.48	.48
NATURAL GAS (4.6 MM BTU/Ton X .40 $/MM BTU)	1.84	1.84
LABOR (7 Positions @ 4.00 $/hr)	33.60	11.20
FURNACE REPAIRS (1.2%)	5.82	3.52
FRINGE BENEFITS AND SUPERVISION	33.60	11.20
GENERAL MAINTENANCE AND SUPPLIES (1.25%)	6.08	3.67
TOTAL DIRECT COSTS	112.62	63.11
INDIRECT COSTS		
DEPRECIATION(@ 10%)	48.48	29.29
INTEREST (AT 7%, 20% DEBT)	6.74	4.10
LOCAL TAXES AND INSURANCE (@ 3%)	14.54	8.79
PLANT AND LABOR OVERHEAD	40.37	13.44
TOTAL INDIRECT COSTS	110.13	55.62
TOTAL AVERAGE COST[2] ($/Net Ton)	222.75	118.73
GENERAL AND SALES EXPENSES($/Ton)	4.46 $/Net Ton	2.37 $/Net Ton
F.O.B. COST ($/Net Ton)	227.21 $/Net Ton	121.10 $/Net Ton
AVERAGE PRODUCT REVENUE ($/Ton)	340.00 $/Ton	340.00 $/Net Ton
AVERAGE PROFIT AFTER TAXES (@ 50%)	56.40 $/Ton	109.45 $/Ton
CASH FLOW ($MM/YR)	0.69 $MM/YR	2.75 $MM/YR
RETURN ON INVESTMENT (%)	14.5%	46.7%

(1)Small tank furnace facility

(2)Transporation charges alter costs substantially

Source: J.M. Robinson, G.I. Gruber, W.D. Lusk and M.J. Santy (19)

There are no significant air or noise pollution problems directly associated with wastewater treatment and control technologies. The wastewaters and sludges are odorless and no nuisance conditions result from their treatment or handling. Incineration is not used in the treatment technologies so no air pollution is caused by this source. Water vapor resulting from the evaporation of reverse osmosis brines is expected to be relatively pure.

A nonwater quality aspect of perhaps greater significance than air pollution is the high energy required for total recycle systems. In view of the limited availability of clean energy sources and the air pollution problems associated with other energy sources, the benefits derived from a total recycle system should also be weighed against the energy required to operate such a system.

TABLE 65: OPAL GLASS PRODUCTION—ESTIMATED ECONOMICS OF CONTROL PROCESS A BASIS—60 TONS PER DAY OF OPAL GLASS PRODUCTION

Capital Cost Estimates ($1000)

Item Number	Description	Equipment F.O.B. Cost	Installation Factor	Equipment Installation Cost
1.	COOLING DUCT, t_{in} = 1500°F, t_{out} = 400°F, carbon steel, 22,500 cfm 500 ft^2 surface area, air cooled, 2 in. W.G. pressure drop	3	1.67	5
2.	WET CYCLONE, 100 gal/min, 22,500 cfm, neoprene lined steel, 2 in. W.G. pressure drop	50	1.55	78
			Capital Subtotal	83.
			Indirects (@ 15%)	12.
			Contingency (@ 20%)	17.
			Total Capital (as of January 1971)	112.

Operating Cost ($ /hr)

Item Number	Power Cost	Maintenance Cost	Equipment Operating Cost
1	0.05	0.10	0.15
2	0.05	0.13	0.18
Subtotal			0.33
Water (100 gpm, 90% recycle)			0.02
Disposal			—
Total Operating Cost			0.35

Total Operating Cost ($/hr)	0.35
Taxes and Insurance (2%, 330 days)	0.28
Capital (7.1%, 330 working days)	1.00
Pollution Control Cost ($/hr)	1.63
Pollution Control Cost ($/ton)	0.65

Source: J.M. Robinson, G.I. Gruber, W.D. Lusk, and M.J. Santy (19)

TABLE 66: OPAL GLASS PRODUCTION–ESTIMATED ECONOMICS OF CONTROL PROCESS B
BASIS–60 TONS PER DAY OF OPAL GLASS PRODUCTION

Capital Cost Estimates ($1000)

Item Number	Description	Equipment F.O.B. Cost	Installation Factor	Equipment Installation Cost
1	COOLING DUCT, t_{in} = 1500°F, t_{out} = 400°F, carbon steel, 22,500 cfm, 500 ft^2 surface area, air cooled, 2 in. W.G. pressure drop	3	1.67	5
2	BAGHOUSE, 22,500 cfm, 2.5 in. W.G. pressure drop, 0.3 lbs solid loading per hour, fabric	20	4.13	83
	Capital Subtotal			88
	Indirects (@ 15%)			13
	Contingency (@ 20%)			18
	Total Capital (as of January 1971)			119

Operating Cost ($ / hr)

Item Number	Power Cost	Maintenance Cost	Equipment Operating Cost
1	.05	0.10	0.15
2	.06	1.26	1.32
Subtotal			1.47
Water			–
Disposal			–
Total Operating Cost			1.47

Total Operating Cost ($/hr)	1.47
Taxes and Insurance (2%, 330 days)	0.30
Capital (7.1%, 330 working days)	1.07
Pollution Control Cost ($/hr)	2.84
Pollution Control Cost ($/ton)	1.14

Source: J.M. Robinson, G.I. Gruber, W.D. Lusk, and M.J. Santy (19)

TABLE 67: OPAL GLASS PRODUCTION—ESTIMATED ECONOMICS OF CONTROL PROCESS C
BASIS—60 TONS PER DAY OF OPAL GLASS PRODUCTION

Capital Cost Estimates ($1000)

Item Number	Description	Equipment F.O.B. Cost	Installation Factor	Equipment Installation Cost
1	BAGHOUSE, 7500 cfm, 2 lb solid loading per hour, fabric	10.	4.00	40.
			Capital Subtotal	40
			Indirects (@ 15%)	6
			Contingency (@ 20%)	8
			Total Capital (as of January 1971)	54

Operating Cost ($/hr)

Item Number	Power Cost	Maintenance Cost	Equipment Operating Cost
1	0.01	1.26	1.27
Subtotal			1.27
Water			–
Disposal			–
Total Operating Cost			1.27

Total Operating Cost ($/hr)	1.27
Taxes and Insurance (2%, 330 days)	0.14
Capital (7.1%, 330 working days)	0.48
Pollution Control Cost ($/hr)	1.89
Pollution Control Cost ($/ton)	0.76

Source: J.M. Robinson, G.I. Gruber, W.D. Lusk, and M.J. Santy (19)

The incandescent lamp envelope manufacturing subcategory is the only subcategory of the pressed and blown glass industry that may pose an air pollution problem. Ammonia removal by steam stripping is recommended for control of high ammonia discharges from the frosting waste stream. It is possible that the steam and ammonia gas from the stripping unit could be vented to the atmosphere through the furnace exhaust stack.

The ammonia concentration of the combined stack discharge is not expected to exceed 35 mg/cu m (46 ppmv), which is the threshold odor limit for ammonia. Because the ammonia concentration will be below the threshold odor level, steam stripping should not cause a significant air pollution problem.

There are no significant air or noise pollution problems directly associated with the treatment and control technologies of the other subcategories. The wastewaters and sludges are odorless and no nuisance conditions result from their treatment or handling.

Proprietary Control Processes

A process developed by H.R. Swift et al (48) has particular application to the carry-over of batch materials up through the flue system of glass melting furnaces. Accordingly, it is contemplated to coat certain of the checker bricks with an appropriate glass.

The process is carried out by applying a glass frit of the proper softening range to a surface which may or may not be a particular part of original processing equipment. The surface is however in close proximity to the original processing equipment. The softened glass coating acts as an adhesive to which dust particles in the air will cling and thereby be eliminated from the process.

This adhesive surface may be formed by coating furnace or machine parts with a glass of the proper softening range; or, special collector plates, to which the glass coating is applied, may be positioned in proper proximity to the source of dust for collection purposes.

A process developed by G.A. Bowman (49) is a process for making glass in which, by a combination of integrated steps, waste chemicals in the form of vapor or particulates now lost up the glass furnace stack to atmosphere are recovered and recycled into either agglomerated or unagglomerated glass batch.

Also chemicals normally discharged to the sewer in a wastewater stream are recovered and recycled into either agglomerated or unagglomerated glass batch. The process also enables recovery and recycling into the glass batch of a substantial percentage of the Btu content of the fuel supplied to melt the glass batch in a standard furnace. The process enables compliance with most regulations regarding air and stream pollution.

Figure 62 shows a suitable form of apparatus for the conduct of the process. Referring to the drawing, there is illustrated a batching bin **10** delivering powdered raw materials to a mixer **11** which thoroughly mixes the powdered raw materials and delivers them to a holding bin **12** by means of elevator **13**. The mix is then fed using an accurate weigh feeder **14** to an agglomerating unit such as the disc pelletizer **15** where water is added. The pelletized particles may be of any size or shape that fits the batch mixer and equipment used.

The agglomerated particles are fed to a gas dryer **16** which may be any conventional type, preferably compartmentalized, normally waste gas heated unit, wherein the agglomerated particle is placed on moving belts to a depth of at least 4". The actual bed depth will depend on the residence time in the dryer, temperature of input hot gas, the temperature desired for the exhaust gas and the pressure drop that can be tolerated across the sectionalized moving beds. Gas flow is preferably directed up and down through the bed as it enters different compartments. The gas flow is counterflow to the movement of the agglomerated particles on the belt. The hot gases normally will be the waste flow from glass furnaces which have been passed through another higher temperature waste heat exchanger stack furnace **17**.

FIGURE 62: INTEGRATED PROCESS FOR AIR AND WATER POLLUTION CONTROL AND HEAT RECOVERY IN GLASS MANUFACTURE

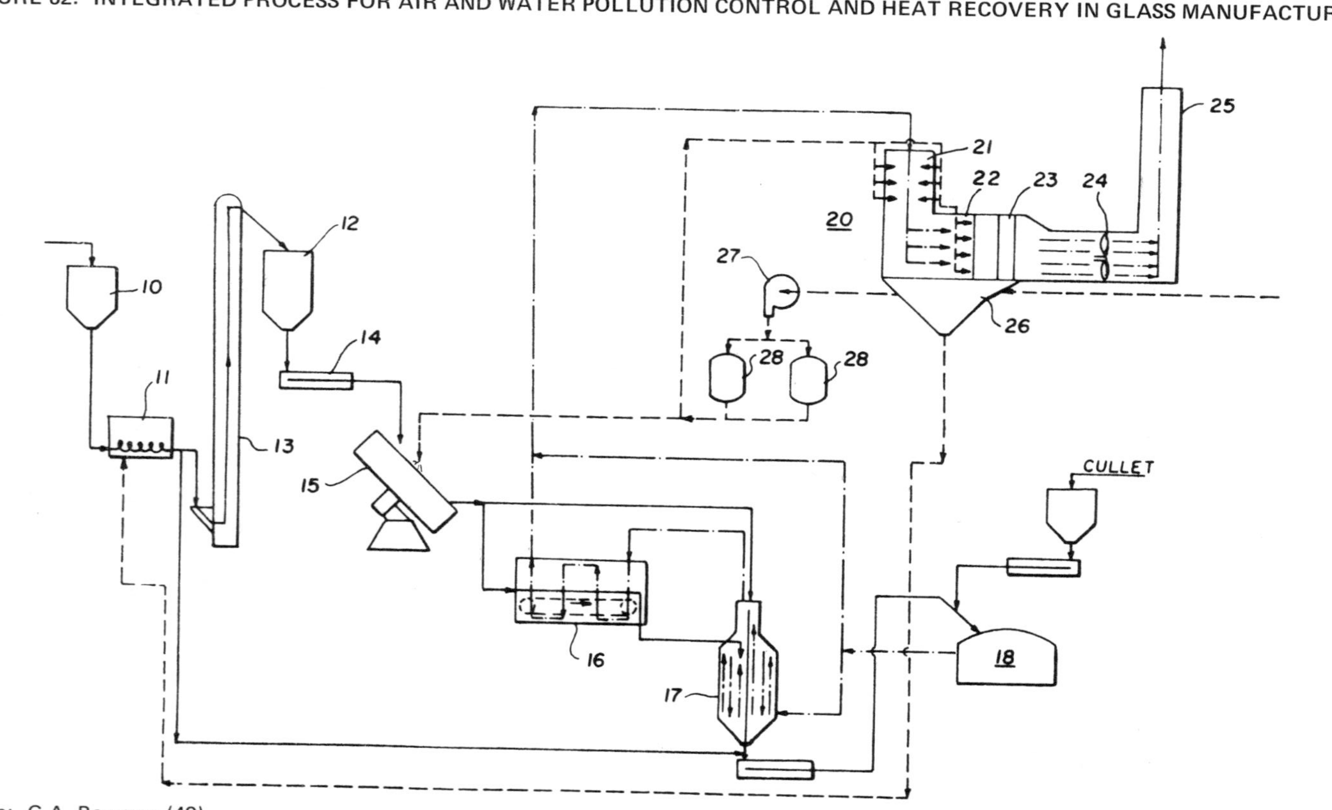

Source: G.A. Bowman (49)

The stack furnace **17** receives the dried agglomerates from the dryer **16** where they come in contact with the hotter gases from the glass melting furnace. Here more heat is absorbed in the agglomerated particle, increasing the temperature of the particle and cooling the waste gas. The transfer of heat depends on the Btu content and temperature of the incoming gas, residence time of gas and particles in contact with each other, specific heat and temperature of the incoming particle. The flow of the particles is vertically downward, countercurrent to the upward flow of waste gas.

Under some conditions of operations, the compartmentalized dryer may be by-passed or eliminated. In this event, the top portion in the stack of the stack furnace is used as the dryer for incoming agglomerated particles. Care must be taken to make certain that moist particles are not dried too rapidly, which would cause steam to form inside the particle.

Cracking would follow as the pressure relieves itself. After severe cracking, particles may tend to disintegrate before melting as they move through the remainder of the process. In addition cracked smaller chips from originally larger particles will fill voids between agglomerates thereby increasing the pressure drop in the gas flow.

From the stack furnace **17** the heated pellets are weighed or measured and conveyed to the glass melting furnace **18** at a selected rate. The cullet is also carefully weighed and fed at a controlled rate to the glass furnace at this point in the process. The temperature of the heated particle of agglomerated glass batch as it enters the glass furnace should be from 100° to 200°F below the temperature of the hot gas exiting from the glass melting furnace. The preheated particles of glass batch feed are now heated to reaction temperature and vitrification takes place forming glass.

The waste gases normally leave the checkers of the glass melting furnace at about 1000° to 1200°F. As they proceed upward through the stack furnace **17** heat is transferred to the particles coming down the stack. The gas leaves the stack furnace and enters the dryer **16** if one is used. In the dryer, the gas passes through the bed of particles on the compartmentalized belt and heat is absorbed by the particles from the gas. The gas normally leaves the dryer at a temperature of between 200° and 300°F.

The gas leaving dryer **16** is conducted to a wet scrubbing system **20** consisting of a wet quenching elbow **21**, scrubber **22**, moisture eliminator or separator **23**, fan **24** and motor, chimney stack **25**, thickener **26**, pumps **27** and water filters **28**. In the wet elbow **21**, the hot gases are quenched to adiabatic saturation. The scrubber will be selected to remove vapors, and all particulate matter to meet the codes of regulatory authorities at the plant site.

The water flow through the scrubber is collected in a thickener located under the scrubber. The moisture separator removes condensation as droplets, also collecting them in the thickener. The fan provides suction for transfer of exhaust gases from the glass melting furnace through the stack furnace, insulated ducts, dryer, wet elbow, scrubber, moisture separator and out the chimney stack to atmosphere. The pressure drop across the fan will vary between 15" to 50" water gauge depending on the system designed and especially the type of scrubber selected.

Sodium sulfate is one of the most prevalent chemicals in the waste gas stream. This chemical leaves the glass furnace as a vapor. When the gas stream temperature cools to about 400°F, the sodium sulfate begins to condense out of the gas stream as very fine solid particles. As part of this process, some of these fine chemical particles adhere to the agglomerated briquette or pellet, thereby causing the chemicals to be removed from the gas stream and remaining in the bed of agglomerated particles.

The remainder of the chemicals are carried into the scrubber **20** where they are wetted by the sprays, nozzles or other wetting procedures by which the selected scrubber operates. Here the chemicals are dissolved by the scrubbing solution. As the solution is recycled through the scrubber, it becomes more and more concentrated as the soluble chemicals dissolve. Sodium sulfate, one of the most prevalent chemicals is very soluble, and readily dissolves into the water solution. The scrubber also removes nondissolving chemicals as solids at this point. Thus, substantially all of the polluting chemicals are removed from the waste gas stream going to the atmosphere.

As part of this process, a thickener **26** is placed under the scrubber. It acts as a gravity receiver of the chemical solution and nondissolved suspended solids coming out of the scrubber as a slurry. This latter class of materials consists of such things as silica sand, lime cores, insoluble or slow dissolving sulfates. By properly sizing the thickener with regards to flow and temperature, clarified water rises to a weir at the top and solids settle to the bottom.

The solids on the bottom are removed as blowdown. The blowdown solids slurry will be piped back to the batch mixer where water is desirable for improved mixing procedures. In this manner, the chemicals in the blowdown from the thickener are restored to the batch and are recycled through the glass-making process. There is no loss or pollution to the environment.

The overflow of clarified water, from the thickener containing dissolved chemicals is piped through filters **28** to sprays **31** at the pelletizer or briquetting machine. Here the water and chemicals are absorbed in the batch. In this manner, the chemicals dissolved in the water solution are restored to the batch and are recycled through the glass-making process, without loss or pollution to the environment.

The overflow from the thickener consisting of settled water is filtered to remove any particles remaining in this clarified water flow. While any of several types of filters may be used, the preferred type is one containing deep sand beds which is insensitive to thickener upsets caused by variation in water temperature. Therefore, filters are used containing deep sand beds. They filter in vertical cross section of the filter bed, and use air to loosen and scour the solids from the carefully sized relatively large sand grain media.

The air scour takes place simultaneously with an upward backwash rinse flow which carries the solids up and out of the filter tank. As part of this process, the loosened solids, having been agglomerated in the filter is piped to the thickener inlet where it now settles to the thickener bottom and is removed with the other solids in blowdown. It is recycled to the mixer, combining here with the glass batch. Here, too, there is no leakage or loss to the environment. Filtrate from the filters **28** is used on the pelletizer disc sprays or on the nozzles in certain types of scrubbers. Clean filtrate will help give trouble-free nozzle spray operation.

As part of this process, in the event it is desired only to recycle chemicals collected in the scrubber water and not recover heat, the water from the thickener containing these chemicals in the blowdown is piped to the mixer. In this manner, the solids collected from the gas stream by the scrubber are recycled to the glass batch.

If this partial installation is made, nonagglomerated batch may continue to be fed to glass-making furnace as is now normally done. However, a conventional cooling tower is required to reject the heat accumulated in the scrubber water circuit as it is picked up from the hot waste gas exhaust stream.

A process developed by W.P. Mahoney (50) involves scrubbing furnace emissions, and particularly glass furnace emissions, from furnace exhausts, by spraying an aqueous solution of sodium silicate into the hot exhaust gases as the exhaust gases are being exhausted through the glass furnace exhaust stack, collecting precipitated prill and recycling the precipitate back into the glass batch materials.

WATER POLLUTION

The major constituents requiring treatment for primary flat glass manufacturing and automotive fabrication are suspended solids and oil. The treatment methods presently employed have been developed for this purpose.

No process wastewater and, therefore, no treatment is required for the rolled and sheet subcategories. In all cases, polyelectrolyte addition with lagoon sedimentation is practiced for plate glass manufacturing. Upgrading the lagoon system and partial recycle are methods of reducing waste loads from the plate process. Float wastewater is of high quality and presently is not treated.

Solid tempered automotive glass wastewater is also not treated but oil and suspended solids must be reduced. Flotation and centrifugation are used to reduce the oil discharged by the windshield fabrication process. Additional treatment will further reduce and assure low discharge levels in the flat glass industry. In some cases, treatment technologies developed for other industries will have to be used.

The primary pollutants from the pressed and blown glass industry are oil, fluoride, ammonia, lead, and suspended solids. Oil is contributed to the wastewater from all subcategories except hand-pressed and blown glass manufacturing. Fluoride and lead are added by the finishing steps for television picture tube envelopes and hand-pressed and blown glass. Fluoride and ammonia are carried over into the wastewater following frosting of incandescent lamp envelopes. Suspended solids are a result of grinding, acid treatment, and cullet quenching.

The industry is currently treating its wastewaters to reduce or eliminate most of the pollutants. Oil is reduced by using gravity separators such as belt skimmers and API separators. Treating for fluoride and lead involves adding lime, rapid mixing, flocculation, and sedimentation of the resulting reaction products. Several glass container plants recycle noncontact cooling and cullet quench water. Treatment for ammonia removal is presently not practiced in the industry.

Flat Glass Manufacture

Manufacture of the basic sheet of flat glass from sand and other raw materials is defined as primary flat glass manufacturing. Sheet, rolled, plate, and float glass are primary flat glass products. The primary glass sheets may be used directly or may be fabricated into glass products as indicated in Figure 63. Among the many fabricated products are mirrors and other coated glass, automotive and architectural tempered glass, windshields, and numerous speciality products such as bulletproof glass, basketball backboards, and glass hot plates. Tempered automobile glass and windshields are the only fabricated products covered by this study.

Flat glass is manufactured by melting sand together with other inorganic materials and then forming the molten material to a flat sheet. Within the primary flat glass industry, several distinct methods are used to make flat glass. These are the float, plate, sheet, and rolled processes. Although the raw materials and the melting operations are essentially the same, each process uses a different method for forming the molten glass into a flat sheet.

In the float process, the glass is drawn across a molten tin bath while in the plate process, rolls control the initial thickness with the final thickness determined by grinding and polishing. The glass is formed by a vertical drawing process in the sheet process. Finally, texturizing rolls are used to impart various surface textures in the rolled process.

All primary glass may be used as architectural glass. Plate glass is used in preference to sheet where a distortion-free glass is desired. Rolled glass is used where a decorative or translucent surface is desired. Float glass is used for the same purposes as plate glass, both of which can serve as the raw material for automobile window glass. Fabricated glass is produced using all of the primary glass types.

Automobile window glass fabrication is divided into two processes. Windshield laminating consists of bonding two layers of glass to an inner layer of vinyl plastic. The major unit of equipment in the process is the autoclave which is used to complete the bonding

FIGURE 63: FLAT GLASS INDUSTRY

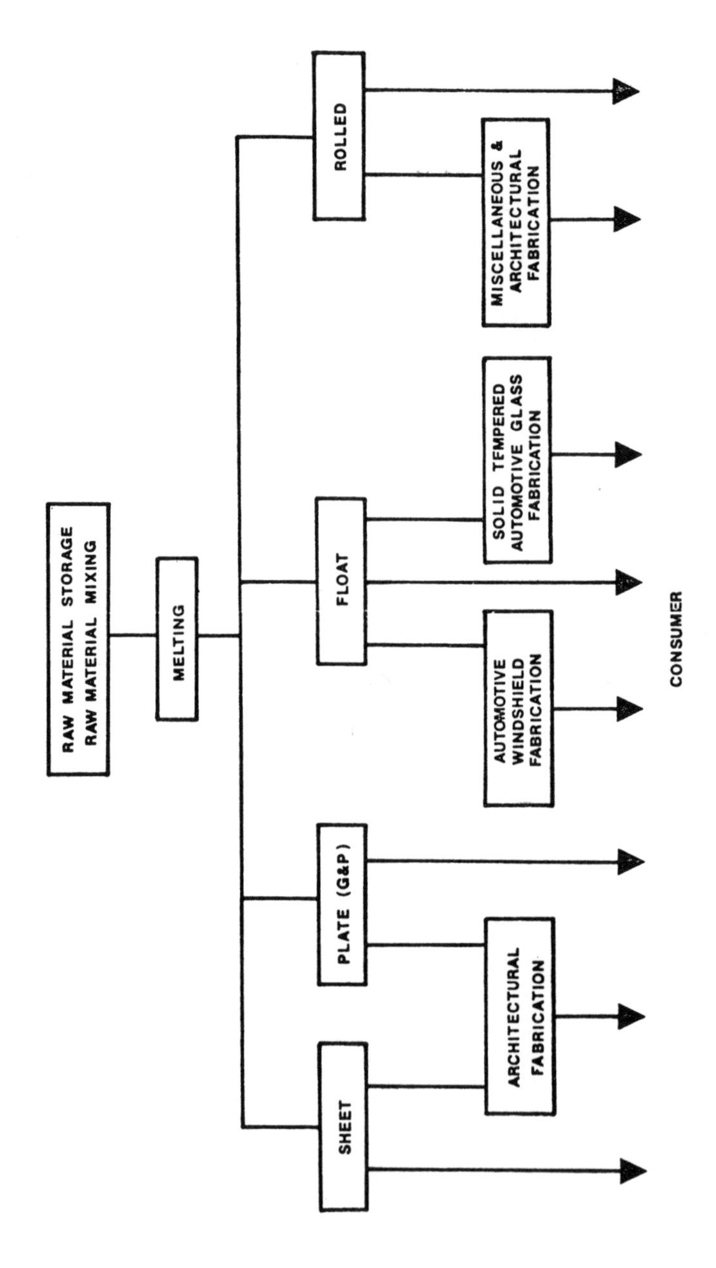

Source: EPA (46)

operation under conditions of high temperature and pressure. The purpose of laminating is to make the glass shatter-resistant. Unlike windshields, side and back lights (windows) are fabricated from solid pieces of glass. The process includes edge grinding, bending and tempering. The purpose of tempering which consists of heating, followed by rapid cooling is to increase the strength of the glass over that of ordinary annealed glass to a considerable degee so that, in the event the glass is broken, it will shatter into small, rounded pieces.

No process wastewater is produced by the sheet and rolled glass subcategories. The manufacturing processes are dry with process water used only in the batch for dust control. Both subcategories have significant cooling requirements and use substantial quantities of cooling water. Although cooling systems are not specifically covered in this report, one system related to water pollution control will be discussed briefly.

Several sheet glass plants are eliminating a pollution problem by disposing of chromium-treated cooling tower blowdown in the batch. Approximately 42 liters/metric ton (10 gallons/short ton) can be disposed of by this method. This is especially interesting in view of the adverse affect of chromium on glass quality. At a low concentration, which has not been defined, chromium causes stones or imperfections in the glass. Apparently, this concentration is not exceeded in sheet glass or, more likely, the imperfections are not significant because the glass is thinner and of lower quality than other types of primary glass.

Discussions with glass industry personnel have indicated great reluctance to consider disposal of cooling tower blowdown, especially chromium-treated, in the batch for plate, float, or rolled glass. These glasses are thicker and of higher quality and it is thought that noticeable imperfections in the glass will result.

An unsuccessful attempt at batch disposal could be quite costly. Glass melting is a continuous process with a large volume of melted glass contained in the furnace. If an undesirable concentration of some contaminant is introduced into the glass furnace, it might result in a week or two of production of unusable glass to dilute the furnace contents to an acceptable level.

Disposal of other glass plant auxiliary wastes in the batch such as boiler blowdown and softener and deionizer regenerants is also within the realm of possibility, although none of these has as yet been demonstrated. Wash water, especially from the float process, should also be amenable to batch disposal. Although batch disposal is not a cure-all for primary glass wastewater disposal [the volume of process and auxiliary wastewater discharge far exceeds the 42 liters/metric ton (10 gallons/short ton) maximum that can be accepted by the batch], experimentation within the industry should be encouraged.

Batch disposal of cooling tower blowdown for sheet glass manufacturing has been demonstrated and should be applicable at many plants. It is too early, however, to predict universal applicability and each process should be considered on an individual basis.

Sheet Glass Manufacture

Sheet glass manufacturing operations may be defined as the processing of raw materials to form thin glass sheets of saleable size. Figure 64 is a flow diagram indicating water usage with respect to the manufacturing steps. Noncontact cooling water is used, but no process wastewater is produced by this subcategory.

The only water used in the process is 42 liters/metric ton (10 gallons/short ton) added to the raw materials for dust suppression. This is evaporated in the melting tank. No process water is used and, therefore, no process wastewater is produced by the sheet-glass subcategory. However, it should be noted that architectural tempering or other wastewater-producing fabrication steps may be operated in the same facility in which sheet glass is produced. The effects of fabrication steps must be considered when analyzing the total effluent from a sheet glass facility.

FIGURE 64: SHEET GLASS MANUFACTURING

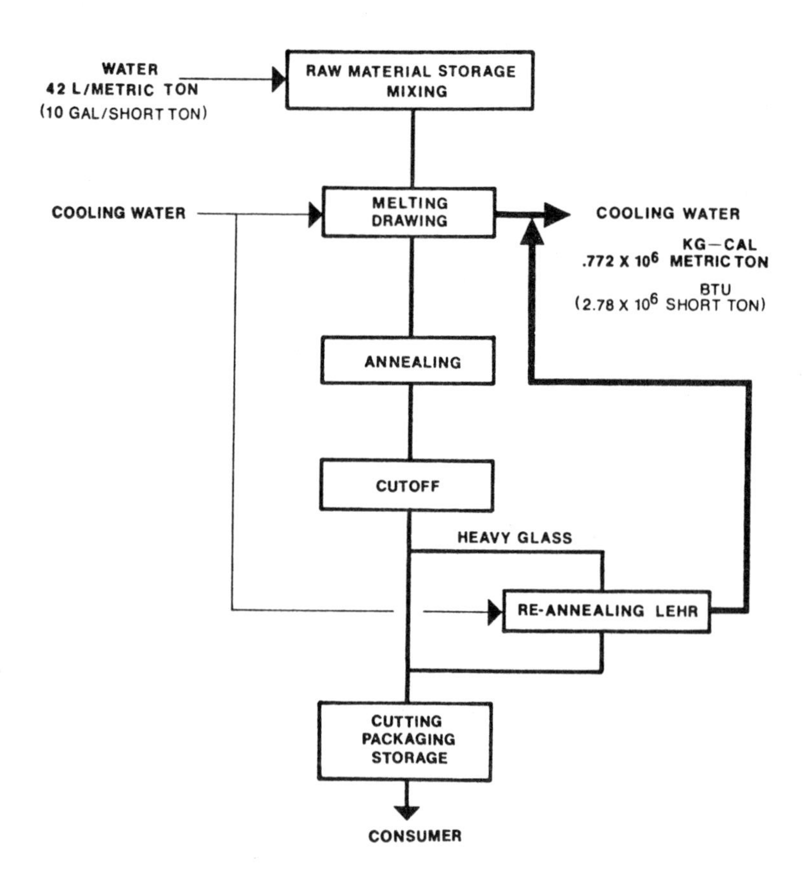

Source: EPA (46)

In the sheet glass manufacturing process, cooling water is required for the melting tank, drawing kiln, compressors and the reannealing lehr. Average heat rejection for plants using the Pennvernon process is 772,000 kg-cal/metric ton (2,780,000 Btu/short ton) with a range of 741,000 kg-cal/metric ton (2,670,000 Btu/short ton) to 877,000 kg-cal/metric ton (1,260,000 Btu/short ton). Another sheet glass plant, using the Fourcault process, reports heat rejection at 350,000 kg-cal/metric ton (1,260,000 Btu/short ton). The reason for the difference in heat rejection between the Pennvernon process and the Fourcault process is the relative proximities of the drawing kiln to the melting tank. In the Fourcault process the molten glass flows in a canal to the drawing kiln which is not as close to the melting tank as in the Pennvernon process.

By traveling the longer distance (by way of canals), the glass has an opportunity to cool so that not as much cooling water is required in the drawing kiln. The Libbey-Owens process is also used for making sheet glass. No heat rejection data were available from these plants.

Rolled Glass Manufacture

Rolled glass manufacturing consists of melting raw materials and drawing the molten glass through rollers to form a glass sheet. The major process steps and points of water usage are listed in Figure 65. Noncontact cooling water is used, but no process wastewater is produced by this subcategory.

Approximately 42 liters/metric ton (10 gallons/short ton) of water are added to the raw materials for dust suppression. This water is evaporated in the melting tank. No process water is used and, therefore, no process wastewater is produced by the rolled glass subcategory.

FIGURE 65: ROLLED GLASS MANUFACTURING

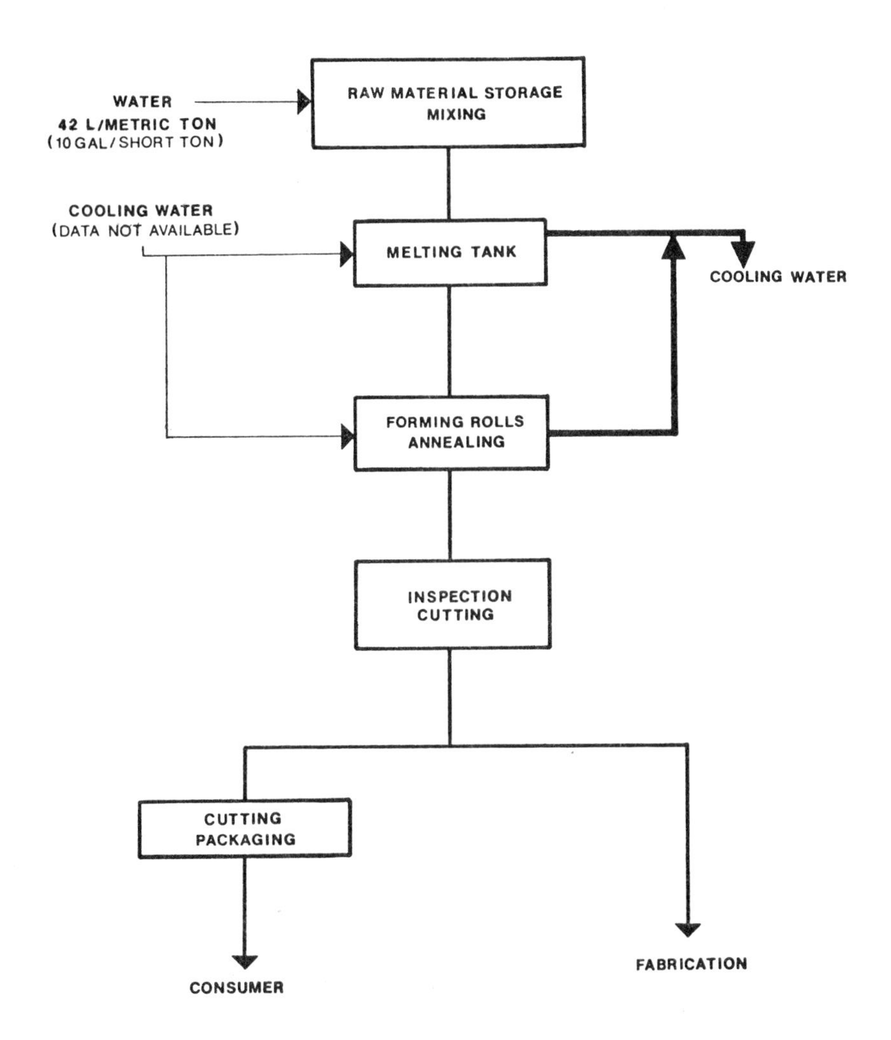

Source: EPA (46)

Fabricating operations generally occur in conjunction with rolled glass manufacturing. It should be noted that numerous and highly variable wastewater streams may result. Although primary rolled-glass production is a dry process, wastewaters may be generated by a rolled-glass facility because of fabrication wastewater.

In the rolled-glass manufacturing process, cooling water is required for the melting tank, forming rolls, annealing lehr and compressors. Although no heat-rejection data are available for rolled glass plants, it is expected that heat-rejection requirements are similar to the plate glass process because of similarities in process configuration.

Plate Glass Manufacture

Plate glass manufacturing is the production from raw materials of a high-quality thick glass sheet. This subcategory has historically been the greatest source of waste in the industry since large volumes of high-suspended-solids wastewater are produced.

Owing to high production costs and related water pollution problems, plate glass is being replaced by float glass. Only three plate glass plants remain in the United States. The major process steps and points of water usage are shown in Figure 66.

FIGURE 66: PLATE GLASS MANUFACTURING

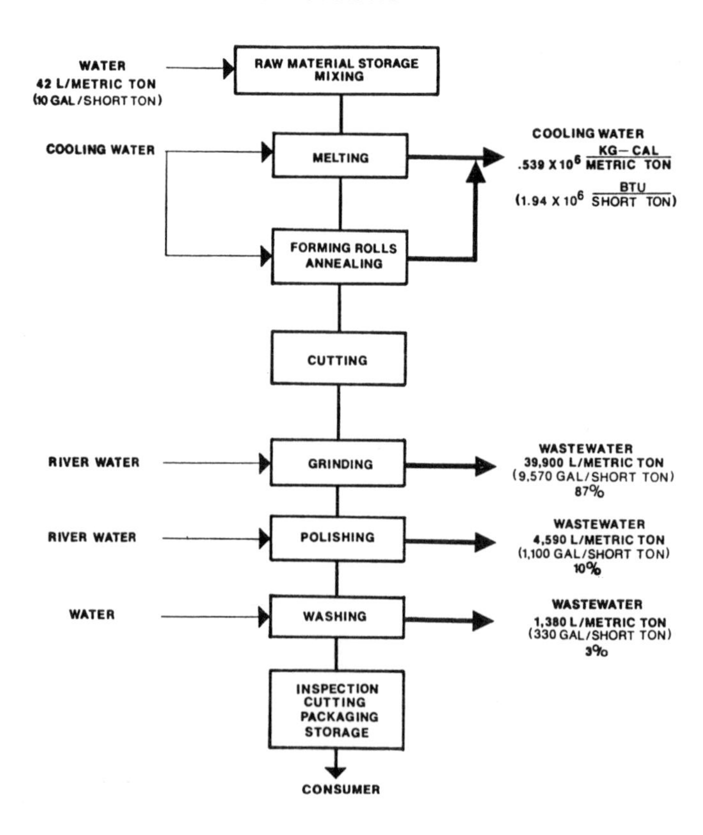

Source: EPA (46)

The typical plate glass manufacturing plant may be located in any part of the country and is at least 12 years old. Advanced plate glass manufacturing technology is used, but this has not been improved since the early 1960's when the advantages of the float process became apparent. Production is continuous seven days a week.

Process water is used in the batch, grinding, polishing, and washing operations. Approximately 42 liters per metric ton (10 gallons per short ton) of water are added to the raw materials for dust suppression. This water is evaporated in the melting tank. Wastewater results from grinding, polishing, and washing the glass. River water is generally used for grinding and polishing, but city or treated water is required for final washing and rinsing.

Grinding is the first step in the process to transform the rough glass sheet into the finished plate glass product. A sand slurry is used in conjunction with large iron grinding wheels to actually grind down the glass surface. Relatively coarse sand is used initially, with progressively finer sand used as the glass proceeds down the grinding line. Sand slurry is recycled from a gravity classifier.

All of the return from the grinders enters one end of the classifier, the sand particles settle according to size, and the wastewater overflows at the other end. The grinding slurry is drawn from progressive segments of the bottom of the tank. Sand classification is regulated by the velocity of water passing through the tank and is controlled by both the slurry drawoff rate and the tank overflow.

Particles too small to settle are removed in the overflow. This wastewater stream is very high in suspended solids consisting of fine sand, glass, and iron particles. About 87% of the flat glass wastewater is contributed by the grinding process.

Polishing is similar to grinding except that smaller particles are used. Rouge (ferric oxide) has generally been used as the polishing medium, but at least one company uses cerium oxide. Neither grinding medium has an apparent advantage in terms of raw wastewater characteristics. Felt pads are used to apply the polishing medium to the glass surface and, therefore, contribute some organic matter to the wastewater stream.

The glass is ground first on one side and then on the other. A bedding medium is required to evenly support the glass. Plaster of Paris is traditionally used for bedding; however, proprietary methods using a reusable medium have been developed for the newer polishing lines. Polishing contributes about 10% of the plate glass wastewater volume. The major constituents include, rouge or cerium oxide, glass particles, felt, and calcium sulfate if plaster of Paris bedding is used.

The residue resulting from grinding and polishing is removed by a series of washing steps. River water is generally used for the first rinse, followed by an acid wash and a final rinsing with city water. The city water rinse may be followed by a deionized water rinse. Washing contributes about 3% of the plate glass wastewater volume. The water is clean as compared with the grinding and polishing wastewater. The initial wash contains significant suspended solids, but the final wash is very clean. Acid carry-over is quickly neutralized by the other waste streams which tend to be basic.

Typical characteristics for the combined wastewater stream are listed in Table 68. In all cases except for pH, the values listed are the quantities added to the water as a result of the plate glass process; concentrations in the influent water have been subtracted. A variable volume of process water is used for plate glass manufacturing.

Flows range from 14,600 to the typical flow of 45,900 l/metric ton (3,500 to 11,000 gallons per short ton) or 4,920 to 18,200 cu m/day (1.3 to 4.8 mgd). Water usage is related to the type and age of the equipment used, being highest at the oldest plants which require extensive modifications to significantly reduce usage. At plants built before water conservation and pollution control were widely practiced, open channels were provided for flushing away any wastes or spillage. Large quantities of water are needed to maintain sufficient velocity to prevent settling.

TABLE 68: RAW WASTEWATER*–PLATE GLASS MANUFACTURING PROCESS

Flow	45,900 l/metric ton	11,000 gal/short ton	
pH	9		
Temperature**	2.8°C	6°F	
Suspended solids	690 kg/metric ton	1,375 lb/short ton	15,000 mg/l
COD**	4.6 kg/metric ton	9.2 lb/short ton	100 mg/l
Dissolved solids**	8.0 kg/metric ton	16.1 lb/short ton	175 mg/l

*Represents typical plate glass process wastewater prior to treatment. Absolute value given for pH, increase over plant influent level given for other parameters.
**Indication of approximate level only; insufficient data are available to define actual level.

Source: EPA (46)

Suspended solids are the major wastewater constituents. Available data show a wide variation in concentration, but good correlation in terms of pounds per ton. About 690 kg per metric ton (1,375 lb/short ton) of suspended solids are discharged. The major wastewater source is the grinding operation, with lesser quantities from polishing and washing. Although sufficient information is not available to definitely establish the dissolved solids, BOD, COD, and temperature levels, the data indicate these are insignificant as compared with suspended solids. Detergents are not used, no increase in phosphorus should occur. While some lubricating oil dripping can be expected from the process equipment, it cannot be detected in the large volume of wastewater.

Wastewater flows from plate glass manufacture, a continuous operation, are relatively constant, being only slightly reduced when polishing is not on line and the suspended solids loadings are consequently lower. Wastewater flows and characteristics vary little during start-up or shutdown. When the plate furnace is drained every three to five years for rebuilding, the glass is drained into a quench tank and cooled with water, which evaporates. No discharge occurs. There are no known toxic materials in wastewater from plate glass manufacture.

In the plate glass manufacturing process, cooling water is required for the melting tank forming rolls, annealing lehr and compressors. Two of the three plate glass plants in the United States reported heat-rejection data. They are 311,000 kg-cal/metric ton (1,120,000 Btu/short ton) and 766,000 kg-cal/metric ton (2,760,000 Btu/short ton). The wide variation in the two values cannot be explained. The larger value is probably more representative of actual plate glass heat-rejection requirements.

Plate glass manufacturing produces a large volume of wastewater, high in suspended solids with lesser concentrations of dissolved solids, BOD, and COD. Plate glass manufacturing is rapidly being replaced by the float process. Float glass is of similar quality, but is less expensive to produce and process wastewaters are insignificant compared to the plate process. Only three plate glass plants remain in this country and at least two of these may be closed by 1977.

Owing to the high operating costs and pollution load, no new plate glass facilities are anticipated. The industry trend of replacing plate glass with float glass has shown that plate glass manufacturing can be successfully eliminated, therefore, only treatment technologies for reducing pollutant loadings from existing plate glass plants will be discussed.

No apparent in-plant modifications of pollutional significance have been developed for the plate glass manufacturing process. The three remaining plants are relatively modern by plate standards. Plate glass technology development ended with the advent of the float process. Sand and rouge recovery systems are based on the latest technology.

In one case, cerium oxide rather than iron oxide is used as the polishing medium. This plant also has the most efficient wastewater treatment of all of the plate glass plants indicating the possible beneficial effects of cerium oxide. Cerium oxide has a higher specific gravity than rouge which may account for better settling ability.

The switch from rouge to cerium oxide was made for reasons other than pollution control and comparative effluent data before and after the change do not exist. Although insufficient information is available to conclude that cerium oxide is more easily removed from plate wastewater, this method might be considered where problems are experienced with iron oxide removal.

Each of the three remaining plate glass plants use lagoon treatment with polyelectrolyte added to the influent wastewater. The typical flow rate is 45,900 liters/metric ton (11,000 gallons/short ton) or 18,168 cubic meters/day (4.8 mgd) and the suspended solids loading is 690 kg/metric ton (1,375 lb/short ton) or 15,000 mg/l. A cationic polyelectrolyte is added to the influent sewer with mixing accomplished through the natural turbulence of the water. The typical lagoon is square, has an area of approximately 5.26 ha (13 acres) and a working depth of 2.44 m (8 feet).

Available data indicate the highest efficiency presently available using this system is 99.6% suspended solids reduction to produce an effluent concentration of 2.5 kg/metric ton (5 lb/short ton) or 54 mg/l. The COD is reduced approximately 90% to 0.45 kg/metric ton (0.9 lb/short ton) or 10 mg/l. Additional treatment methods can be employed to further reduce effluent suspended solids levels.

Several methods for upgrading existing plate glass lagoon systems to increase suspended solids removal efficiency are apparent. Improved polyelectrolyte addition and a two-stage lagoon system should reduce suspended solids to 30 mg/l. An additional reduction to less than 5 mg/l can be accomplished by sand filtration of the lagoon effluent. The filter volume can in turn be reduced by recycling the lagoon effluent back to the grinding and polishing process. These methods of treatment are illustrated in Figure 67.

It should be possible to reduce lagoon effluent suspended solids to 30 mg/l or 1.38 kg/metric ton (2.75 lb/short ton) by improving coagulant mixing and using a two-stage lagoon system. The maximum daily discharge from this system will be 60 mg/l or 2.76 kg/metric ton (5.5 lb/short ton). A mixing tank is added at the lagoon influent to assure proper polyelectrolyte dispersion. The mixing is for one minute or less and is of sufficient velocity that essentially instantaneous mixing of the polyelectrolyte is assured.

The lagoon is divided into two stages by constructing an additional levee. This will produce two lagoons of 2.43 ha (6 acres) each in the typical plant. The two-stage system will reduce the effects of wind action which is a major cause of low effluent quality. A second mixing tank with provision for adding additional polyelectrolyte between the lagoon segments is provided. It is not certain that the second polyelectrolyte addition step will be necessary, but the equipment is included for cost estimating purposes.

The minimum allowable lagoon surface area and detention time is not known. The data from existing one-cell lagoon systems indicate no correlation between surface area or detention time and suspended solids removal. This may be due to poor design, lack of solids removal, or other factors. The lowest effluent concentration was produced in the lagoon with the shortest detention time. This lagoon was used as the basis for the recommended improvements. The other lagoon systems, having a longer detention time, if properly operated, should have no trouble achieving the same effluent levels.

Many polyelectrolytes are on the market, but laboratory testing is required to determine the most efficient one for each application. Based on current practice in glass plants, a liquid cationic polyelectrolyte is most effective, although some inorganic coagulants may also be effective. The latter should be avoided, if possible, where recycle is considered because of the dissolved solids increase. Coagulation and sedimentation are widely employed for both water and wastewater treatment. An effluent concentration much less than 30 mg/l suspended solids is achieved in many systems. Although most conventional systems are operated in specially designed tanks, there is no evidence to indicate that a lagoon system with sufficient protection against short circuiting and wind cannot achieve an average effluent of 30 mg/l suspended solids.

FIGURE 67: WASTEWATER TREATMENT–PLATE GLASS PROCESS

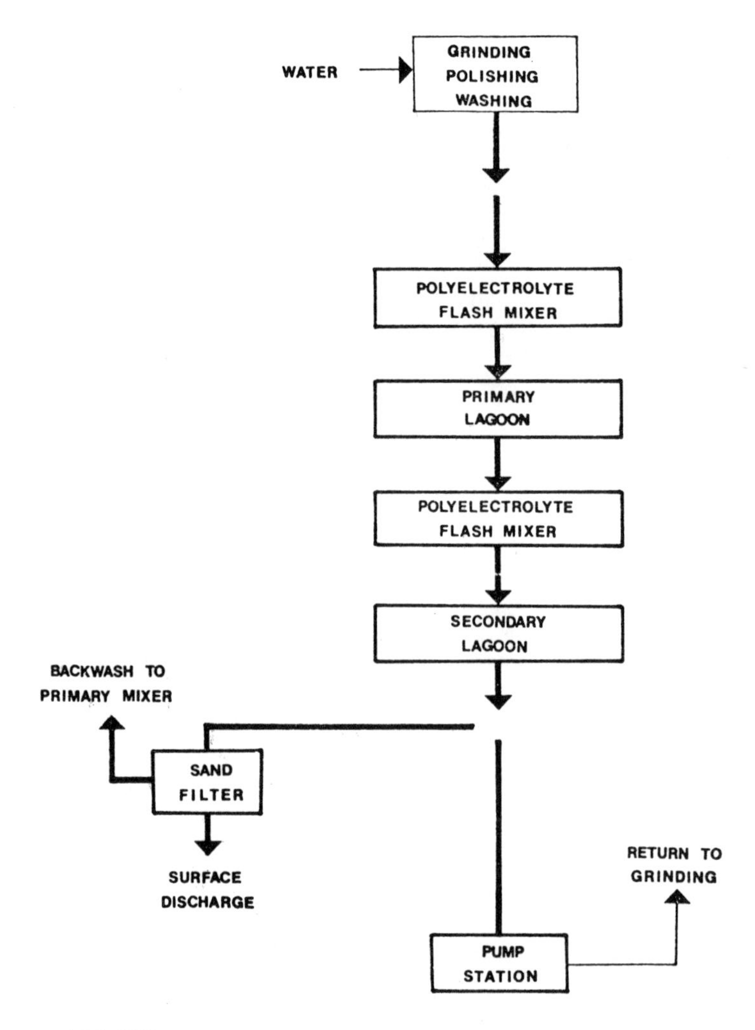

Source: EPA (46)

Lagoon effluent suspended solids can be further reduced to less than 0.23 kg/metric ton (0.46 lb/short ton) or 5 mg/l by rapid sand filtration. The entire lagoon effluent is filtered through a standard gravity sand filter at an assumed 163 l/min/sq m (4 gpm/sq ft). The filter backwash is recycled to the head of the lagoon for suspended solids removal.

Rapid sand filtration is a widely used and thoroughly proven technology. Such filters are used extensively in water treatment plants following coagulation and sedimentation. Suspended solids levels substantially below 5 mg/l are almost the rule in the water treatment industry and similar values have also been achieved for the filtration of secondary sewage effluent. Other filters such as mixed media, pressure, and upflow filters are also available and may be more desirable in some cases, but rapid sand filters are chosen for illustrative purposes because more background information on cost and treatment efficiency is available. The volume of water requiring filtration can be substantially reduced by recycling the lagoon effluent back to the grinding process. Recycle is not presently employed in the industry, but there is adequate reason to believe that it is feasible especially if lagoon effluent suspended solids are reduced to 30 mg/l. In most cases, this is a lower concentration than the raw river water presently being used.

A liberal 20% blowdown from the recycle system is assumed to allow for any unforeseen dissolved solids problems. It is likely that a lower blowdown rate can and will be achieved to reduce filtration requirements. The filtered effluent suspended solids concentration will still be 5 mg/l or less but, owing to the 80% volume reduction, the effluent loading will be reduced to 0.045 kg/metric ton (0.09 lb/short ton). COD will be reduced to 0.09 kg/metric ton (0.18 lb/short ton).

The typical plate glass manufacturing plant may be located in any part of the country and is at least 12 years old. Advanced plate glass manufacturing technology is used but this has not been improved since the early 1960's when the advantages of the float process became apparent. Annual production at the plant is approximately 150,000 metric tons. Costs and effluent quality for the four treatment alternatives are summarized in Table 69.

Alternative A – Lagoon with Polyelectrolyte Addition

Alternative A is the treatment universally practiced at plate glass plants and includes polyelectrolyte addition to the raw wastewater followed by sedimentation in a one-cell lagoon.

Costs – No additional cost

Reduction Benefits – Suspended solids are reduced 99.6% and COD is reduced 90%.

Alternative B – Lagoon Improvements

Alternative B consists of partition of the existing one-cell lagoons into two cells in series with polyelectrolyte addition at the entrance to each.

Costs – Incremental investment costs are $57,000 and total annual costs are $32,800 over Alternative A.

Reduction Benefits – The incremental reduction of suspended solids compared to Alternative A is 70%. Total reduction of suspended solids is 99.8%.

Alternative C – Filtration

Alternative C is sand filtration of the lagoon effluent resulting from Alternative B.

Costs – Incremental investment costs are $472,000 and total annual costs are $142,500 over Alternative B.

Reduction Benefits – The incremental reduction of suspended solids compared to Alternative B is 83%. Total reduction of suspended solids is almost 100%.

Alternative D – Filtration and Recycle

Alternative D is recycle of the lagoon effluent resulting from Alternative B to the plate glass grinders and sand filtration of a 20% blow-down prior to discharge to the receiving stream. Owing to the lower operating cost for Alternative D, the annual cost for this system is less than for Alternative C.

Costs – Incremental investment costs are $127,000 over Alternative C but total annual costs are $36,800 less than Alternative C.

Reduction Benefits – Incremental reductions are 80% for suspended solids and COD compared to Alternative C. Total reductions are essentially 100% for suspended solids and 98% for COD.

TABLE 69: WATER EFFLUENT TREATMENT FOR MANUFACTURING PLATE GLASS

Alternative Treatment or Control Technologies:	($1,000)			
	A	B	C	D
Investment	0	57.	529.	656.
Annual Costs:				
Capital Costs	0	4.6	42.3	52.5
Depreciation	0	2.9	26.5	32.8
Operating and Maintenance Costs (excluding energy and power costs)	0	22.7	99.7	49.7
Energy and Power Costs	0	2.6	6.8	3.5
Total Annual Cost	0	32.8	175.3	138.5

Effluent Quality:

Effluent Constituents		Raw Waste Load	Resulting Effluent Levels			
Flow	(l/metric ton)	45,900	45,900	45,900	45,900	9,200
Suspended Solids	(kg/metric ton)	690.	2.5	1.38	.23	.045
COD	(kg/metric ton)	4.6	.45	.45	.45	.09
Flow	(l/sec)	210	210	210	210	40
Suspended Solids	(mg/l)	15,000	54	30	5	5
COD	(mg/l)	100	10	10	10	10

Source: EPA (46)

In Table 69 the cost data include the traditional expenditures for equipment purchase, installation, and operation and where necessary, include solid waste disposal. No significant production losses due to the installation of water pollution control equipment are anticipated. The costs are based on a typical plant for subcategories where no treatment is practiced and on an exemplary plant where treatment is employed. In some cases production rates and wastewater volume are adjusted to be more representative of the industry subcategory.

Investment costs include all the equipment, excavations, foundations, buildings, etc. necessary for the pollution control system. Land costs are not included because the small additional area required is readily available at existing plants. In all cases, the lagoon systems used for plate glass wastewater treatment are already in operation and no additional land costs are required.

Costs have been expressed as August 1971 dollars and have been adjusted using the national average Water Quality Office—Sewage Treatment Plant Cost Index. The cost of capital was assumed to be 8% and is based on information collected from several sources including the Federal Reserve Bank. Depreciation is assumed to be 20-year straight-line or 5% of the investment cost. Operating costs include labor, material, maintenance, etc., exclusive of power costs. Energy and power costs are listed separately. August 1971 energy costs were assumed to be $0.018 per kwh for electricity and $1/million Btu for the steam required for brine evaporation.

Three plate glass plants remain in operation in the United States. All of these plants practice Alternative A treatment but none of the other alternatives are practiced at present. There is no apparent benefit for phasing costs within an alternative. However, where one alternative includes a previous alternative, the earlier alternatives may be built first.

The cost of Alternative B is not expected to vary significantly between plants. The cost of Alternative C and D will vary somewhat depending on the volume of water filtered. A reduction in plant water usage, although theoretically possible, is not practical because extensive inplant modifications will be required. For this reason, costs are based on the plant having the highest flow rate and will be somewhat less for other plants. Another unknown factor is the amount of blowdown required for the recycle system. A liberal 20% blowdown is assumed. Filtration costs will be reduced if the allowable blowdown can be reduced.

The age of equipment and process employed do not significantly affect costs. No process changes are required and significant engineering or nonwater quality environmental impact problems are not anticipated.

Float Glass Manufacture

The float process may be considered the replacement for plate glass manufacturing. Float glass production is substantially less expensive and process wastewater has all but been eliminated. The major process steps and points of water usage are illustrated in Figure 68.

FIGURE 68: FLOAT GLASS MANUFACTURING

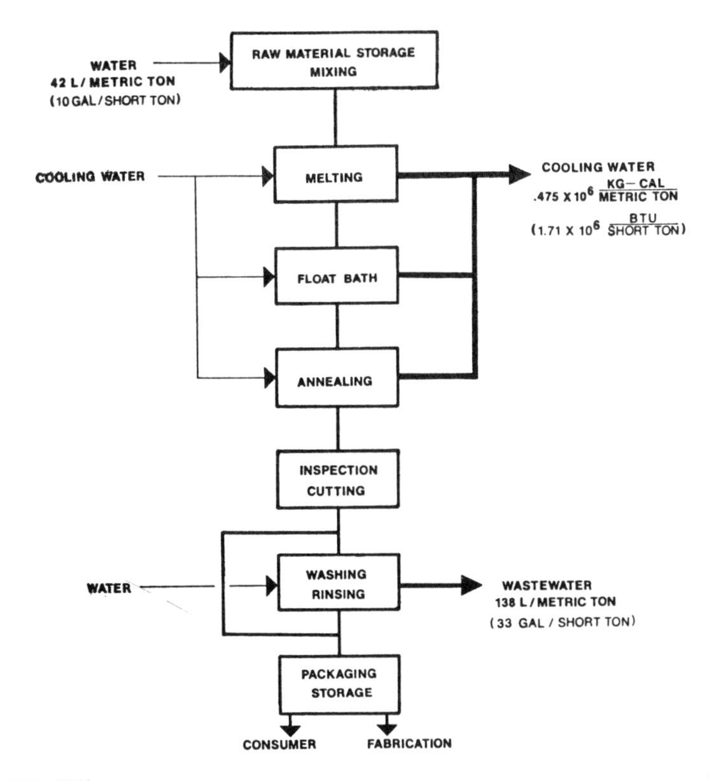

Source: EPA (46)

The typical float glass plant may be located in any part of the country and has been built since 1960. Both float and mirror washing are practiced so that flows are approximately 30% higher than if float washing alone is practiced. Float production is continuous seven days a week, but the mirror washer is operated only as required.

Process water is used in the batch and in some cases for washing. Approximately 42 liters per metric ton (10 gal/short ton) of water are added to the raw materials for dust suppression. This water is evaporated in the melting tank. Some plants wash the glass prior to packing, and this constitutes the only wastewater stream for this subcategory.

Sulfur dioxide is sprayed on the underside of the glass sheet soon after forming to develop a protective coating. Sodium sulfate is formed which, in high enough quantity, will show up as a visible film on the glass which may be removed by washing.

Some plants wash as part of the float process and others do not. Generally, where they do not, the glass is to be fabricated in the same facility, and washing is the first step in the fabrication process. Glass to be used for mirror manufacture is always washed in a special washer not directly connected to the float line. The available data do not distinguish between regular washing and mirror washing, so both types are considered to be part of the float process in this report. The mirror washer effluent is probably of higher quality than the float wash water, but the differences are not significant for this study.

Two basic types of washing systems are used. Most plants presently use a one-or two-stage wash of city-water quality followed by a deionized water rinse. The water is heated to 52° to 65°C (125° to 150°F) to prevent glass breakage and to enhance dissolution of the soluble film. Maximum recycle is practiced, with blowdown governed by dissolved solids buildup. This system is typical of the industry.

An older three-stage system using detergents is still used at some plants. The first stage is a recycled detergent wash followed by a recycled city-water rinse and a recycled deionized water rinse. Blowdown is governed by dissolved solids and detergent buildup.

Some typical characteristics of float glass wash water blowdown are listed in Table 70. In all cases except for pH, the values listed are the quantities added to the water as a result of the float glass process, and concentrations in the influent water have been subtracted.

The volume of wash water discharge is influenced by make-up water characteristics and mirror washing requirements. Flows range from 88 to 138 l/metric ton (21 to 33 gallons per short ton) or 34 to 136 cu m/day (0.009 to 0.036 mgd), with the highest volumes recorded for plants that are washing mirror glass. The typical flow is 138 l/metric ton (33 gallons per ton).

The volume of wastewater discharged depends upon the dissolved solids content of the makeup water. Blowdown rates are manually adjusted and are generally held constant even though the square meters of glass washed may vary considerably. The flow is set so that acceptable dissolved solids concentrations are maintained at the highest washing rate. Dissolved solids in the wash prior to the deionized rinse are generally limited to 300 to 400 mg/l.

Blowdown from the float washer is of fairly high quality, as can be seen from Table 70. The most significant increase of 14 g/metric ton (0.028 lb/short ton) or 100 mg/l is noted for dissolved solids. COD and suspended solids show increases of only 2 g/metric ton (0.0041 lb/short ton). Trace quantities of BOD and oil are also present.

The available data indicate a pH range of 7.4 to 8.2 with a typical pH of 8. Only one temperature reading was available; this gives an indication of the water temperature, but should not be taken as typical. No information was available on the phosphorus content of float washer effluent where detergents are used, but the use of detergent is not typical.

TABLE 70: RAW WASTEWATER*–FLOAT GLASS MANUFACTURING PROCESS

Flow	138 l/metric ton	33 gal/ton	
pH	8		
Temperature**	37°C	98°F	
Suspended Solids	2 g/metric ton	0.0041 lb/short ton	15 mg/l
Oil	0.7 g/metric ton	0.0014 lb/short ton	5 mg/l
COD	2 g/metric ton	0.0041 lb/short ton	15 mg/l
BOD	0.25 g/metric ton	0.0005 lb/short ton	2 mg/l
Phosphorus	***		
Dissolved Solids	14 g/metric ton	0.028 lb/short ton	100 mg/l

*Representative of typical float glass process wastewater. Absolute value given for pH and temperature, increase over plant influent level given for other parameters.

**Indication of approximate level only; insufficient data are available to define actual level.

***No information is available on wastewater containing phosphorus.

Source: EPA (46)

Deionizer regeneration is not considered process wastewater and, therefore, was not included in the characterization. Process wastewater from the float subcategory is of fairly high quality and is disposed of only because of the dissolved solids concentration. There is no significant change in wastewater characteristics during start-up or shutdown. The float furnace is drained every 3 to 5 years for cleaning. Molten glass is drained into a quench tank and cooled with a water spray. The cooling water evaporates and no discharge occurs.

In the float glass manufacturing process, cooling water is required for the melting tank, float bath, annealing lehr and the compressors. Average heat rejection is 475,000 kilogram-calories/metric ton (1,710,000 Btu/short ton) removed from the melting tank with a range of 400,000 kilogram-calories/metric ton (1,440,000 Btu/short ton) to 561,000 kilogram-calories/metric ton (2,020,000 Btu/short ton).

Float glass manufacturing is rapidly replacing the plate glass process as the method for producing high-quality thick glass sheets. Conversion to the float process has drastically reduced pollution loadings related to the manufacture of this type of glass. Washing is required for some types of glass and this is the only process wastewater resulting from float glass production.

Raw waste suspended solids loadings are reduced from 690 kg/metric ton (1,375 lb/short ton) for plate glass to 2 g/metric ton (0.0041 lb/short ton) for float glass. The wastewater loading for other parameters is equal to or less than that for suspended solids. The typical flow of raw wastewater amounts to only 138 l/metric ton (33 gal/short ton) or 136 cu m/day (0.036 mgd). Owing to the high quality, float wash water is presently not treated.

Until several years ago, detergents were used in the float washer. In an effort to reduce phosphorus discharge and prevent foaming in the receiving body of water, most plants have now found that sufficient washing can be accomplished without detergents. Non-detergent washing is now typical. There is no evidence to indicate that elimination of detergents in the float wash is detrimental to the product or the process. Elimination of detergents in the float wash is believed possible in all cases.

Recycling washer systems are typical for the industry. Recycling, although having no effect on the quantity of pollutants discharged, does conserve water and should be encouraged. A typical system involves one or two stages of city-water washing and a final, totally recycled deionized water rinse. Dissolved solids are removed in the first washer stages and any residual that might cause spotting is removed by the deionized water rinse.

Deionizer regeneration requirements are governed by the buildup of dissolved solids in the preceding wash. The more dissolved solids carried over into the final rinse, the more they were picked up and thus removed by the deionizer.

Wastewater phosphorus loadings can be eliminated by discontinuing the detergent wash and all effluent loadings can be eliminated by recycling the wash water to other processes. Dissolved solids is the limiting factor governing discharge from a recycling float washer system. Dissolved solids removal is required if the water is recycled for washing. Where no detergents are added, the wash water is of high quality and can be recycled to the batch and cooling tower. These systems are illustrated in Figure 69.

FIGURE 69: WASTEWATER TREATMENT FLOAT PROCESS

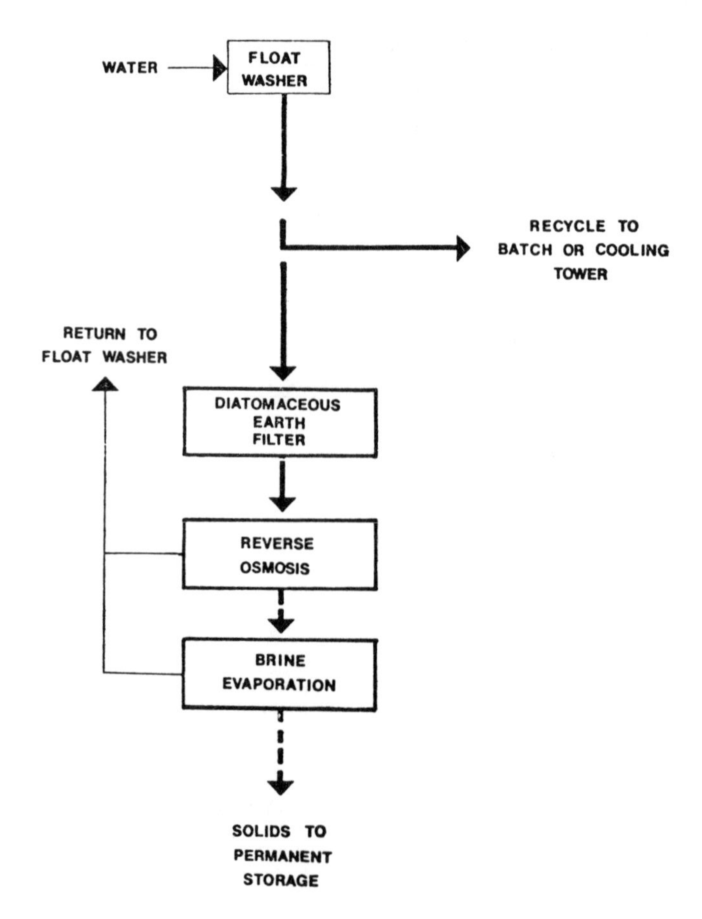

Source: EPA (46)

The use of detergents for float glass washing can be eliminated without any adverse effects on the manufacturing process as discussed above. Although this is an in-plant modification, by reducing phosphorus, it has the same effect as treatment and is considered as such for the sake of continuity in this report.

Elimination of detergent essentially eliminates phosphorus from float process wastewater as no other source is known. No data are available on the quality of phosphorus presently discharged, but elimination of detergents will achieve essentially 100% removal. With credit given for evaporation, trace phosphorus, and analytical error, a typical plant can achieve an effluent phosphorus concentration of 0.05 g/metric ton (0.0001 lb/short ton) or 0.5 mg/l.

Float glass wash water, where no detergents are used, is of high quality and can be recycled as batch water or cooling tower makeup. The data indicate very low increases of all contaminants result from washing. The dissolved solids will average 300 to 400 mg/l and the concentration of other constituents will be less than 15 mg/l. The exact temperature is not known, but in one case was measured at 37°C (98°F). With the exception of temperature, these characteristics are not significantly different from those of the city water presently being used in the batch or as cooling tower makeup.

Wash water can be collected and pumped through overhead piping to the batch house or cooling tower. The maximum flow acceptable for the batch is 42 l/metric ton (10 gallons per short ton). The remaining 96 l/metric ton (23 gallons per short ton) may be pumped to the cooling tower. Batch and cooling tower disposal of float wash water has not been demonstrated, but owing to the high quality of this wastewater, there are no apparent reasons why this disposal method should not be implemented.

It is theoretically possible to recycle the wash water back to the washer following dissolved solids removal. Three dissolved solids removal systems are sufficiently developed at present to be considered currently available. These are ion exchange, electrodialysis, and reverse osmosis. Ion exchange is already used extensively for final rinse water treatment; however, this process significantly increases the total dissolved solids loadings when regeneration wastes are considered.

Current research and development effort in dissolved solids removal technology center on reverse osmosis. Significant improvement and future development of this process are anticipated. For these reasons, reverse osmosis is selected for dissolved solids removal in this report.

It is assumed that reverse osmosis will concentrate the dissolved solids approximately five times and produce a wastewater flow rate of 20% of the initial volume treated. This wastewater stream must be disposed of if any net pollution reduction is to be achieved. It may be possible to discharge this wastewater to the batch, but this has not been demonstrated. The proven method of evaporation to dryness will be assumed in this report.

A complete recycle system using reverse osmosis might be set up as follows. The wash water discharge will first pass through a diatomaceous earth filter with an oil absorptive media to reduce both oil and suspended solids to less than 5 mg/l. Both of these constituents have an adverse effect on the reverse osmosis membranes. The filter is a dry discharge type, and spent diatomaceous earth is discharged at approximately 15% dry solids content, suitable for land disposal.

Following filtration, dissolved solids are removed by reverse osmosis. The water is forced at high pressure through a semipermeable membrane that retains most of the dissolved ions. Product water is returned to the washer and the waste brine is evaporated. The steam is condensed and also returned to the washer and the salt residue must be stored permanently in lined basins to prevent ground water contamination.

No total recycle systems have been demonstrated or contemplated in the flat glass industry. At the present time, reverse osmosis is used mainly for boiler water treatment, generally in competition with ion exchange. With the present state of the art, it is impossible to accurately predict the feasibility of the system without pilot plant data. Even if technically feasible, the cost/benefit ratio will be high. Capital and operating costs are high, relatively large amounts of energy are required, and two types of solid waste must be disposed of on land. The untreated wash water contains only 300 to 400 mg/l dissolved solids and less than 15 mg/l of other constituents. In most cases these concentrations will not significantly affect the receiving stream.

The typical float glass manufacturing plant may be located in any part of the country and has been built since 1960. Annual production is approximately 360,000 metric tons. Three alternative methods of treatment are discussed. Costs and effluent quality are summarized in Table 71.

Alternative A – No Waste Treatment or Control

Alternative A is the elimination of detergent usage in the float washer. As can be seen, the wastewater is of high quality and all plants presently discharge this water untreated.

Costs – None.
Reduction Benefits – Close to 100% phosphorus reduction.

Alternative B – Recycle to Batch and Cooling Tower.

Alternative B includes recycle of the float wash water to the batch and cooling tower. Process wastewater discharge is eliminated. The waste load recycled to the batch will become part of the glass and the waste load recycled to the cooling tower will constitute a portion of the cooling tower blowdown.

Costs – Incremental investment costs are $7,000 and total annual costs are $3,100 over Alternative A.
Reduction Benefits – Elimination of process wastewater discharge.

Alternative C – Total Recycle

Alternative C is the total recycle of wastewater back to the process following treatment using diatomaceous earth filtration for suspended solids removal and reverse osmosis for dissolved solids removal. Waste brine is evaporated to dryness and residual salt permanently stored. Sufficient suspended and dissolved solids are removed so that the water can be reused for float washing. No liquid wastes are discharged.

Costs – Incremental investment costs are $127,000 and total annual costs are $55,500 over Alternative B.
Reduction Benefits – Wastewater discharge is totally eliminated. Reduction of suspended solids, COD, phosphorus and all other pollutant constituents of 100%.

About half of the float glass plants produce process wastewater in the form of wash water. Washing is not required at the other plants and no process wastewater is produced. Washing is necessary where practiced and cannot be eliminated on the basis of the information gathered for this study.

No cost is associated with Alternative A. The evidence gathered indicates that detergent can simply be eliminated from the process. The cost of Alternatives B and C will depend upon the quantity of glass produced and the allowable dissolved solids build-up. The typical plant is one of the largest float plants so that the costs should be somewhat conservative with respect to the entire subcategory.

The high cost of Alternative C is the result of dissolved solids removal and waste brine disposal. These costs could be only roughly estimated since no system of this type is presently in operation. Each of the alternatives is a separate system and there is no benefit to be derived from cost phasing.

The age of equipment and the process employed do not significantly affect costs. No process changes are required and no significant engineering or nonwater quality environmental impact problems are anticipated.

TABLE 71: WATER EFFLUENT TREATMENT COSTS FOR MANUFACTURING FLOAT GLASS

Alternative Treatment or Control Technologies:	($1,000) A	B	C
Investment	0	7.	134.
Annual Costs:			
Capital Costs	0	.6	11.
Depreciation	0	.4	6.7
Operating and Maintenance Costs (excluding energy and power costs)	0	2.	28.4
Energy and Power Costs	0	.1	12.5
Total Annual Cost	0	3.1	58.6

Effluent Quality:

Effluent Constituents		Raw Waste Load	Resulting Effluent Levels		
Flow	(l/metric ton)	138	138	No Discharge	No Discharge
Suspended Solids	(g/metric ton)	2	2		
Dissolved Solids	(g/metric ton)	14	14		
COD	(g/metric ton)	2	2		
Flow	(l/sec)	1.6	1.6	No Discharge	No Discharge
Suspended Solids	(mg/l)	15	15		
Dissolved Solids	(mg/l)	100	100		
COD	(mg/l)	15	15		

Source: EPA (46)

Solid Tempered Automotive Glass Manufacture

Solid tempered automotive fabrication is the fabrication from glass blanks of automobile backlights (back windows) and sidelights (side windows). The major process steps and points of water usage are illustrated in Figure 70. The typical solid tempered automotive glass fabrication plant may be located in any part of the country and uses process equipment that has been modified within the last 10 to 15 years.

FIGURE 70: SOLID TEMPERED AUTOMOTIVE GLASS FABRICATION

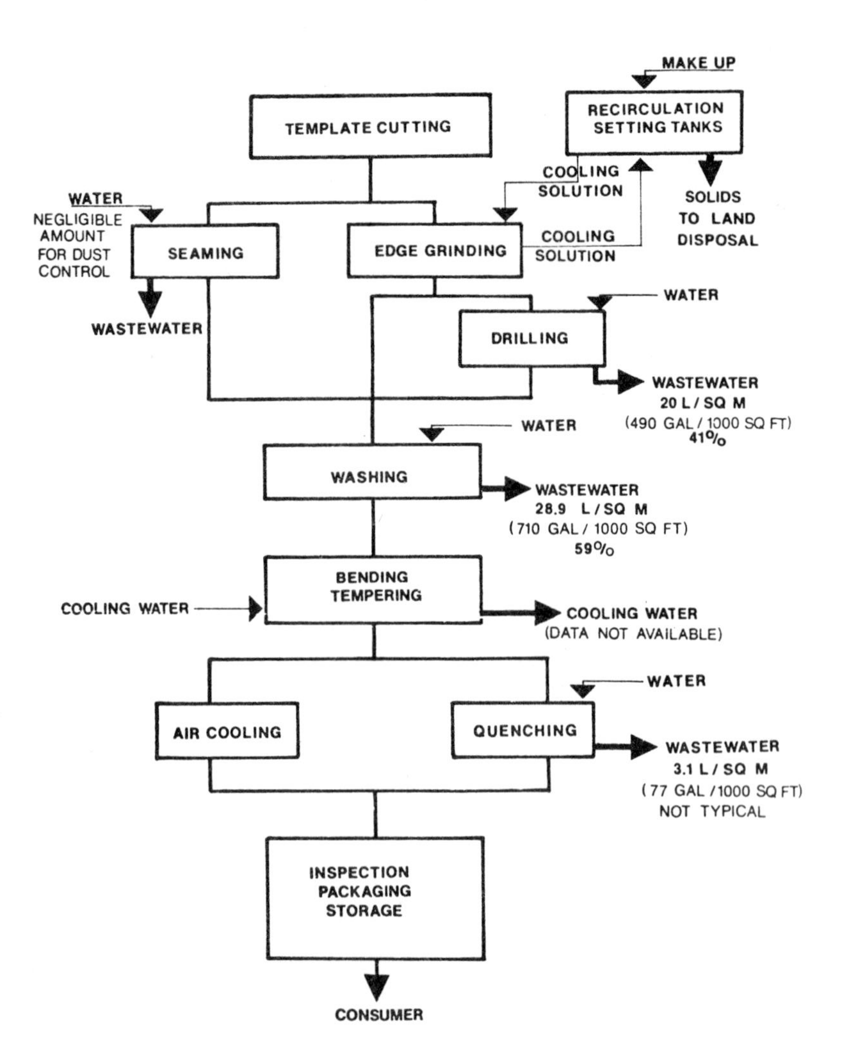

Source: EPA (46)

Production schedules are variable, but in many cases the plant is operated five or six days a week for 24 hours a day. Water is used in solid tempered automotive fabrication for seaming, grinding, drilling, quenching, cooking, and washing. The wash water is the major source of contaminated wastewater. Seaming is a light grinding to remove the sharp edges on backlights. In some cases, a fine spray of water is used to hold down the dust.

Edge grinding is used to form the smooth rounded edge on the exposed surfaces of sidelights. An oil-water emulsion collant-solution is used which also serves to flush away the glass particles. All plants recycle the coolant through a gravity sedimentation chamber where the glass particles settle and are removed along with free floating oil and scum.

The coolant is continuously recycled and the only blowdown from the system is the carry-over that remains on the glass. In the typical plant, the settled sludge and skimmings are collected for disposal as landfill; a few plants, however, discharge this waste to the wastewater system. About 11.2 g/sq m (2.3 lb/1,000 sq ft) of dry sludge is produced.

Holes are dilled in sidelights for window handles and brackets. Water is used in this process to cool the drill and flush away the glass particles. The typical flow is 20 l/sq m (490 gal/1,000 sq ft).

Washing is required to remove residual coolant and glass particles. One or two washing steps may be used before the bending furnace, depending on the plant set-up. Where the plant is set up on a production-line basis, the glass goes directly from edge grinding through drilling and washing to tempering, and only one washer is used. The edging or drilling and seaming lines may also operate independently of the tempering line, in which case washing occurs following drilling and seaming and again before tempering. More water is used in the two-stage process, but the pollutant loadings are not significantly different.

Both once-through and recycling washers are used; two or more stages may be used with each recycling from its own reservoir. Make-up water is added to the last stage and wastewater is discharged from the first stage. The recycle systems reduce the water usage, but the quantity of waste products is not reduced.

The wash water is heated to accelerate removal of oily residues. Recycling is limited by the build-up of oil and suspended solids. A typical plant uses one or two wash steps, with some recycling. The typical flow is 28.9 l/sq m (710 gal/1,000 sq ft).

Rapid cooling is required by the tempering process. Air cooling is typical, but quenching is also done with a water spray. Quench water is considered a process waste because the water comes in direct contact with the glass. Very little, if any, contaminants are picked up. The only apparent benefit of water quenching is that less space is required than for air cooling. About 3.1 l/sq m (77 gal/1,000 sq ft) is used where quenching is employed.

Some typical characteristics of the combined wastewater resulting from solid tempered automotive fabrication are listed in Table 72. In all cases except for pH, the values listed are the quantities added to the water as a result of solid tempered automotive fabrication. The background level in the influent water has been subtracted. The significant parameters are BOD, suspended solids, and oil.

Process wastewater flows vary significantly, ranging from 40.7 to 105 l/sq m (1,000 to 2,600 gal/1,000 sq ft) or 492 to 1,551 cu m/day (0.13 to 0.41 mgd). The typical flow is considered to be 49 l/sq m (1,200 gal/1,000 sq ft). As stated above, the wastewater flow rates are influenced both by the number of washing steps employed and by recycling. The high flow rate is indicative of a plant which does not recycle water.

Suspended solids are added to the waste stream in the form of glass particles resulting from seaming, grinding, and drilling. A typical plant generates 4.9 sq m (1 lb/1,000 sq ft). Some decrease in suspended solids loading may be expected if dry seaming is practiced, but a quantitative estimate of the reduction is not available.

Almost all the oil is contributed by the grinding solution carry-over, with trace quantities added by miscellaneous machine lubricants. Typical plant wastewater contains 0.64 g/sq m (0.13 lb/1,000 sq ft).

A small quantity of BOD is contributed to the wastewater by the oil in the coolant solution carry-over and to a much lesser extent by traces of oil entering the wastewater stream as a result of machinery lubrication. The typical raw wastewater loading is 0.73 g/sq m (0.15 lb/1,000 sq ft). Some information is also available on pH, temperature, COD, and dissolved solids. Limited data are available for temperature and COD, but 8°C (17°F) and 1.22 g/sq m (0.25 lb/1,000 sq ft) are indicative of the increases to be expected.

TABLE 72: RAW WASTEWATER[a]–SOLID TEMPERED AUTOMOTIVE GLASS FABRICATION

Flow	49	l/sq m	1200	gal/1000 sq ft		
pH	7					
Temperature (b)	8 C		17 F			
Suspended Solids	4.9	g/sq m	1	lb/1000 sq ft	100	mg/l
Oil	.64	g/sq m	.13	lb/1000 sq ft	13	mg/l
COD (b)	1.22	g/sq m	.25	lb/1000 sq ft	25	mg/l
BOD	.73	g/sq m	.15	lb/1000 sq ft	15	mg/l
Dissolved Solids	4.9	g/sq m	1	lb/1000 sq ft	100	mg/l

(a) Representative of typical solid tempered automotive process wastewater. Absolute value given for pH, increase over plant influent level given for other parameters.

(b) Indication of approximate level only; insufficient data is available to define actual level.

Source: EPA (46)

A pH of nearly 7 was recorded in all cases, indicating that pH is not a problem in solid tempered automotive glass fabrication. The dissolved solids increase of 4.9 g/sq m (1 lb per 1,000 sq ft) is higher than was expected. Water treatment regenerants and boiler blowdown (which are combined with the process wastewater stream for much of the sample data) are assumed to have contributed at least in part to the dissolved solids increase.

No significant variations in wastewater volume or characteristics are experienced during plant start-up or shutdown, and there are no known toxic materials in wastewater from the solid tempered automotive glass manufacturing process.

Cooling water is required at some solid tempered automotive glass plants for the tempering hearth and quenching. Although no data are available, heat rejection for these operations can be expected to be low with respect to the other subcategories.

Solid tempered automotive glass fabrication produces a wastewater with significant quantities of suspended solids, and lesser quantities of oil and BOD. The BOD is the result of oil contamination. Typical raw wastewater characteristics are: 100 mg/l suspended solids, 13 mg/l oil and 15 mg/l BOD.

The typical flow rate is 49 l/sq m (1,200 gal/1,000 sq ft). None of the plants studied presently treat solid tempered automotive wastewater. In-plant modifications may reduce wastewater volume and loading. Most plants presently collect the sludge removed from the coolant recycle system for disposal as landfill; however, in a few cases this is discharged to the sewer system imparting an unnecessary load on the treatment system.

The method of collection and dewatering used by most plants is a chain-driven scraper system which scrapes the sediment to discharge at one end of the tank and skims the floating material for discharge at the other end. The combined sludge is collected in a portable container for hauling to landfill. The sludge has an approximate moisture content of 90% and is well suited for land disposal.

In some plants, cooling water is sprayed directly onto the glass. Although little contamination is picked up in this quenching process, the water is in contact with the glass and is, therefore, a process wastewater. Quenching may be replaced by air cooling thereby reducing the volume of wastewater requiring treatment.

Wastewater volumes, but not the quantity of pollutants discharged, can be reduced by using recycling washers. Generally older washers tend to be of the once-through type, while new equipment is generally recycling with a two-stage system most common. Water is pumped over the glass from two separate reservoirs, and make-up water is added to the second wash tank. Overflow from second tank goes to the first wash tank and overflow from this tank is discharged to the sewer.

Sufficient water pressure and volume is required for the washer sprays to dislodge and flush away glass particles, oil, or dirt that might be on the glass. Recycling does not significantly affect these requirements until the concentration of contaminants increases to the point where residue is left on the glass. Some recycling can be employed in all plants, even where dissolved solids are high. Only one recycle will cut the wastewater flow to half that required for a once-through system.

The extent of recycle is limited by oil and suspended solids buildup. It is theoretically possible to remove these contaminants using a diatomaceous earth filter with oil absorption media. This type equipment is discussed below in more detail. Complete recycle is limited by dissolved solids buildup, with at least one company believing that 300 to 400 milligrams per liter is the allowable maximum concentration.

The major contaminants to be removed from solid tempered automotive glass are suspended solids and oil. Treatment may be accomplished at the individual washer or at end of pipe. It may be beneficial to consider individual treatment for new sources, but owing to limited floor space, end of pipe treatment is most practical for existing plants. Location of the treatment system does not influence the degree of pollutant reduction. Coagulation-sedimentation and filtration are common methods for reducing suspended solids and oil that are applicable to solid tempered wastewater. These treatment methods and a recycle system using reverse osmosis will be discussed and are illustrated in Figure 71.

FIGURE 71: WASTEWATER TREATMENT—SOLID TEMPERED AUTOMOTIVE GLASS FABRICATION

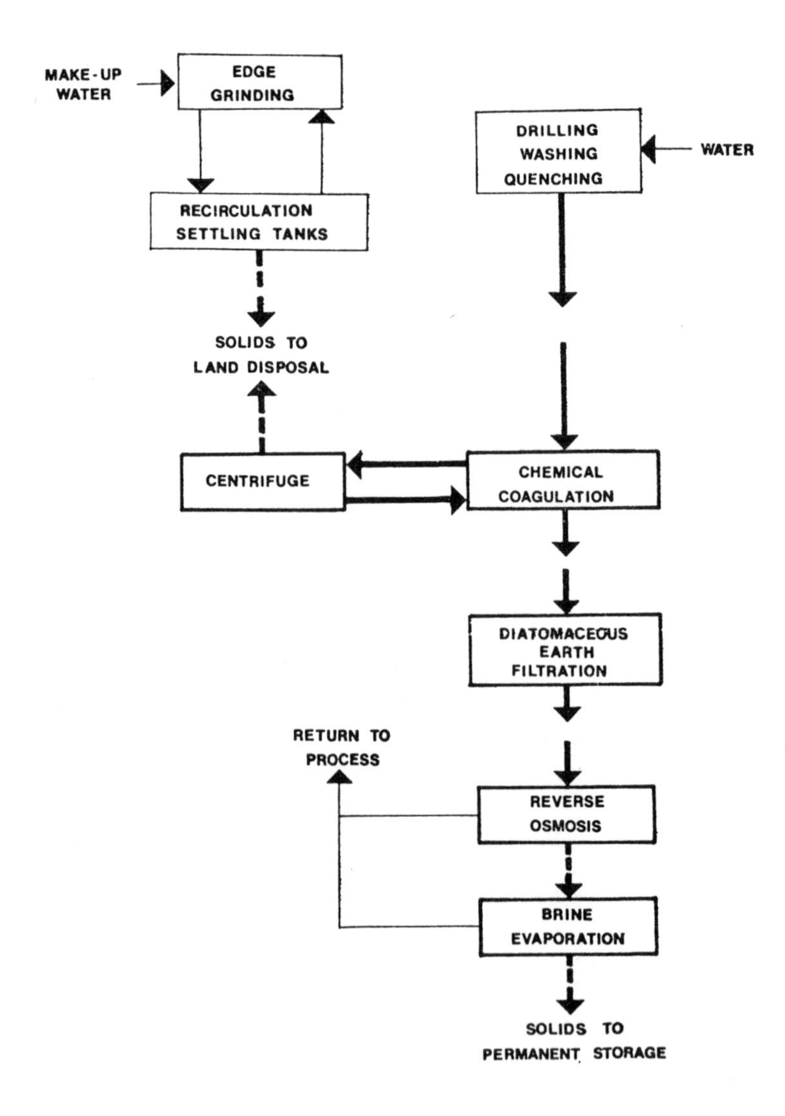

Source: EPA (46)

For cost estimating purposes, no wastewater treatment is considered to be treatment Alternative A. Coagulation and sedimentation is commonly used in the water industry for suspended solids removal. Solid tempered automotive glass wastewater is not unlike some of the river water commonly treated except for the higher oil content. It should be possible, using a properly designed system and the correct coagulant to achieve an effluent suspended solids concentration of 25 mg/l.

A solids contact coagulation-sedimentation system with sludge dewatering by centrifugation is assumed. Solids contact differs from conventional coagulation-sedimentation in that a portion of the sludge is returned to provide more surface area for trapping the newly coagulated particles. Numerous organic and inorganic flocculants and flocculant aids are available and individual testing will be required in each case to determine the optimum chemicals and addition rate.

Polyelectrolytes are preferable to inorganic flocculants because they do not contribute dissolved solids. Owing to the nature of the wastewater, however, it is likely that an inorganic flocculant such as alum or a coagulant aid such as bentonite clay will be required. Design parameters cannot be accurately predicted without at least laboratory scale studies. Conventional design rates can be assumed.

Sludge will be dewatered by centrifugation. It is difficult to accurately predict the sludge volume or moisture content without experimental data. A conservative estimate of the volume expressed in terms of production is 21 cu cm/sq m (0.07 cu ft/1,000 sq ft) with an 80% moisture content. Sufficient capacity for all equipment will be required so that effluent quality is maintained when portions of the equipment are down for maintenance.

Coagulation-sedimentation for suspended solids removal is a well-established process that can be successfully applied to solid tempered automotive glass wastewater. An effluent of 25 mg/l suspended solids or 1.22 g/sq m (0.25 lb/1,000 sq ft) should be readily achieved and the maximum daily concentration should not exceed 40 mg/l or 1.95 g/sq m (0.4 lb per 1,000 sq ft).

It is likely that oil and, therefore, BOD will also be removed, especially if inorganic coagulants are used, but lacking substantiating evidence, no credit is given for oil and BOD removal. Dissolved solids will be increased somewhat if inorganic coagulants are used.

A further decrease in suspended solids and oil can be achieved by filtering the settled effluent through a diatomaceous earth filter with a media especially treated for oil removal. This type of filter is commonly used to remove oil from boiler condensate.

The diatomaceous earth filtration system will consist of the filter, precoat tank, and a slurry tank for continuously feeding diatomaceous earth. The filter will be of the dry discharge type so that the sludge will not require dewatering. Sufficient units will be provided so that the system will continue to function with one unit down for cleaning or maintenance.

Experimentation on solid tempered automotive wastewater will be required to develop exact design parameters, but the approximate filter rate will be 20.4 to 40.7 l/min/sq m (0.5 to 1 gpm/sq ft) and approximately 0.9 kg (2 lb) of diatomaceous earth is required per 0.45 kg (1 lb) of oil removed. Oil, rather than suspended solids, is expected to be limiting; therefore, approximately 1.28 g/sq m (0.26 lb/1,000 sq ft) of diatomaceous earth will be required.

Effluent oil and suspended solids should be reduced to well below 5 mg/l by diatomaceous earth filtration. The BOD reduction resulting from oil removal can only be estimated. An effluent BOD of 10 mg/l is assumed although actual values will probably be lower. These loadings expressed in terms of typical plant production are as follows: 0.24 g/sq m (0.05 lb/1,000 sq ft) suspended solids, 0.24 g/sq m (0.05 lb/1,000 sq ft) oil and 0.49 g per sq m (0.1 lb/1,000 sq ft) BOD. COD will probably be reduced in equal or greater proportion than BOD.

Equivalent effluent levels can also be achieved using sand filtration, as described for plate glass (wastewater treatment), if sufficient oil is removed in the coagulation-sedimentation process. Inorganic coagulants such as alum will absorb oil. Trace quantities of oil are commonly removed by coagulation-sedimentation in water treatment plants.

The quantity of oil removed and the conditions for removal cannot be accurately stated without experimental data. Another consideration is oil fouling of the sand media. Oil will tend to coat the sand particles, and if sufficient quantities reach the filter, special cleaning procedures may be required. Due to the unknown factors related to sand filtration, a diatomaceous earth filtration system is used for cost estimating.

As in all cases for the flat glass industry, it is theoretically possible to completely recycle the treated effluent following dissolved solids removal. No such system is presently employed in the industry and only very general assumptions on the type of equipment required and the treatment efficiency can be made.

Filtered effluent can be passed through a reverse osmosis unit with 80% of the flow returned to the plant. The other 20%, consisting of waste brine, is evaporated with the steam condensed and returned to the plant and the salt permanently stored in a lined lagoon. The reverse osmosis system is similar for all flat glass applications and is discussed more fully in the float glass treatment section.

Dissolved solids data from the solid tempered automotive subcategory is limited and difficult to interpret because high dissolved solids from auxiliary waste streams are included. The maximum allowable dissolved solids concentration is also unknown. It is certain, however, that maximum possible recycle will be practiced prior to reverse osmosis. For cost estimating purposes, a conservative estimate of half of the typical flow or 24.4 l per sq m (600 gal/1,000 sq ft) will be assumed to be treated by reverse osmosis.

Only limited benefit, in terms of pollution reduction, will be achieved by going to a complete recycle system. Parameters other than dissolved solids have essentially been eliminated by prior treatment. Available data indicate a 100 mg/l or 4.9 g/sq m (1 lb/1,000 sq ft) dissolved solids increase at present water usage, which may be considered insignificant. Relatively large capital, operating, and power costs are required for reverse osmosis and an expensive landfill is needed for salt storage.

The typical solid tempered automotive glass fabrication plant may be located in any part of the country and uses process equipment that has been modified within the last 10 to 15 years. Annual production is 3.5 million square meters. Cost and effluent quality for the four treatment alternatives discussed are summarized in Table 73.

Alternative A – No Wastewater Treatment or Control

Alternative A is no wastewater treatment or control. The wastewater is of relatively high quality except for suspended solids. At the present time, no plants treat solid tempered automotive wastewater. Land disposal of coolant sludge is assumed, as this is almost universally practiced in the industry.

Costs – None
Reduction Benefits – None

Alternative B – Coagulation-Sedimentation

Alternative B is solids contact coagulation-sedimentation of all process wastewater, centrifugation of waste sludge and land disposal of dewatered waste solids.

Costs – Incremental investment costs are $81,000 and total annual costs are $24,100 over Alternative A.
Reduction Benefits – Effluent suspended solids are reduced 75%.

Alternative C – Filtration

Alternative C is oil absorptive diatomaceous earth filtration of the effluent from Alternative B. The spent diatomaceous earth is also disposed of as landfill.

Costs – Incremental investment costs are $68,000 and total annual costs are $18,000 over Alternative B.

Reduction Benefits – Incremental reduction of suspended solids is 80%. Total reductions of suspended solids, oil and BOD are 95, 62 and 33% respectively.

Alternative D – Total Recycle

Alternative D is the further treatment of the effluent from Alternative C using reverse osmosis. The waste brine is evaporated and the residual salt permanently stored. Sufficient suspended and dissolved solids are removed so that the water can be reused in the manufacturing process. No liquid wastes are discharged.

Costs – Incremental investment costs are $215,000 and total annual costs are $84,300 over Alternative C.

Reduction Benefits – Reduction of suspended solids, oil, BOD and all other pollutant constituents of 100%.

TABLE 73: WATER EFFLUENT TREATMENT COSTS FOR MANUFACTURING SOLID TEMPERED AUTOMOTIVE GLASS

Alternative Treatment or Control Technologies:	($1,000) A	B	C	D
Investment	0	81.	149.	364.
Annual Costs:				
Capital Costs	0	6.5	11.9	29.1
Depreciation	0	4.1	7.5	18.2
Operating and Maintenance Costs (excluding energy and power costs)	0	11.7	17.9	53.4
Energy and Power Costs	0	1.8	4.8	25.7
Total Annual Cost	0	24.1	42.1	126.4

Effluent Quality:

Effluent Constituents		Raw Waste Load	Resulting Effluent Levels			
Flow	(1/sq m)	49	49	49	49	No Discharge
BOD	(g/sq m)	.73	.73	.73	.49	
Suspended Solids	(g/sq m)	4.9	4.9	1.22	.24	
Oil	(g/sq m)	.64	.64	.64	.24	
Flow	(1/sec)	7.9	7.9	7.9	7.9	No Discharge
BOD	(mg/1)	15	15	15	10	
Suspended Solids	(mg/1)	100	100	25	5	
Oil	(mg/1)	13	13	13	5	

Source: EPA (46)

The volume of water to be treated depends on the amount of recycling practiced. More extensive recycling at the typical plant is representative of the better plants in this subcategory. A further reduction in water usage may be possible but is not assumed in the cost estimate. For those plants presently using more water than the typical plant, higher cost may be required for increased treatment cost or for in-plant modifications to reduce water usage.

The costs recorded here are representative of an above-average-size plant with moderate water recycling and reuse practices. A flow reduction of 50% prior to the reverse osmosis system in Alternative D is assumed.

None of the treatment methods is presently practiced in the flat glass industry. The technology is transferred from other industries and for this reason the cost estimates may be somewhat rough. This is especially true for the reverse osmosis system in Alternative D, where many unknowns had to be assumed. There is no apparent benefit for phasing costs within an alternative; however, where one alternative includes other alternatives, the earlier alternatives may be built first.

The age of equipment and the process employed do not significantly affect costs. No process changes are required and no significant engineering or nonwater quality environmental impact problems are anticipated.

Windshield Fabrication

Windshield fabrication is the manufacturing of laminated windshields from glass blanks and vinyl plastic. Oil resulting from oil autoclaving is the major constituent in this wastewater. The major process steps and points of water usage are illustrated in Figure 72.

The typical windshield fabrication plant may be located in any part of the country and uses oil autoclaves. Air autoclaves have been installed at some new plants, but oil autoclaves are still used for 90% of the windshields produced. The production schedule is variable and ranges from an eight-hour five-day week to a 24-hour six-day week.

Water is used in windshield fabrication for cooling, seaming, and washing. Three or four washes are required when oil autoclaves are used. Initial, vinyl, and postlamination washes are required in all cases. The prelamination wash has been eliminated by some plants.

Wet or dry seaming may be used in the windshield fabrication process. With wet seaming, a small volume of water is used for dust control and to flush away the glass particles produced. About 8.2 l/sq m (200 gal/1,000 sq ft) of finished windshields is used.

The first wash occurs early in the manufacturing process, following cutting and seaming. Traces of cutting oil, residual glass particles, and any dust which may have accumulated on the glass while in storage is removed. Only water, which is generally heated, is used. No detergents or other cleaning compounds are required.

Various types of washers are used. In some cases, once-through wash water is discharged directly to the sewer. Newer plants generally use recycling washers to reduce water usage. The wastewater flows vary from 28.5 to 138 l/sq m (700 to 3,400 gal/1,000 sq ft) of finished windshields. The typical flow is 81.5 l/sq m (2,000 gal/1,000 sq ft).

The two pieces of glass used to form a windshield are bent as one unit, and a parting material is used to prevent the two pieces from fusing during the bending process. The parting material is usually washed before the vinyl sheet is inserted, but in some cases a material is used that does not require washing. The exact nature of the parting materials used and the details of their application and removal are considered proprietary by the industry. The prelamination washer also serves to clean the glass surface of any dirt or spots since they cannot be removed following lamination. A three-stage washer is usually used. The first stage is a detergent wash followed by a city-water rinse and a final demineralized water rinse.

FIGURE 72: WINDSHIELD FABRICATION

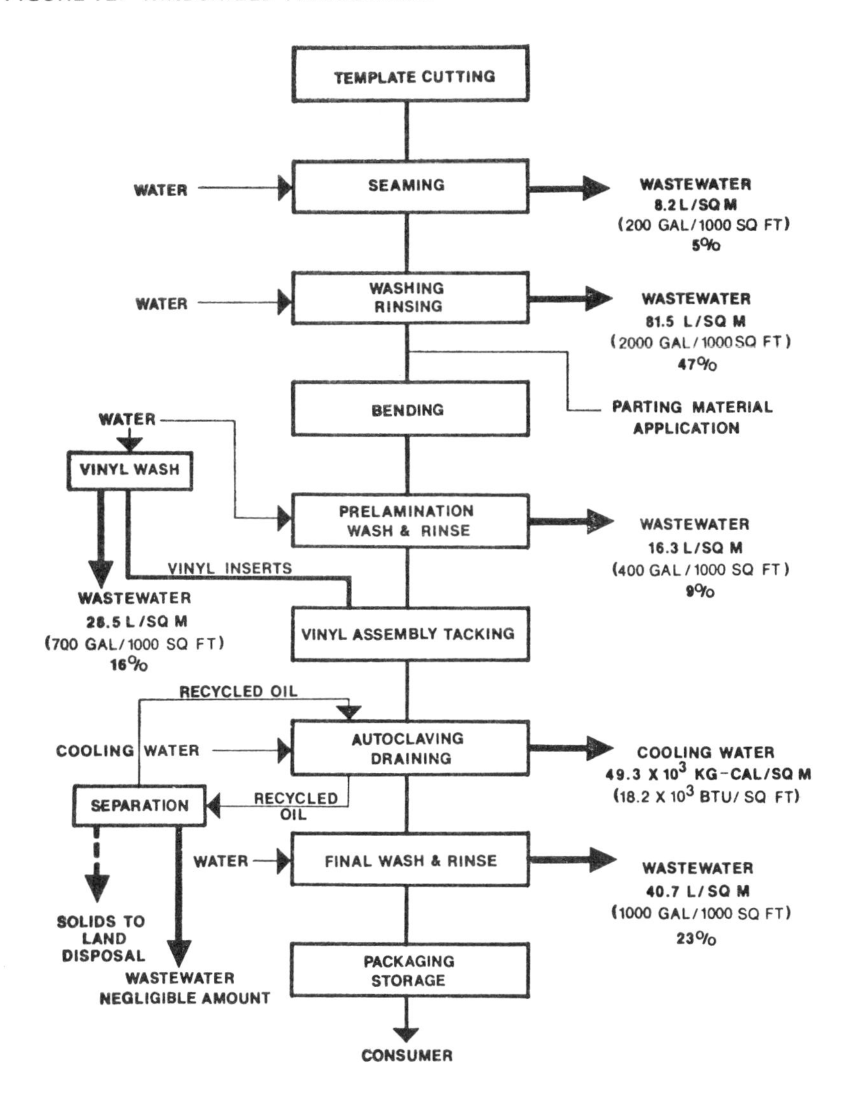

Source: EPA (46)

Deionized rinse water is the only makeup to the system. It is recycled through the stages and discharged as blowdown from the detergent wash. All stages are heated. Water usage is about 16.3 l/sq m (400 gal/1,000 sq ft) of windshields produced. The limited data available indicate a hot wastewater with relatively high phosphorus, moderate dissolved solids, and low organic and suspended solids increases. The plastic used for laminating is shipped from the manufacturer in rolls. Sodium bicarbonate is used as a parting material to keep the plastic from sticking and is removed in a two- or three-stage washer.

The three-stage system uses two city-water washes in series followed by a deionized-water rinse. The two-stage system is used where relatively low dissolved solids water is available and consists of two city-water rinses in series. Highly variable quantities of water are used for washing plastic, ranging from 12.2 to 285 l/sq m (300 to 7,000 gal/1,000 sq ft). The typical flow rate is 28.5 l/sq m (700 gal/1,000 sq ft). The wastewater is high in dissolved solids because of the sodium bicarbonate. The data also indicate a COD of 100 mg/l or 2.8 g/sq m (0.58 lb/1,000 sq ft) based on the typical flow. The high COD is unexpected and has not been explained.

Residual oil from the laminating autoclaves is removed in a series washing operation. Two basic systems are employed in the industry. In one case only washing is accomplished. In the second case, the washing is done in two stages with dry seaming in between washing steps. The wastewater characteristics are similar for both systems. For the purposes of this report, both systems will be grouped and discussed as one process.

Wash water for each stage is recycled out of a reservoir. In some cases, flows are countercurrent with blowdown from the following stage serving as makeup for the preceding stage. The old method of postlamination washing, still used at some plants, is to use a detergent wash as the first stage followed by two city-water rinses and possibly a final deionized-water rinse. Large quantities of detergent are required with this system, and very oily emulsified wastewater is produced.

Using a hot-water rinse before the detergent wash has been found to cut detergent usage by up to 95%. Most of the oil is removed by the hot water, and proportionately less detergent is required to emulsify the residual oil. Although the same quantity of oil remains in the wastewater stream, the majority is free oil and is more readily removed than emulsified oil.

The wastewater flows are the same for both methods. The typical postlamination washer flow is 40.7 l/sq m (1,000 gal/1,000 sq ft) of windshields produced. Wastewater characteristics are also similar for both methods, except higher phosphorus concentrations resulting from higher detergent usage are expected where an initial detergent wash is used.

Small amounts of water are picked up by the autoclave oil from condensation, cooling water leaks, and other sources. The oil and water separate in the oil storage tanks and the water is removed to a second tank where further gravity separation takes place. The oil is recycled to the autoclaves and the water is either discharged to the sewer or to the autoclave washwater treatment system. The stream accounts for only 1 to 2% of the total wastewater flow.

Some typical characteristics of the combined wastewater stream resulting from windshield fabrication are listed in Table 74. In all cases except for pH, the values listed are the quantities added to the water as a result of windshield fabrication. The influent water background levels have been subtracted. These data apply to a plant where an initial hot water rinse is used for the postlamination wash. No information is available for plants using an initial detergent wash.

Wastewater flow rates from plants considered typical of the windshield fabrication process vary from 52.9 to 492 l/sq m (1,300 to 12,100 gal/1,000 sq ft) of windshields produced. This corresponds to 454 to 2,195 cu m/day (0.12 to 0.58 mgd). The variability is due to the type of washers used (once-through as opposed to recycling) and to the dissolved solids content of the plant water. Less recycling can be practiced where influent dissolved solids are high. The typical flow is 175 l/sq m (4,300 gal/1,000 sq ft).

Suspended solids are contributed to windshield fabrication wastewater as a result of seaming. Carry-over results, even when dry seaming is used. The data indicate a typical reported value of 137 g/sq m (28 lb/1,000 sq ft) or 780 mg/l. This figure is much higher than the actual suspended-solids level because of oil interference since free oil tends to collect on the filter used in the suspended solids determination, causing high readings.

The actual typical suspended solids are estimated at 4.4 g/sq m (0.9 lb/1,000 sq ft) or 25 mg/l. Almost all the oil is contributed by the laminating process, with trace amounts resulting from machinery lubrication. The typical loading is 298 g/sq m (61 lb/1,000 sq ft).

A significant COD is noted as a result of the high oil content from the postlamination wash. Almost the entire loading of 298 g/sq m (61 lb/1,000 sq ft) may be attributed to oil. As indicated above, some COD is also contributed by the vinyl wash water.

The pH for all of the plants for which data was received ranged between 7 and 8. Sodium bicarbonate, removed from the vinyl, is the only constituent added which would be expected to significantly affect pH. Sufficient dilution is provided by the other wastewaters so that little effect is noted.

Phosphorus results from detergents used in the preassembly and postlamination washes. The available information on phosphorus loading shows substantial variation, indicating variable detergent usage. No basis is available for defining phosphorus or detergent limitations; therefore, the typical phosphorus value is based on the plants with high phosphorus loadings. The typical value is 0.98 g/sq m (0.2 lb/1,000 sq ft).

TABLE 74: RAW WASTEWATER*–WINDSHIELD FABRICATION USING OIL AUTOCLAVES

Flow	175 l/sq m	4,300 gal/1,000 sq ft	
pH	7-8		
Temperature	18.9°C	40°F	
Suspended Solids	4.4 g/sq m	0.9 lb/1,000 sq ft	25 mg/l
Oil	298 g/sq m	61 lb/1,000 sq ft	1,700 mg/l
COD	298 g/sq m	61 lb/1,000 sq ft	1,700 mg/l
BOD	5.9 g/sq m	1.2 lb/1,000 sq ft	33 mg/l
Total Phosphorus	0.98 g/sq m	0.2 lb/1,000 sq ft	5.6 mg/l

*Representative of typical process wastewater from the fabrication of windshields using oil autoclaves. Absolute values are listed for pH; the increase over plant influent level is given for other parameters.

Limited information is available on BOD and temperature characteristics for raw windshield lamination wastewater. The data indicate a BOD loading of 5.9 g/sq m (1.2 lb per 1,000 sq ft) or 33 mg/l. As with COD, the BOD can be attributed to the oily wastewater temperatures. These show an average discharge temperature immediately following the process of 38.9°C (102°F or 18.9°C (40°F) over the influent temperature. Sufficient data are not available to give an indication of the dissolved-solids levels.

No significant variations in wastewater volume or characteristics are experienced during plant start-up or shutdown, and there are no known toxic materials in wastewater from the windshield fabrication process.

Cooling water is required for autoclave operations and the compressors. Data are available from two plants. Values are 40,100 kg-cal/sq m (14,800 Btu/sq ft) and 58,600 kg-cal per sq m (21,600 Btu/sq ft) of fabricated automotive glass. The average heat rejection is 49,300 kg-cal/sq m (18,200 Btu/sq ft).

Oil is the major contaminant to be removed from windshield lamination wastewater. The oil contributes to a high organic loading as measured by COD. Lesser quantities of suspended solids and phosphorus are contributed as a result of seaming and detergent washing. Typical concentrations of these parameters are as follows: 1,700 mg/l oil, 1,700 mg per l COD, 25 mg/l suspended solids and 5.6 mg/l phosphorus. The typical flow rate is 175 l/sq m (4,300 gal/1,000 sq ft).

A combination of in-plant modification and end-of-pipe treatment will most efficiently reduce pollutant concentrations. Oil and phosphorus concentrations can be significantly reduced by modifying washing techniques. Residual oil and suspended solids can be reduced by filtration. An alternate to much of the oil removal equipment is the use of air autoclaves. In theory, zero discharge can be accomplished by using reverse osmosis for dissolved solids removal.

In-plant modifications can significantly contribute to a reduction of wastewater volume and to the quantity of oil and phosphorus discharged. Reduction of wastewater volume through recycling and reuse, though not reducing the quantity of pollutants discharged, will reduce the size of required treatment units. Windshield fabrication wastewater is almost entirely the result of washing operations. Three of four washes are required, depending on the production process, but the number of washes does not significantly affect wastewater volume.

Of much greater significance is the extent of wash water recycle. The same general considerations govern windshield wash water recycle as govern solid tempered wash water recycle. Older washers tend to be of the once-through type and some type of recycling is generally provided on new equipment. The typical plant employs some recycling, but water usage has not been minimized. Recycle is limited by factors, such as the manufacturing process and background dissolved solids concentration, that vary from plant to plant and cannot be generalized. It is probable that maximum recycle will be practiced whereever possible to minimize treatment costs.

It is now typical in the industry to use an initial hot water rinse in the postlamination wash to reduce detergent usage and to eliminate the large volume of emulsified oil that is produced when an initial detergent wash is used. This practice should become standard and is assumed as part of all treatment methods.

The limited available data on effluent phosphorus concentrations show increases from near zero to 0.98 g/sq m (0.2 lb/1,000 sq ft) indicating significant variation in detergent usage. Insufficient information was available to define the reasons for variable detergent usage, but it is apparent that some plants are producing acceptable windshields with much lower phosphorus discharges than others. Two exemplary plants are discharging less than 0.2 g per sq m (0.04 lb/1,000 sq ft) and, therefore, it can be assumed that other plants can develop the technology to reduce phosphorus to this level.

Another method for reducing oil contamination to trace levels is to use air rather than oil autoclaves. Air autoclaves are now used for windshield lamination by several small manufacturers, but are not typical of the industry. Greater handling problems and apparently more manpower are required for air autoclaves.

It was impossible to obtain the background data necessary to determine the relative cost of the two systems. However, one company has indicated that its analysis has shown a trade-off for new plants between the cost of extra handling using air autoclaves and treatment requirements using oil autoclaves. It is possible to reduce oil to trace levels by using diatomaceous earth filtration as indicated below.

Replacement of existing oil autoclaves would be expensive in terms of the investment required and the loss of production during the change-over. It is likely, however, that owing to the reduction in water usage and elimination of a potential pollution problem, air autoclaves will be installed in new plants.

Primary methods of windshield fabrication washwater treatment involve removal of the oil from postlamination wash water. Most of the oil can be removed by centrifugation, plain flotation in an American Petroleum Institute (API) separator, or dissolved air flotation. Suspended solids and residual oil can be removed by oil absorptive media filtration. In theory, it is possible to go to complete recycle by removing dissolved solids. The progression of treatment methods is illustrated in Figure 73. Phosphorus concentrations will be lowered by reducing detergent usage as indicated above. An initial hot water rinse is assumed for all treatment methods. For cost estimating purposes, no wastewater treatment is considered to be treatment Alternative A. When an initial hot water rinse is used, the oil removed collects in the initial wash reservoir. The oil is not emulsified since no detergents are used and can readily be removed by gravity separation. A cream separator-type centrifuge is used for oil removal at one exemplary plant and this method is the most efficient and economical of those observed.

Oil and water are drawn from the surface of the hot water rinse reservoir and passed through a centrifuge commonly used in the dairy industry for cream separation. Concentrations of up to 50% oil are reduced to less than 50 mg/l. The oil is sufficiently free of water to be returned to the autoclaves and the water is returned to the hot water rinse reservoir. A cartridge filter is used prior to the centrifuge to minimize cleanouts due to solids build-up, but this feature is optional as the suspended solids content is low.

Sufficient oil is removed so that the only blowdown from the initial hot water rinse is the residual carried over on the glass. So little oil reaches the second stage detergent wash that carry-over on the glass is also the only blowdown or loss from the detergent wash. As a result, the only wastewater from this exemplary postlamination washer is blowdown from the third-stage recycle rinse tank and once-through final rinse water.

The rinse water passes through an API separator, but little removal takes place in this unit. An API separator is a good safety feature, however, for trapping any oil that might accidently be discharged and will be included in cost estimates for this system. Oil and COD levels, for the typical total plant effluent, can be reduced to 1.76 g/sq m (0.36 lb per 1,000 sq ft) and 4.9 g/sq m (1 lb/1,000 sq ft), respectively, or a reduction of over 98% in both cases.

No credit is taken for phosphorus and suspended solids removal with this system. Similar effluent quality can be obtained by treating with dissolved air flotation although at higher cost because more sophisticated equipment and chemicals are required. With this system, presently in operation at another exemplary plant, oil is not removed at the initial hot water rinse tank, but blowdown from this and all the wash and rinse tanks is collected and treated by dissolved air flotation. Free oil is removed by belt skimmers prior to the flotation unit.

The raw wastewater is treated with polyelectrolyte to break any emulsion and combined with a portion of the treated effluent that has been pressurized and saturated with air. This mixture is discharged into the flotation cell. When the pressure is released, small bubbles are formed which cause the oil to float. A disadvantage of this system is that both skimmings and sediment are produced. These are not suitable for reuse and are disposed of as landfill. No information is available on the characteristics of this sludge.

A continuously recycling initial hot water rinse with oil removal has been successfully demonstrated. This system or the alternate dissolved air flotation system can be implemented throughout the industry. The equipment is readily available and can be installed on existing equipment without any interruption of normal operations.

Residual oil and suspended solids can be reduced to trace quantities by filtration in either of two systems that are available. The entire windshield fabrication wastewater stream may be filtered through oil-absorptive diatomaceous earth or only the laminating wash water may be filtered through oil-absorptive diatomaceous earth and the total wastewater stream filtered through sand or an equivalent medium.

The diatomaceous earth filters are the same type discussed for solid tempered automotive glass treatment. Sand filters are discussed in the section on plate glass treatment. More process steps are required for sand filters because the filter backwash must be dewatered. It is assumed that the backwash would be treated by batch coagulation-sedimentation and that the resulting sludge would be dewatered by centrifuge. Approximately a 20% solids sludge would be obtained by this method.

FIGURE 73: WASTEWATER TREATMENT—WINDSHIELD FABRICATION

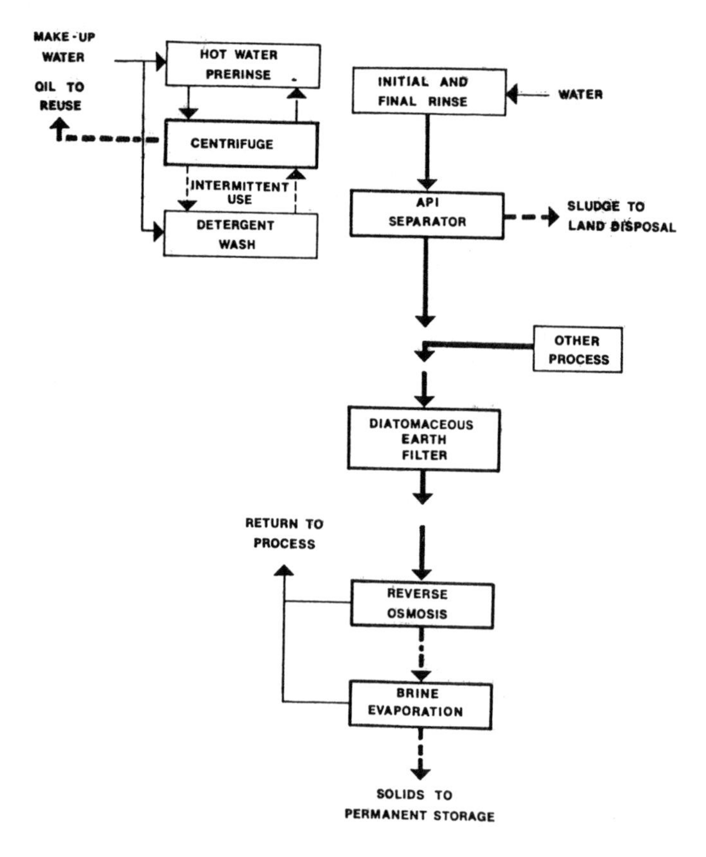

Source: EPA (46)

No additional equipment is required with the diatomaceous earth filters as these discharge a dry cake that is suitable for land disposal. The diatomaceous earth filtration system is somewhat less expensive and is used for cost-estimating purposes.

Wastewater effluent quality is assumed to be the same for both systems. Oil is reduced at least 50% compared to the above discharge and 99+% compared to the raw wastewater for a residual loading of 0.88 g/sq m (0.18 gal/1,000 sq ft) or 5 mg/l based on the typical flow rate. Suspended solids are reduced at least 80% compared to the raw wastewater for a typical effluent concentration of 5 mg/l and, therefore, have the same residual loading as for oil.

The effluent loadings are conservatively estimated because neither system has been demonstrated in the flat glass industry. No credit is taken for COD reduction as most of the residual at this point is assumed to be contributed by the vinyl wash water and not by oil. This technology has been demonstrated in other industries and can be successfully employed for windshield fabrication wastewater treatments. As with the other subcategories, it is theoretically possible to totally recycle windshield fabrication wastewater following reverse osmosis.

No system of this type has been demonstrated or anticipated by the industry for windshield fabrication wastewater. Many factors are related to the feasibility of a reverse osmosis system and it can only be assumed that such a system is technically feasible for windshield lamination effluent.

The anticipated system is similar to those indicated for float glass and solid tempered automotive glass fabrication. Maximum recycle would be achieved prior to reverse osmosis, but this is assumed to account for only a 33% reduction because significant recycle is already practiced in the windshield fabrication process. The reverse osmosis product water will be recycled to the manufacturing process. Waste brine will be evaporated with the steam returned to the process and the residual salt permanently stored in a lined lagoon.

Capital and operating costs for a dissolved solids removal system will be high and land will be permanently wasted for salt storage. Little benefit in terms of pollution reduction will be achieved because wastewater dissolved solids concentrations are low.

The typical windshield fabrication plant may be located in any part of the country and uses oil autoclaves. Annual production is 750,000 square meters. Cost and effluent quality for the four treatment alternatives discussed are summarized in Table 75.

Alternative A – No Wastewater Treatment or Control

Alternative A is no wastewater treatment or control.

Costs – None.
Reduction Benefits – None.

Alternative B – Lamination Wash Water Treatment

Alternative B is modification of the postlamination washer sequence to provide a continuously recycling initial hot water rinse, oil removal by centrifugation of the recirculating hot rinse water, recycle of oil back to the process, and treatment of other postlamination rinse waters by gravity oil separation. Other process waters are not treated. Negligible waste solids are produced.

Costs – Incremental investment costs are $32,000 and total annual costs are $14,600 over Alternative A.
Reduction Benefits – Oil is reduced by 99.4% and COD is reduced by 98.4%.

Alternative C – Filtration

Alternative C includes oil absorptive diatomaceous earth filtration of all process wastewater in addition to the treatment system described for Alternative B. The spent diatomaceous earth is disposed of as landfill. Phosphorus is reduced by more vigorous in-plant detergent control and improved washing techniques.

Costs – Incremental investment costs are $83,000 and total annual costs are $18,200 over Alternative B.
Reduction Benefits – Incremental reductions are 50% for oil and 80% for suspended solids. Total reductions are 99.7% for oil, and 80% for suspended solids and phosphorus.

Alternative D – Total Recycle

Alternative D is total recycle and reuse of the water following reverse osmosis treatment for dissolved solids removal. The waste brine is evaporated and the residual salt is permanently stored.

Costs – Incremental investment costs are $202,000 and the total annual costs are $90,000 over Alternative C.
Reduction Benefits – Reduction of oil, COD, suspended solids, phosphorus and all other pollution constituents of 100%.

As with the other subcategories, the volume of water treated and, therefore, the cost of treatment is related to the amount of recycling that can be practiced.

TABLE 75: WATER EFFLUENT TREATMENT COSTS—FLAT GLASS MANUFACTURING WINDSHIELD FABRICATION

Alternative Treatment or Control Technologies	A	B	C	D
	($1,000)			
Investment	0	32.	115.	317.
Annual Costs:				
Capital Costs	0	2.6	9.2	25.4
Depreciation	0	1.6	5.8	15.8
Operating and Maintenance Costs (excluding energy and power costs)	0	8.	13.6	48.5
Energy and Power Costs	0	2.4	4.2	33.1
Total Annual Cost	0	14.6	32.8	122.8

Effluent Quality:

Effluent Constituents		Raw Waste Load	Resulting Effluent Levels			
Flow	(l/sq m)	175	175	175	175	No Discharge
Oil	(g/sq m)	298	298	1.76	.88	
COD	(g/sq m)	298	298	4.9	4.9	
Suspended Solids	(g/sq m)	4.4	4.4	4.4	.88	
Phosphorus	(g/sq m)	.98	.98	.98	.2	
Flow	(l/sec)	6	6	6	6	No Discharge
Oil	(mg/l)	1700	1700	10	5	
COD	(mg/l)	1700	1700	28	28	
Suspended Solids	(mg/l)	25	25	25	5	
Phosphorus	(mg/l)	5.6	5.6	5.6	1	

Source: EPA (46)

Approximately the same absolute l/sq m are required at all plants, but the quantity discharged can be reduced by using recycling washers. Relatively more recycle is presently practiced for windshield fabrication than for other subcategories but it may still be possible by using recycling washers in all cases and by carefully controlling flow to further reduce usage. The typical plant is of less than average size and practices moderate water recycling and reuse. Costs may be as much as four times higher for the larger plants because of the higher water volume. A flow reduction of 33% prior to the reverse osmosis system in Alternative D is assumed.

New plants will probably use air rather than oil autoclaves. This will reduce the wastewater flow rate by approximately 23% and eliminate the need for Alternative B treatment. The technology for Alternatives C and D was transferred from other industries and is presently not practiced in the flat glass industry. The cost estimates for these alternatives may be somewhat rough because of the unknowns involved. This is especially true for the reverse osmosis system in Alternative D. There is no apparent benefit for phasing costs within an alternative; however, where one alternative includes other alternatives, the earlier alternatives may be built first.

It is possible, but not likely that some modification of the washers may be required to effect the detergent reduction indicated for Alternative C. No equipment modification was required at the exemplary plant where this technology is used but it is possible that modification will be required if another type of washer is used. Other considerations such as the age of equipment or the process employed do not significantly affect cost and no significant engineering of nonwater quality environmental impact problems are anticipated.

Glass Container Manufacturing

Glass container manufacturing consists of melting raw materials and then forming the molten glass using a blow-mold technique. The major process steps and points of water usage are shown in Figure 74.

Process water is used for cullet quenching and noncontact cooling of the batch feeders, melting furnaces, forming machines, and other auxiliary equipment. At some plants, a small amount of water is also added to the batch to control dust. The volume discharged depends on the quantity of once-through cooling water and on the water conservation procedures employed at the glass container plant. The typical flow is representative of a plant using some once-through cooling water and practicing reasonable water conservation.

Water is added to the batch for dust suppression at some plants in all of the subcategories covered by this study, but the practice is not considered typical for the industry. When water is added, it is generally at a rate of about 11.5 l/metric ton (2.75 gal/ton).

Noncontact cooling water is used to cool batch feeders, melting furnaces, forming machines, and other auxiliary equipment. The typical flow of cooling water is 1,380 l/metric ton (330 gal/ton). This represents 47% of the total flow. Reported and calculated heat-rejection rates vary from 361,000 kg-cal/metric ton (1,300,000 Btu/ton) to 13,900 kg-cal per metric ton (50,000 Btu/ton). Owing to the wide variation and absence of sufficient information to explain the differences, it is not possible to define a typical heat-rejection value. The average value is 97,300 kg-cal/metric ton (350,000 Btu/ton).

Cullet quench water is required to dissipate the heat of molten glass that is intentionally wasted or discharged during production interruptions, or to quench hot pieces which are imperfect. Some plants use noncontact cooling water for the dual purposes of furnace and equipment cooling and cullet quenching. The typical cullet quench water flow is 1,540 l/metric ton (370 gal/ton) or 53% of the total flow.

Repair and maintenance departments are required in all glass container plants. Wastewater is produced in the maintenance departments from the cleaning of production machinery. The machinery is inspected, cleaned, and repaired at specific intervals. The cleaning operation includes steam cleaning of large parts and caustic batch cleaning of items such as molds. The wastewater from the maintenance department is of very low volume and is primarily occasional rinse water from the cleaning operations.

Several glass container plants have corrugator facilities to manufacture boxes. Wastes developed from the corrugator facilities are of low volume and include cleanup water from the gluing and ink labeling equipment, lubricating oil, and steam condensate. The wastes are usually contained at the plant site and treated or discharged to a municipal sewer system. The corrugator box manufacturing operation is not covered in the SIC codes under study in this report. Typical characteristics for the combined noncontact cooling and cullet quench wastewater streams for a glass container plant are listed in Table 76.

FIGURE 74: GLASS CONTAINER MANUFACTURING

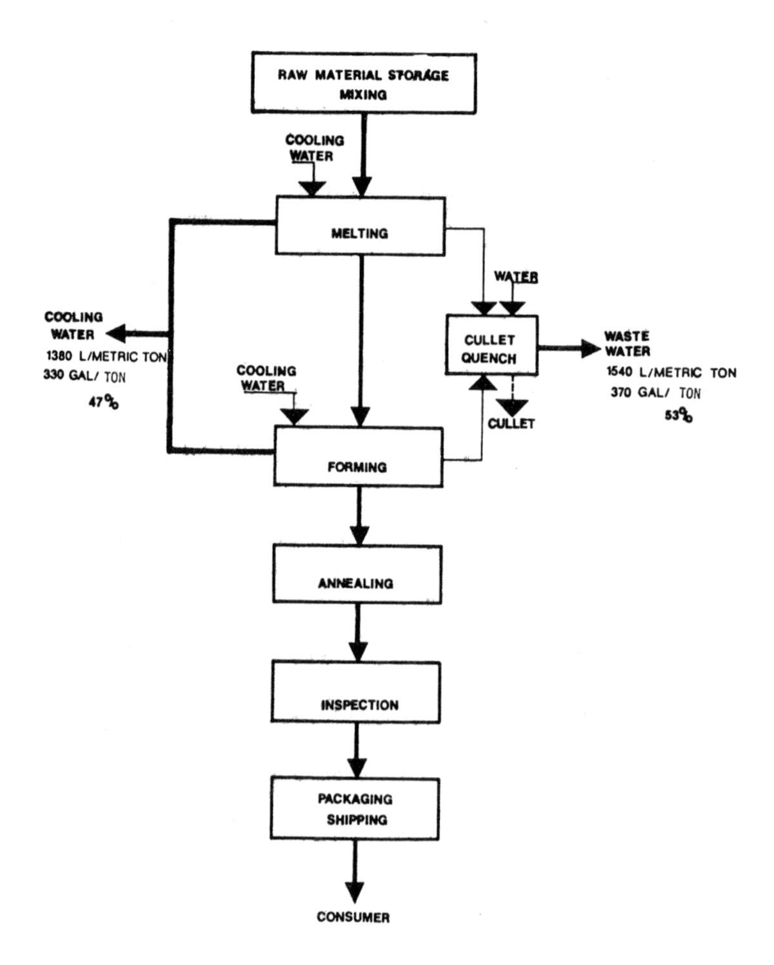

Source: EPA (47)

In all cases, except for pH, the values listed are the quantities added to the water as a result of glass container manufacturing; concentrations in the influent water have been subtracted. The significant parameters are oil and suspended solids. BOD and COD are a result of oil in the wastewater; control of oil therefore controls oxygen demand.

The quantity of wastewater produced in the manufacture of glass containers is highly variable. Flows range from near zero to 6,250 l/metric ton (1,500 gal/ton) or from near zero to 2,460 cu m/day (0.65 mgd). Some plants have indicated no discharge, but are apparently discharging an unknown quantity of blowdown. This blowdown may be in the form of water carried with the cullet and fed to the furnace during batching.

The typical flow is 2,920 l/metric ton (700 gal/ton). The amount of water usage depends, to a certain extent, on the raw water source and age of the plant.

TABLE 76: RAW WASTEWATER[a]–GLASS CONTAINER MANUFACTURING

Flow	2920	l/metric ton	700	gal/ton	
Temperature	6°C		11°F		
pH	7.5				
BOD	0.0145	kg/metric ton	0.029	lb/ton	5 mg/l
COD	0.145	kg/metric ton	0.29	lb/ton	50 mg/l
Suspended Solids	0.07	kg/metric ton	0.14	lb/ton	24 mg/l
Oil	0.03	kg/metric ton	0.06	lb/ton	10 mg/l

(a) Representative of typical glass container manufacturing waste water. Absolute value given for pH, increase over plant influent level given for other parameters.

Source: EPA (47)

Glass container plants receive water from various sources including plant-owned wells, surface water, and municipal water systems. The amount of water conservation and recirculation is considerably greater at plants that use water from a municipal system. Plant age is another factor which may affect water usage. Newer plants may use somewhat less water because of more attention to water conservation.

A small amount of BOD is added to the wastewater as shear spray or lubricating oil. Shear spray is an oil-water emulsion used to cool and lubricate the shears and the chutes that convey the glass to the IS machine. Many plants now use a synthetic biodegradable shear spray to reduce the effects of oil on the receiving stream.

Excess shear spray eventually finds its way into the cullet quench water. Another potential source of BOD is leakage of lubricating oils into the cooling water system. The typical raw wastewater loading is 0.0145 kg/metric ton (0.029 lb/ton) of BOD_5. The COD is contributed by the same sources that contribute BOD, namely shear spray oil and lubricating oil. The typical plant wastewater contains 0.145 kg/metric ton (0.29 lb/ton) of COD.

Suspended solids enter the plant wastewater as the result of cullet quenchine and plant cleanup. The cullet quench water picks up fine glass particles; additional suspended solids are added during cleanup of the IS machine area. A typical plant generates 0.07 kg/metric ton (0.14 lb/ton) of suspended solids.

Oil is added to the plant wastewater as shear spray oil and leaking lubricants. The typical oil loading is 0.03 kg/metric ton (0.06 lb/ton). Some information is available on the temperature and pH of glass container plant wastewaters. The average rise in temperature over the plant influent water is 6°C (11°F). The typical pH of the wastewater is 7.5 and reported values range from 6.5 to 8.6.

Glass container plant operation is continuous (24 hr/day, 7 day/week); and therefore, wastewater flows are relatively constant. No significant variations in wastewater volume or characteristics occur during plant startup or shutdown, and there are no known toxic materials in the wastewater. The melting tanks must be drained every three to five years for rebuilding and excessive quantities of cullet quench water are produced for one or two days during this period. In larger plants with several furnaces, this discharge may occur several times a year. The very limited data available indicate that temperature is the only significant parameter and that receiving stream standards may necessitate cooling of the quench water in some cases. As was stated above, wastewater results from forming and cullet quenching in the manufacture of glass containers. In most plants, these wastewaters are combined with noncontact cooling water prior to discharge, and in some cases a portion of the cooling water is used for cullet quenching.

Oil and suspended solids are the only significant parameters contained in this wastewater. The quantity of pollutants discharged may be reduced by recycling the cullet quench water and treating the blowdown using dissolved air flotation followed by diatomaceous earth filtration as illustrated in Figure 75.

FIGURE 75: WASTEWATER TREATMENT—GLASS CONTAINER MANUFACTURING MACHINE PRESSED AND BLOWN GLASS MANUFACTURING

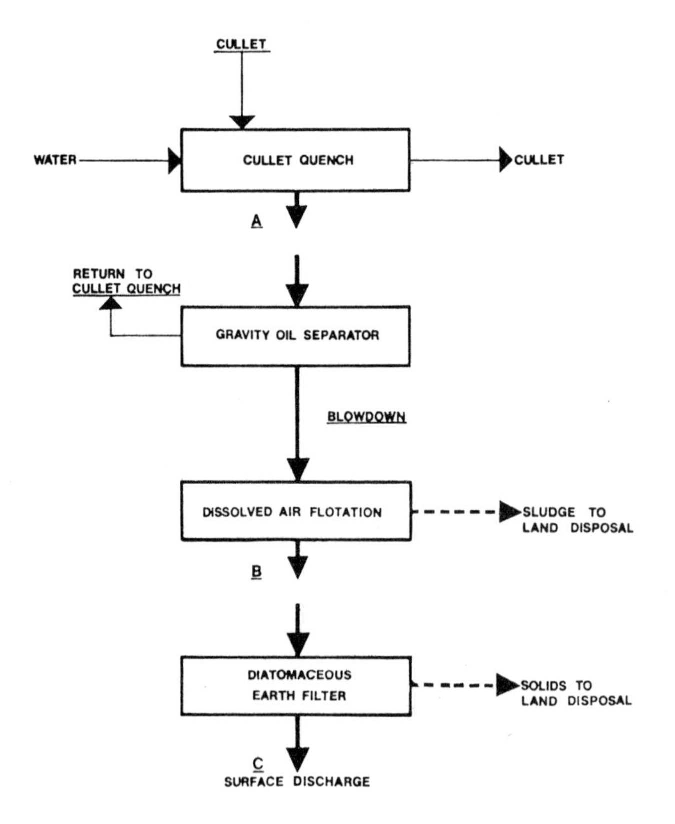

Source: EPA (47)

Existing Treatment and Control (Alternative A): Both in-plant techniques and end-of-pipe methods have been employed to reduce pollutant discharge. Many plants have achieved low effluent levels with only in-plant methods, and the presence of end-of-pipe treatment systems has not necessarily assured a high-quality effluent. A number of plants without end-of-pipe treatment are achieving low discharge levels while several plants with treatment are discharging at rather high levels. The typical combined cooling and cullet quench water flow for a glass container plant is 2,920 l/metric ton (700 gal/ton) with 53% of the total flow being process water. Suspended solids discharges of 70 g/metric ton (0.14 lb per ton) and oil discharges of 30 g/metric ton (0.06 lb/ton) are presently being achieved by 70% of the 40 plants for which data are available. These values correspond to 24 mg/l for suspended solids and 10 mg/l for oil at the typical flow.

These effluent levels should be readily achievable by all plants with a minimum of in-plant modification or end-of-pipe treatment. In-plant modifications to reduce pollutant discharge include shear spray collection of forming machine shop oil, or modified cleanup procedures. Many plants collect and recycle shear spray. Pans are placed around the shears to collect as much of the excess spray as possible.

The collected material is filtered and returned to the shear spray make-up tank. Approximately 70% of the shear spray can be recovered in this manner. In some plants, troughs are built around the forming machines to collect the oily runoff resulting from excess lubrication and leaks. The oily waste flows by gravity to a storage tank and is periodically hauled away for reclamation or disposal.

It is the practice of many plants to periodically hose down the area around the forming machines. It may be possible to use a dry removal method for at least part of the cleanup. One of the oil-adsorptive sweeping compounds might be used and disposed of as solid waste, thereby eliminating some of the oil discharged into the wastewater system.

End-of-pipe treatment might involve some type of sedimentation system with oil removal capability. This will serve to reduce suspended solids and free oil but will not significantly reduce emulsified oil. Although end-of-pipe treatment systems are presently being used by some plants, it would appear that in-plant techniques will be more effective and less expensive to achieve the suggested effluent levels.

Recycle with Dissolved Air Flotation of Blowdown (Alternative B): Effluent levels can be further reduced by segregating the cullet quench water from the cooling water system, recycling the cullet quench water through a gravity separator and treating the blowdown using dissolved air flotation. Suspended solids and oil will build up in the recriculation system, but the dissolved solids concentration will probably be limiting.

A conservative value of 5% blowdown is assumed, based on the operating dissolved solids level of approximately 1,700 mg/l in an existing recycle system. Dissolved solids levels of 4,000 to 5,000 mg/l are probably acceptable, but supportive data are not presently available. A cooling tower is not considered necessary. Existing recycle systems use only a tank that serves as the recycling pump wet well and from which oil is removed using a belt skimmer.

Segregation of the noncontact cooling water and recycle with a 5% blowdown will reduce the typical contact water discharge flow to 77 l/metric ton (18.5 gal/ton). Blowdown from the recycle system can be treated to 2 g/metric ton (0.004 lb/ton) using dissolved air flotation. COD is expected to be reduced in proportion to the oil removed. The sludge production is approximately 1,900 l/day (500 gal/day) at 3% solids concentration.

Several glass container plants are presently recycling cullet quench water to conserve water, but none are treating the blowdown using dissolved air flotation. This technology is practiced in the flat glass industry and should be readily transferable to the pressed and blown glass industry.

At least one plant has recently begun treating a portion of the recycling quench water using diatomaceous earth filtration. Long-term operating data for this system are not available. It is possible that diatomaceous earth filtration will not be effective for the high oil concentrations in a recycling system or that excessive diatomaceous earth usage will be required. For this reason, the proposed model system includes dissolved air flotation.

Diatomaceous Earth Filtration (Alternative C): Diatomaceous earth filtration may be employed to further reduce the oil and suspended solids in the dissolved air flotation discharge stream to less than 5 mg/l or 0.4 g/metric ton (0.0008 lb/ton). Approximately 50 l/day (13 gal/day) of 15% solids sludge is produced.

This technology has been commonly employed for steam condensate treatment and should be readily transferable to the pressed and blown glass industry. As stated above, at least one plant is presently employing the diatomaceous earth filtration technology.

Rather than treat to such a low effluent level, it may be feasible for a plant to consider complete recycle or discharge of the blowdown into the batch. Several container manufacturers are investigating this possibility and a number of plants have achieved nearly complete recycle. More data than were available for this study will be necessary to evaluate the feasibility of zero discharge through application of this technique.

The typical glass container manufacturing plant may be located in any part of the country and may be 50 or more years old. The daily production is approximately 454 metric tons (500 tons). Cullet quenching and noncontact cooling water are not segregated. Costs and effluent quality for the three treatment alternatives are summarized in Table 77.

TABLE 77: WATER EFFLUENT TREATMENT COSTS–GLASS CONTAINER MANUFACTURING

Alternative Treatment or Control	($1000)		
	A	B	C
Investment	0	285.	312.
Annual Costs:			
Capital Costs	0	22.8	25.
Depreciation	0	14.3	15.7
Operating and Maintenance Costs (excluding energy and power costs)	0	17.2	23.
Energy and Power Costs	0	1.8	3.2
Total Annual Cost	0	56.1	66.9

Effluent Quality:

Effluent Constituents	Raw Waste Load	Resulting Effluent Levels		
Flow (l/metric ton)	2920	2920	77	77
Oil (g/metric ton)	30	30	2	0.4
Suspended Solids (g/metric ton)	70	70	2	0.4
Flow (l/sec)	15.3	15.3	.41	.41
Oil (mg/l)	10	10	25	5
Suspended Solids (mg/l)	24	24	25	5

Source: EPA (47)

Alternative A – Existing Treatment and Control

Alternative A involves no additional treatment. These effluent levels are readily achievable by all plants within this subcategory through normal maintenance and clean-up operations within the plant and represent the raw waste loadings expected from a glass container plant. Improved housekeeping techniques may be required at some plants to achieve the typical effluent levels, while others may elect to provide end-of-pipe treatment in the form of some type of sedimentation system with oil removal capabilities. It is felt however, that in-plant techniques will be a more effective and a less expensive means of achieving effluent levels.

Costs – No additional cost.
Reduction Benefits – Upgrading of all effluent discharges to this level.

Alternative B – Recycle with Dissolved Air Flotation of Blowdown

Alternative B involves segregation of noncontact cooling water from cullet quench water. The cullet quench water is recycled back to the cullet quench process through a gravity oil separator, and blowdown is treated using dissolved air flotation. The blowdown is 5% of the total cullet quench water flow.

Costs – Incremental investment costs are $285,000 and total annual costs are $56,100 over Alternative A.
Reduction Benefits – The incremental reductions of oil and suspended solids compared to Alternative A are 93% and 97%, respectively.

Alternative C – Diatomaceous Earth Filtration

Alternative C provides further treatment of the effluent from Alternative B by diatomaceous earth filtration.

Costs – Incremental investment costs are $27,000 and total annual costs are $10,800 over Alternative B.
Reduction Benefits – The incremental reductions of oil and suspended solids compared to Alternative B are 80%. Total reductions of oil and suspended solids are 98.7 and 99.4%, respectively.

Machine Pressed and Blown Glass Manufacturing

Machine pressed and blown glass manufacturing consists of melting raw materials and then forming the molten glass using presses or other techniques to manufacture tableware, lenses, reflectors, sealed headlamp glass parts, and other products not covered in the other subcategories. The major process steps and points of water usage are listed in Figure 76.

Water is used in the manufacturing of machine pressed and blown products primarily for noncontact cooling and cullet quenching. Cullet quenching is the cooling of molten glass or hot rejects with water. Some plants use a portion of the noncontact cooling water for cullet quenching. Water may also be added to the batch for dust suppression and an oil-water emulsion is used for shear spraying.

Noncontact cooling water is required to cool batch feeders, melting furnaces, presses, and other auxiliary equipment. The typical flow of noncontact cooling water is 2,710 l/metric ton (650 gal/short ton). Noncontact cooling water is 48% of the combined flow from this category. Although no heat-rejection data are available for machine pressed and blown glass plants, it is expected that heat-rejection requirements are similar to those of glass container plants.

Quench water is required at all machine pressed and blown glass plants to cool intentionally wasted molten glass during production interruptions, and to quench hot pieces that are wasted or rejected because of imperfections. The configuration of equipment is similar to a glass container plant. Quench water and waste glass are discharged into chutes and flow to a cart located in the furnace basement. Excess quench water overflows the cart and is discharged to the sewer.

The typical quantity of water used for cullet quenching is 2,920 l/metric ton (700 gal/ton). This is 52% of the total flow. Some machine pressed and blown glass plants have small plating shops where molds are periodically cleaned and chrome-plated. Low volumes of rinse waters are periodically discharged, but no evidence of chromium contamination was found in the data collected during this study. Chromium discharges should be regulated by the effluent limitations developed for plating wastes.

Finishing may be employed at some machine pressed and blown glass plants, but most of the finishing techniques produce no wastewater. The great majority of the finishing steps can be classified as decorating and involve painting or coating and reannealing. Other finishing steps may produce small quantities of wastewater, but these are not covered in this study. It is recommended that where treatment is required, the technology developed for hand pressed and blown glass finishing be applied.

FIGURE 76: MACHINE PRESSED AND BLOWN GLASS MANUFACTURING

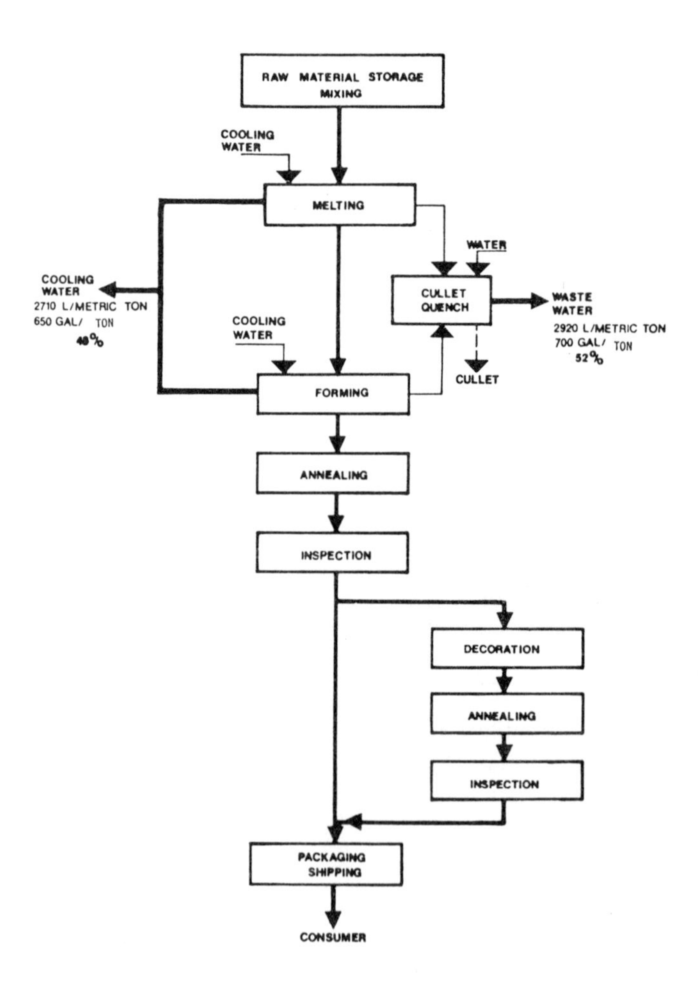

Source: EPA (47)

Typical characteristics of the combined noncontact cooling and cullet quench wastewater streams are listed in Table 78. In all cases, except for pH, the values listed are the quantities added to the water as a result of machine pressed and blown glassware manufacturing. Background concentrations in the influent water have been subtracted. Oil and suspended solids are the significant parameters. The COD is contributed by the oil.

TABLE 78: RAW WASTEWATER[a]–MACHINE PRESSED AND BLOWN GLASS MANUFACTURING

Flow	5630	l/metric ton	1350	gal/ton	
Temperature	10°C		18°F		
pH	7.8				
BOD	0.028	kg/metric ton	0.056	lb/ton	5 mg/l
COD	0.28	kg/metric ton	0.56	lb/ton	50 mg/l
Suspended Solids	0.14	kg/metric ton	0.28	lb/ton	25 mg/l
Oil	0.056	kg/metric ton	0.11	lb/ton	10 mg/l

(a) Representative of typical machine pressed and blown glass manufacturing waste water. Absolute value given for pH, increase over plant influent level given for other parameters.

Source: EPA (46)

A variable volume of water is used for machine pressed and blown glass manufacturing. Flows range from 2,210 to 27,500 l/metric ton (530 to 6,600 gal/ton) or 87 to 2,650 cu m/day (0.023 to 0.7 mgd). The typical combined flow of noncontact cooling water and cullet quench water is 5,630 l/metric ton (1,350 gal/ton). The variation in water usage depends on the amount of once-through noncontact cooling used and also on the water conservation practiced at the various machine pressed and blown glass plants. Cullet quench water and noncontact cooling water are generally combined prior to discharge.

The typical COD added to the wastewater is 0.28 kg/metric ton (0.56 lb/ton). The COD results primarily from shear spray and lubricating oil leaks. The suspended solids are fine glass particles picked up by the cullet quench water. The typical suspended solids loading is 0.14 kg/metric ton (0.28 lb/ton).

Oil is added to the wastewater as shear spray and lubricating oil. Water-soluble oil is used to lubricate the gob shear cutters and the glass gob chute. The shear spray oil flows from the gob chute and enters the cullet quench water. Lubricating oil leaks may also contaminate the cooling water and cullet quench water. The typical quantity of oil discharged is 0.056 kg/metric ton (0.11 lb/ton).

Some information is also available on the BOD, pH, and temperature. The typical pH is 7.8 and the typical temperature rise is 10°C (18°F). The temperature increase resulting from cullet quenching alone is not known. These appear in Table 78.

Machine pressed and blown plants operate at various schedules, some continuously, while others operate for an 8 hr/day, 5 day/week. Continuous operation is desirable because furnace heat must be maintained. Some glass must be wasted during the off periods to maintain the flow of glass through the furnace; therefore, cullet quench water is always required. No significant variations in waste volume or characteristics are experienced during plant start-up or shutdown, and there are no known toxic materials in wastewater from the manufacture of machine pressed and blown glassware. Excessive quench water volumes will be produced when a tank is drained for rebuilding or for a change in the composition of the glass. Heat is the only significant pollutant parameter contained in this wastewater source.

Owing to similar manufacturing techniques, wastewater resulting from machine pressing and blowing of glass is similar to glass container manufacturing wastewater. Oil and suspended solids are the significant pollutant parameters and their discharge may be reduced by recycling the cullet quench water and then treating the blowdown using dissolved air flotation followed by diatomaceous earth filtration as illustrated in Figure 75.

Existing Treatment and Control (Alternative A): The in-plant and end-of-pipe pollution control methods employed in the glass container industry are also used for machine pressed and blown glass manufacturing. The typical combined cooling and cullet quench wastewater flow is 5,630 l/metric ton (1,350 gal/ton) with 52% of the total flow being process water.

Suspended solids discharges of 140 g/metric ton (0.28 lb/ton), equivalent to 25 mg/l at the typical flow, are presently being achieved by three of the nine plants for which data are available. Oil discharges of 60 g/metric ton (0.12 lb/ton) are being achieved by four of six plants. This corresponds to a concentration of 10 mg/l at the typical flow.

These effluent levels should be readily achievable by all plants with a minimum of in-plant modification or end-of-pipe treatment. Possible methods of treatment are described earlier in the glass container manufacturing subsection.

Recycle with Dissolved Air Flotation of Blowdown (Alternative B): Effluent levels can be further reduced by segregating the cullet quench water from the cooling water system, recycling the cullet quench water through a gravity separator, and treating the blowdown using dissolved air flotation. Suspended solids and oil concentrations will increase in the recirculation system, but the dissolved solids level will again be limiting. No machine pressed and blown glass plants are presently recycling cullet quench water, but the same technology used for container plants should apply.

Using a maximum dissolved solids level of 1,700 mg/l, only eight cycles of concentration are possible as compared to 20 cycles for container manufacturing. The decrease in allowable cycles results from the higher dissolved solids increase reported per cycle in the data with regard to the machine pressed and blown glass manufacturing subcategory. Eight cycles correspond to a blowdown rate of 370 l/metric ton (88 gal/ton).

Further study by the industry will probably indicate that substantially higher dissolved solids levels are acceptable, thereby substantially reducing blowdown requirements. Effluent concentrations of 25 mg/l for both oil and suspended solids can be achieved. This is equivalent to 9 g/metric ton (0.018 lb/ton) at the typical blowdown. Approximately 720 liters/day (190 gal/day) of 3% sludge is produced.

Diatomaceous Earth Filtration (Alternative C): Diatomaceous earth filtration may be employed to further reduce the oil and suspended solids in the dissolved air flotation discharge to less than 5 mg/l. This is equivalent to 1.8 g/metric ton (0.0036 lb/ton). Sludge production is approximately 50 l/day (13 gal/day) at 15% solids.

The typical machine pressed and blown glass manufacturing plant may be located in any part of the country and may be 50 or more years old. The daily production is 90.9 metric tons (100 tons). Cullet quenching and noncontact cooling waters are not segregated. Costs and effluent quality for three treatment alternatives are summarized in Table 79.

TABLE 79: WATER EFFLUENT TREATMENT COSTS–MACHINE PRESSED AND BLOWN GLASS MANUFACTURING

Alternative Treatment or Control Technologies:	A	B ($1000)	C
Investment	0	187.	214.
Annual Costs:			
Capital Costs	0	15.	17.2
Depreciation	0	9.4	10.8
Operating and Maintenance Costs (excluding energy and power costs)	0	16.2	22.6
Energy and Power Costs	0	1.2	2.6
Total Annual Cost	0	41.8	53.2

Effluent Quality:

Effluent Constituents	Raw Waste Load	Resulting Effluent Levels		
Flow (l/metric ton)	5630	5630	370	370
Oil (g/metric ton)	56	56	9	1.8
Suspended Solids (g/metric ton)	140	140	9	1.8
Flow (l/sec)	5.9	5.9	.39	.39
Oil (mg/l)	10	10	25	5
Suspended Solids (mg/l)	25	25	25	5

Source: EPA (47)

Alternative A – Existing Treatment and Control

Alternative A involves no additional treatment. These effluent levels are readily achievable by all plants within this subcategory through normal maintenance and clean-up operations within the plant and represent the raw waste loadings expected from a typical machine pressed or blown ware plant. Improved housekeeping may be required at some plants to achieve the typical effluent levels, while others may elect to provide end-of-pipe treatment in the form of some type of sedimentation system with oil removal capabilities. It is felt however, that in-plant techniques will be a more effective and a less expensive means of achieving these effluent levels.

Costs – No additional cost.
Reduction Benefits – Upgrading of all effluent discharges to this level.

Alternative B – Recycle with Dissolved Air Flotation of the Blowdown

Alternative B involves the segregation of noncontact cooling water from cullet quench water, recirculation of the cullet quench water following gravity oil separation, and treatment of the 13% cullet quench system blowdown by dissolved air flotation.

Costs – Incremental investment costs are $187,000 and total annual costs are $41,800 over Alternative A.

Reduction Benefits – The incremental reductions of oil and suspended solids over Alternative A are 84% and 94%, respectively.

Alternative C – Diatomaceous Earth Filtration

Alternative C involves the treatment of the effluent from the dissolved air flotation unit by employing diatomaceous earth filtration.

Costs – Incremental investment costs are $27,000 and total annual costs are $11,400 over Alternative B.

Reduction Benefits – The incremental reduction of oil and suspended solids compared to Alternative B is 80%. Total reductions are 97 and 98.7%, respectively.

Glass Tubing Manufacturing

The manufacture of tubing consists of melting raw materials and forming the molten glass on a rotating mandrel or other forming device. The partially formed tubing is then drawn into lengths and cut by scribing or by thermal shock. The major process steps and points of water usage are illustrated in Figure 77.

FIGURE 77: GLASS TUBING MANUFACTURING

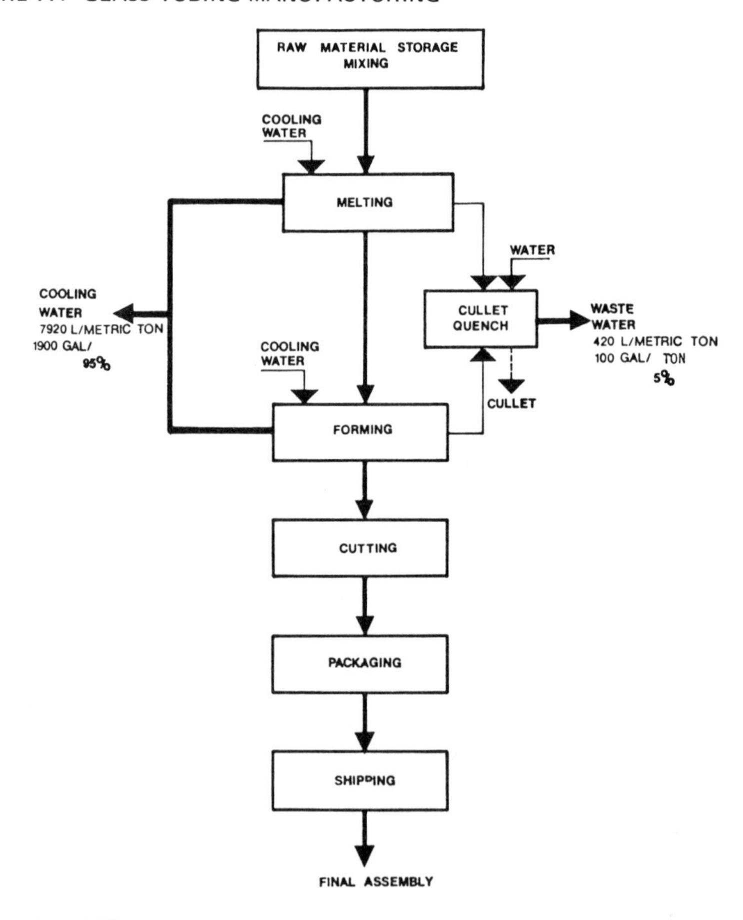

Source: EPA (47)

The only process water used in the manufacturing of glass tubing is for cullet quenching. Cullet quenching is infrequent compared with that amount common to the other subcategories and is done only when a break or disruption occurs in the drawing process. During this period, glass is wasted at the same rate that tubing is drawn so that a constant flow through the furnace is maintained.

Cooling water is primarily used for noncontact cooling of furnace walls and mandrel transmissions. The typical flow of noncontact cooling water is 7,920 l/metric ton (1,900 gal per ton) and accounts for about 95% of total plant water usage.

Cullet quenching is required only when there is a break or disruption in the drawing process. During a stoppage, molten glass continues to run over the mandrel or forming device, but is formed into a ribbon by two rollers that are cooled by a spray of water. The cullet ribbon and quench water drop to a segregated storage area in the melting tank basement. The quenching system is activated only when required. The typical flow is 420 l per metric ton (100 gal/ton) and accounts for 5% of the total typical flow.

Some plants use a cullet quench system similar to that employed at glass container and machine pressed and blown glass plants. A continuous stream of water is discharged into the quench carts so that quench water will be available when necessary. This type of quenching system requires a considerably greater volume of water than the periodic system and is not considered typical. The periodic cullet quench system is more applicable to the tubing manufacturing process because the wasting of cullet is infrequent.

Some typical characteristics of the combined noncontact cooling and cullet quench wastewaters resulting from tubing manufacturing are listed in Table 80. In all cases, except for pH, the values listed are the quantities added to the water as a result of the manufacturing process. Background levels in the influent water have been subtracted. Oil and suspended solids are the significant wastewater parameters. COD is contributed by the oil.

In most plants, noncontact cooling water and cullet quench water streams are discharged as a combined waste stream. Flows range from 3,340 l/metric ton (800 gal/ton) to 9,910 l per metric ton (2,380 gal/ton). The typical flow of 8,340 l/metric ton (2,000 gal/ton). The high flow is due to the use of once-through noncontact cooling water.

Chemical oxygen demand results from oil contamination of the noncontact cooling water. The typical COD is 0.08 kg/metric ton (0.16 lb/ton). This is a concentration of 10 mg/l and is not considered significant.

Suspended solids are added to the wastewater during cullet quenching. Fine glass and miscellaneous solid particles are picked up in the quench tank and discharge trenches leading to the sewer. The typical suspended solids loading is 0.225 kg/metric ton (0.45 lb/ton).

The typical oil loading is 0.085 kg/metric ton (0.17 lb/ton). Oil enters the waste stream from lubricating oil leaks in the noncontact cooling water system. The manufacturing methods used to form tubing do not require shears and, therefore, the oil associated with shear spraying is not a factor in this system.

Some additional information is available on the temperature and pH of glass tubing manufacturing wastewaters. The wastewater temperature increase due to the manufacture of glass tubing is 4.5°C (8°F). The wastewater pH is 7.9 and is in the acceptable range of 6 to 9.

No significant variations in wastewater volume or characteristics are experienced during plant start-up or shutdown, and there are no known toxic materials in the wastewater resulting from glass tubing manufacturing. As with all continuous furnaces, periodic furnace drainage requires large volumes of cullet quench water; however, temperature is the only significant pollutant parameter associated with this wastewater source.

TABLE 80: RAW WASTEWATER[a] GLASS TUBING MANUFACTURING

Flow	8340	l/metric ton	2000	gal/ton	
Temperature	4.5°C		8°F		
pH	7.9				
COD	0.08	kg/metric ton	0.16	lb/ton	10 mg/l
Suspended Solids	0.225	kg/metric ton	0.45	lb/ton	27 mg/l
Oil	0.085	kg/metric ton	0.17	lb/ton	10 mg/l

(a) Representative of typical glass tubing manufacturing waste waters. Absolute value given for pH, increase over plant influent level given for other parameters.

Source: EPA (47)

Process wastewater in the glass tubing manufacturing subcategory results from cullet quenching during periods when normal production has been interrupted. In most plants, cullet quench water is combined with noncontact cooling water prior to discharge. Suspended solids are the only significant pollutant in the quench water along with small quantities of tramp oil. Recycle with treatment of the blowdown, as illustrated in Figure 78, is a feasible method of treatment.

Existing Treatment and Control (Alternative A): Owing to the high quality and erratic discharge of cullet quench water, no plants presently treat this source of wastewater. All four of the plants for which data are available presently achieve a discharge of less than 225 g/metric ton (0.45 lb/ton) of suspended solids, and 85 g/metric ton (0.17 lb/ton) of oil. This corresponds to 27 mg/l of suspended solids and 10 mg/l of oil at the typical combined cullet quench water and noncontact cooling water flow of 8,340 l/metric ton (2,000 gal/ton).

Recycle with Diatomaceous Earth Filtration of Blowdown (Alternative B): Because cullet quench water accounts for only 5% of the combined flow, it is possible to further reduce the oil and suspended solids levels by segregating the cullet quench water from the noncontact cooling water, recycling the quench water through a cooling tower, and treating the blowdown using diatomaceous earth filtration. A flow over the cooling tower of 12.6 liters/second (200 gpm) for the typical plant is assumed.

Assuming a 5% blowdown, the discharge will be 21 l/metric ton (5 gal/ton). The minimum allowable blowdown is unknown because this technology is not presently employed in the glass tubing industry, but 5% is considered a conservative estimate. Suspended solids will probably be limiting because only negligible dissolved solids increases were noted in the available data. It is anticipated that at least a portion of the suspended solids can be removed in a glass trap associated with the collection sump.

It may be possible to use a tank rather than a cooling tower provided sufficient water can be stored to sufficiently dissipate the heat in the glass to be quenched. Information to calculate the required storage volume is not available and, therefore, a cooling tower is assumed.

Blowdown from the recycling system can be treated at a constant rate using diatomaceous earth filtration. Approximately 15 l/day (4 gal/day of 15% solids sludge will be produced. Diatomaceous earth filtration is used to treat boiler condensate and is readily transferable to the glass tubing manufacturing subcategory.

FIGURE 78: WASTEWATER TREATMENT–GLASS TUBING MANUFACTURING

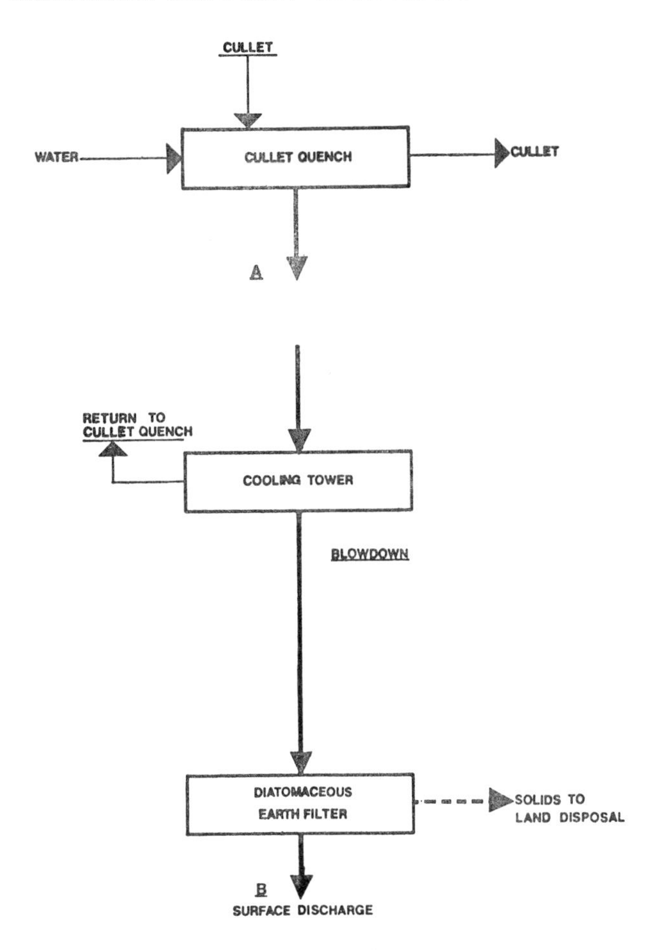

Source: EPA (47)

Disposal of the blowdown in the batch is an alternative to treatment and will allow zero wastewater discharge. The typical blowdown is approximately 1.5% by weight of the furnace fill, well within the range of the 3% of water added when batch wetting is used.

The typical glass tubing manufacturing plant may be located in any part of the country and is at least 10 years old. The daily production at the plant is approximately 90.9 metric tons (100 tons). Costs and effluent quality for the two treatment alternatives are summarized in Table 81.

Alternative A – Existing Treatment and Control

Alternative A involves no additional treatment. There are no plants at present which treat cullet quench water and all plants for which data are available achieve these effluent levels.

Costs – No additional costs.
Reduction Benefits – None.

Alternative B – Recycle with Diatomaceous Earth Filtration of Blowdown

Alternative B involves the recirculation of the cullet quench water stream through a cooling tower. The cooling tower blowdown is treated by diatomaceous earth filtration.

Costs – Incremental investment costs are $97,600 and total annual costs are $22,700 over Alternative A.

Reduction Benefits – Almost complete oil and suspended solids removal is obtained.

TABLE 81: WATER EFFLUENT TREATMENT COSTS–GLASS TUBING MANUFACTURE

Alternative Treatment or Control Technologies:	($1000) A	B
Investment	0	97.6
Annual Costs:		
Capital Costs	0	7.8
Depreciation	0	4.9
Operating and Maintenance Costs (excluding energy and power costs)	0	9.7
Energy and Power Costs	0	0.3
Total Annual Cost	0	22.7

Effluent Quality:

Effluent Constituents	Raw Waste Load	Resulting Effluent Levels	
Flow (l/metric ton)	8340	8340	21
Oil (g/metric ton)	80	80	0.1
Suspended Solids (g/metric ton)	230	230	0.1
Flow (l/sec)	8.8	8.8	.022
Oil (mg/l)	10	10	5
Suspended Solids (mg/l)	27	27	5

Source: EPA (47)

Television Picture Tube Envelope Manufacturing

Television picture tube envelope manufacturing consists of melting the raw materials, forming the screen and funnel sections, adding the components necessary for the final assembly of the picture tube, and polishing the necessary screen and funnel surfaces. The major process steps and points of water usage are illustrated in Figure 79. Water is used in television picture tube manufacturing for cooling, quenching, abrasive polishing, edge grinding, and acid polishing. Noncontact cooling water is required in the forming section of the plant for the batch feeders, furnaces, presses, annealing lehrs, and other auxiliary equipment such as compressors and pumps. Once-through systems are used in all of the plants that submitted data. In most cases, a portion of the water discharged is used as quench water to cool molten glass during manufacturing interruptions or to quench defective pieces from the forming operation. In at least one plant, cooling water is recirculated as rinse water later in the manufacturing process.

FIGURE 79: TELEVISION PICTURE TUBE ENVELOPE MANUFACTURING

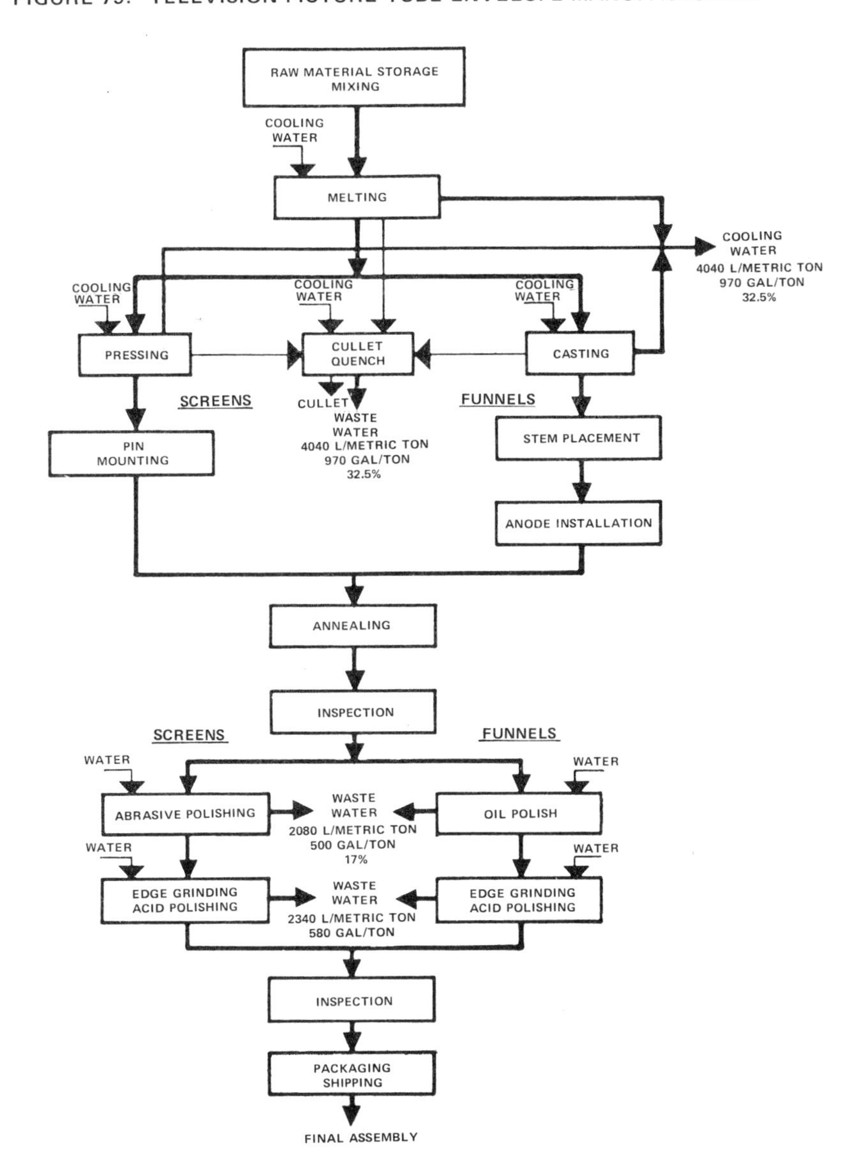

Source: EPA (47)

The typical flow for both the noncontact cooling and cullet quench water streams is 4,040 liters/metric ton (970 gal/ton). Each of these sources accounts for 32.5% of the total plant flow. Reported flows for the combined forming wastewater stream range from 7,230 l/metric ton (1,740 gal/ton) to 14,400 l/metric ton (3,460 gal/ton). The typical flow for a plant practicing reasonable water conservation is 8,080 l/metric ton (1,940 gal per short ton) and accounts for approximately 65% of total water usage. The funnel portion of the picture tube is abrasively polished using a diamond wheel machine with an oil lubricated grinding surface. After grinding, the funnel is rinsed with water.

The outer face of the picture tube screen is also abrasively polished. The screen face plate is polished in a step process using garnet, pumice, and rouge; all of the grinding compounds are in the form of a slurry. Between grindings with the various compounds, each screen is rinsed with water. The grinding compound slurry is recycled and only the blowdown from the slurry system is discharged. After the face has been ground, the connecting edge is ground and then beveled. The typical flow of abrasive wastewater is 2,080 l/metric ton (500 gal/ton) and is 17% of the total flow.

Following abrasive polishing and beveling, the connecting edges of both the funnel and screen are acid polished or fortified. The funnel is dipped into a combination of sulfuric acid and hydrofluoric acid. The two sections are then rinsed with water to remove the residual acid.

Constant overflow-type rinse tanks are generally used. The acid polishing step removes irregularities from the joining surfaces and allows a perfect seal when the screen and funnel are joined. Fume scrubbers are required in the acid polishing area and contribute significant amounts of fluoride to the wastewater. The combined typical wastewater flow for funnel and screen acid polishing is 2,340 l/metric ton (560 gal/ton) or 18% of the total flow.

Typical characteristics for the combined noncontact cooling, cullet quenching, abrasive and acid polishing wastewaters resulting from television picture tube envelope manufacturing are listed in Table 82. In all cases, except for pH, the values listed are the quantities added to the water as a result of the process. Background levels in the influent water have been subtracted. The significant parameters are suspended solids, oil, dissolved solids, fluoride, and lead.

Total wastewater flows, including noncontact cooling water, range from 11,100 to 14,400 liters/metric ton (2,670 to 3,460 gal/ton) or 1,590 to 3,260 cu m/day (0.42 to 0.86 mgd). The typical flow is 12,500 l/metric ton (3,000 gal/ton). The variation in flow rate depends primarily upon the amount of water used for once-through cooling.

TABLE 82: RAW WASTEWATER[(a)]—TV PICTURE TUBE ENVELOPE MANUFACTURE

Flow	12,500 l/metric ton	3000 gal/ton	
Temperature (b)	14°C	25°F	
pH	6-9		
COD	0.435 kg/metric ton	0.87 lb/ton	35 mg/l
Suspended Solids	4.2 kg/metric ton	8.4 lb/ton	335 mg/l
Dissolved Solids	3.25 kg/metric ton	6.5 lb/ton	260 mg/l
Oil	0.125 kg/metric ton	0.25 lb/ton	10 mg/l
Fluoride	1.8 kg/metric ton	3.6 lb/ton	143 mg/l
Lead	0.385 kg/metric ton	0.77 lb/ton	30 mg/l

(a) Represents typical TV picture tube envelope manufacturing process waste water prior to treatment. Absolute value given for pH; increase over plant influent level given for other parameters.

(b) Indication of approximate level only; insufficient data are available to define typical value.

Source: EPA (47)

Suspended solids are added to the wastewater in the form of glass particles and grinding slurry solids from edge grinding and abrasive polishing. Typical plant wastewater contains 4.2 kg/metric ton (8.4 lb/ton) of suspended solids. Dissolved solids are contributed to the wastewater stream from acid polishing and abrasive polishing. The typical loading is 3.25 kg/metric ton (6.5 lb/ton) of dissolved solids.

Fluoride is contributed by the rinse waters following acid polishing, fume scrubbing, and the periodic dumping of the concentrated acid. Hydrofluoric acid is used to polish the edges of both the screen and funnel portions of the picture tube envelope. The typical loading is 1.8 kg/metric ton (3.6 lb/ton) of fluoride.

Lead results from both abrasive and acid polishing. It is not clear if the lead in the abrasive waste stream is truly dissolved or in the form of colloidal particles, but standard analytical procedures show a significant concentration. Lead in the acid waste is assumed to result from the dissolution of the glass. The typical quantity of lead added to the wastewater is 0.385 kg/metric ton (0.77 lb/ton).

Oil is added to the wastewater as shear spray drippage into the quench water during forming operations, as lubrication leaks, and as funnel rinse water. The typical oil loading is 0.125 kg/metric ton (0.25 lb/ton).

Some information is also available on temperature, pH, and COD. The typical increase in COD is 0.435 kg/metric ton (0.87 lb/ton). The low organic content indicated by the COD is not considered significant. Owing to the segregation of the various waste streams, a typical value for the pH of the combined waste streams from a picture tube envelope plant is not available.

Typical pH values for the various process streams are acid polishing, 3.0; abrasive polishing, 9.5; cooling and quenching, 7.6. The cooling and quenching water contributes 65% of the combined plant flow. Owing to this high flow, it is estimated that the raw wastewater pH should be in the range of 6 to 9.

Television picture tube envelope manufacturing plants generally operate continuously, and no significant variations in wastewater volume or characteristics are experienced during plant start-up or shutdown. An additional source of wastewater from a picture tube envelope plant may be chrome-plating wastewater resulting from mold repair.

This is a very low-volume waste and is usually batch-treated at the plant or trucked from the plant for disposal. Available data indicate no chromium is added to the wastewater. Where applicable, the effluent limitations developed for plating wastes should be used.

Wastewaters are produced during both the forming and finishing of television picture tube envelopes. Cullet quench water contains low concentrations of oil and suspended solids and does not require further treatment. Finishing wastewater produced by acid and abrasive polishing of television tube screens and funnels, contains high concentrations of suspended solids, fluoride, and lead.

The acid and abrasive wastes are presently treated using lime precipitation, coagulation and sedimentation, but effluent levels can be further reduced by sand filtration followed by activated alumina adsorption. These treatments are illustrated in Figure 80.

Existing Treatment and Control (Alternative A): Television tube envelope manufacturing plants employ in-plant methods of water conservation and end-of-pipe treatment for fluoride, lead, and suspended solids removal. Because many of the plants have been built within the last 10 years and all within the last 25 years, relatively good water conservation is practiced. Abrasive grinding slurries are recycled to recover usable abrasive material and only the particles too small to be of further value are discharged. Rinse waters are recycled where possible by using countercurrent or overflow-type rinse tanks. Some final rinses are once through because a high-quality water is required to prevent spotting.

FIGURE 80: WASTEWATER TREATMENT–TELEVISION PICTURE TUBE ENVELOPE MANUFACTURING

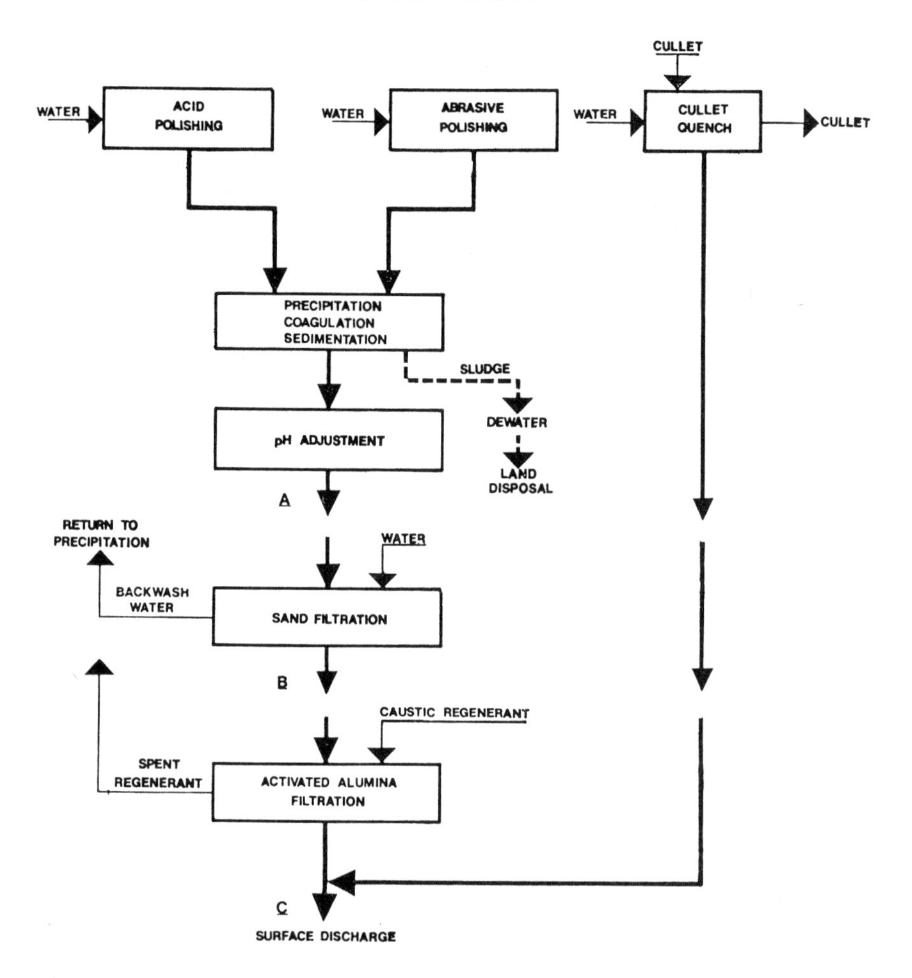

Source: EPA (47)

It may be possible to recycle this water for less critical uses. Spent acid solutions are either bled into the treatment system at a slow rate or returned to the manufacturer for recovery and recycling.

All of the plants, for which information was available, treat abrasive and acid polishing wastewaters by lime precipitation followed by combined coagulation and sedimentation of the calcium fluoride precipitate and abrasive waste suspended solids. The pH is reduced where necessary and sulfuric acid is generally used. Vacuum filtration is the most common method of dewatering, and the sludge is disposed of as landfill.

One plant reports sludge production of 20.9 metric tons/day (23 tons/day). The cullet quench waters are combined with noncontact cooling water prior to discharge and are not treated in any way. The typical combined noncontact cooling and cullet quench water flow has been found to total 8,080 l/metric ton (1,940 gal/ton) while the suspended solids and oil concentrations are below 5 mg/l. When combined with the treatment plant effluent, the total typical flow amounts to 12,500 l/metric ton (3,000 gal/ton).

Effluent levels of 130 g/metric ton (0.25 lb/ton) for suspended solids and oil, 65 g/metric ton (0.13 lb/ton) for fluoride, and 4.5 g/metric ton (0.009 lb/ton) for lead can be achieved using existing treatment methods and equipment. These values are equivalent to concentrations in the treatment plant effluent of 15 mg/l of fluoride and 1 mg/l of lead and concentrations in the combined treated and cullet quench water streams of 10 mg/l for oil and suspended solids.

The fluoride and lead concentrations in the combined flow are 5.2 mg/l and 0.35 mg/l, respectively. Of the four television picture tube envelope manufacturing plants for which data were available, three of the four presently achieve the above discharge level for suspended solids, two of three for oil, three of four for fluoride, and all meet the discharge level for lead. All plants can achieve these levels by upgrading the operation of existing treatment systems and improving housekeeping to minimize pollutant discharge from the forming area.

Sand Filtration (Alternative B): Fluoride and lead precipitates that are not removed during sedimentation may be further reduced by filtering the lime-treated effluent using sand or graded media. The filter backwash can be returned to the head of the lime treatment system and, therefore, no additional sludge handling equipment is required. Filtration will reduce the fluoride to 10 mg/l, the lead to 0.1 mg/l and the suspended solids to less than 5 mg/l.

The total plant discharge, including the treated effluent and the cullet quench water, will be reduced to 60 g/metric ton (0.12 lb/ton) for suspended solids, 45 g/metric ton (0.09 pound/ton) for fluoride, and 0.45 g/metric ton (0.0009 lb/ton) for lead. The concentration of pollutants in the total typical plant discharge for this level of treatment will be 5 milligrams/liter for suspended solids, 10 mg/l for oil, 3.5 mg/l for fluoride, and 0.035 mg per liter for lead.

Filtration of wastewater is not presently practiced in the pressed and blown glass industry, but is a commonly employed treatment method used in the water treatment industry following lime softening.

Activated Alumina Filtration (Alternative C): Reduction of fluoride to 2.0 mg/l can be accomplished by passing the effluent from the sand filter through a bed of activated alumina. The activated alumina may be regenerated with sodium hydroxide; rinsing with sulfuric acid may be necessary to reduce causticity.

The regenerants can be returned to the head of the lime treatment system for removal of the fluoride. With this technology, the fluoride discharge will be reduced to 9 g/metric ton (0.018 lb/ton) and the concentration of fluoride in the total typical plant discharge will be 0.71 mg/l.

Activated alumina is not presently used in the pressed and blown glass segment, but has been successfully used for many years at several potable water treatment plants in the United States. Experiments have indicated that the higher pH associated with lime treatment will not adversely affect the fluoride removal capability. All plants should be able to reduce the average effluent of fluoride wastewaters to 2.0 mg/l using this technology.

The typical television picture tube envelope manufacturing plant can be located in any part of the country and is at least 10 years old. The daily production of the plant is 227 metric tons (250 tons). Costs and effluent quality for the three treatment alternatives are summarized in Table 83. The effluent values are for the combined cullet quench and finishing wastewater streams.

TABLE 83: WATER EFFLUENT TREATMENT COSTS—TELEVISION PICTURE TUBE ENVELOPE MANUFACTURING

Alternative Treatment or Control Technologies:	($1000) A	B	C
Investment	0	67.0	231
Annual Costs:			
Capital Costs	0	5.4	18.5
Depreciation	0	3.4	11.6
Operating and Maintenance Costs (excluding energy and power costs)	0	8.7	37
Energy and Power Costs	0	0.9	1
Total Annual Cost	0	18.4	68.1

Effluent Quality:

Effluent Constituents	Raw Waste Load	Resulting Effluent Levels		
Flow (l/metric ton)	12,500	12,500	12,500	12,500
Oil (g/metric ton)	130	130	130	130
Suspended Solids (g/metric ton)	4200	130	60	60
Fluoride (g/metric ton)	1800	65	45	9
Lead (g/metric ton)	390	4.5	0.45	0.45
Flow (l/sec)	33	33	33	33
Oil (mg/l)	10	10	10	10
Suspended Solids (mg/l)	335	10	5	5
Fluoride (mg/l)	143	5.2	3.5	0.71
Lead (mg/l)	30	0.35	.035	.035

Source: EPA (47)

Alternative A – Existing Treatment and Control

Alternative A involves no additional treatment. Lime addition, precipitation, coagulation, sedimentation, and pH adjustment are presently used throughout the industry for removal of fluoride, lead, and suspended solids from finishing wastes. Cullet quench water is not treated.

Costs — No additonal cost

Reduction Benefits – Total reductions of suspended solids, fluoride, and lead are 97, 96, and 99%, respectively. The wastewater pH is adjusted to neutrality.

Alternative B – Sand Filtration

Alternative B includes sand filtration of the effluent from the lime precipitation system of Alternative A. Filter backwash water is recycled back to the lime precipitation system.

Costs – Incremental investment costs are $67,000 and total annual costs are $15,400 over Alternative A.

Reduction Benefits – The incremental reductions of suspended solids, fluoride, and lead over Alternative A are 54, 31, and 90%, respectively. Total reductions of suspended solids, fluoride, and lead are 98.6, 97.5, and 99.9%, respectively.

Alternative C – Activated Alumina Filtration

Alternative C involves the activated alumina filtration of the effluent from Alternative B. Spent caustic regenerant is recycled back to the lime precipitation system.

Costs – Incremental investment costs are $164,000 and total annual costs are $52,700 over Alternative B.

Reduction Benefits – The incremental reduction of fluoride over Alternative B is 80%. Total reductions of suspended solids, fluoride, and lead are 98.6, 99.5, and 99.9%, respectively.

Incandescent Lamp Envelope Manufacturing

Incandescent lamp envelope manufacturing consists of melting raw materials and forming the molten glass with ribbon machines into clear incandescent lamp envelopes. Many of the clear envelopes are then frosted or etched with a hydrofluoric acid solution. The major process steps and points of water usage are listed in Figure 81.

Process water is used in the manufacturing of incandescent lamp envelopes for cullet quenching and for rinsing frosted bulbs. The frosting wastewater stream is the major source of pollutants and contains high concentrations of both fluoride and ammonia. Cullet quench water is required to cool the wasted molten glass and to quench imperfect lamp envelopes. Quenching practices are similar to those of other pressed and blown glass plants.

Noncontact cooling water from batch feeders, melting furnaces, ribbon machines, and other auxiliary equipment is used as a source of quench water. Additional wastewater is contributed by the emulsified oil solution that is sprayed on the ribbon machine blowpipes and bulb molds. The excess of this oil-water emulsion flows to the cullet quenching area and is discharged with the quench water. Cullet quenching contributes approximately 57% of the total wastewater flow in the typical plant.

Frosting imparts an etched surface inside the lamp envelope that improves the light diffusing capabilities of the light bulb. The frosting solution contains hydrofluoric acid, fluoride compounds, ammonia, and other constituents, but the exact formulation is proprietary. The percentages of lamp envelopes frosted at a given plant range from 40 to 100%.

In the frosting operations, the solution is sprayed on the inside of the bulb and then removed by several countercurrent water rinses. High fluoride and ammonia concentrations in the rinse water result from frosting solution carry-over. Fume scrubbers are required in the frosting area and contribute significant amounts of fluoride and ammonia to the frosting wastewater. Frosting wastewater accounts for approximately 43% of the total flow in a plant where 100% of the envelopes are frosted.

Cullet quenching wastewater and frosting wastewater from an incandescent lamp envelope manufacturing plant must be classified and characterized separately because the percentage of bulbs frosted varies from plant to plant. The discharge from each of these sources must be added to obtain the total plant discharge.

Cullet quenching wastewater is characterized by low concentrations of oil, suspended solids, and COD, while the frosting waste contains high concentrations of fluoride and ammonia. Typical wastewater volumes and characteristics are summarized in Table 84. The typical cullet quenching wastewater flow is 4,500 l/metric ton (1,080 gal/ton) and the typical frosting wastewater flow is 3,420 l/metric ton frosted (820 gal/ton frosted).

FIGURE 81: INCANDESCENT LAMP GLASS MANUFACTURING

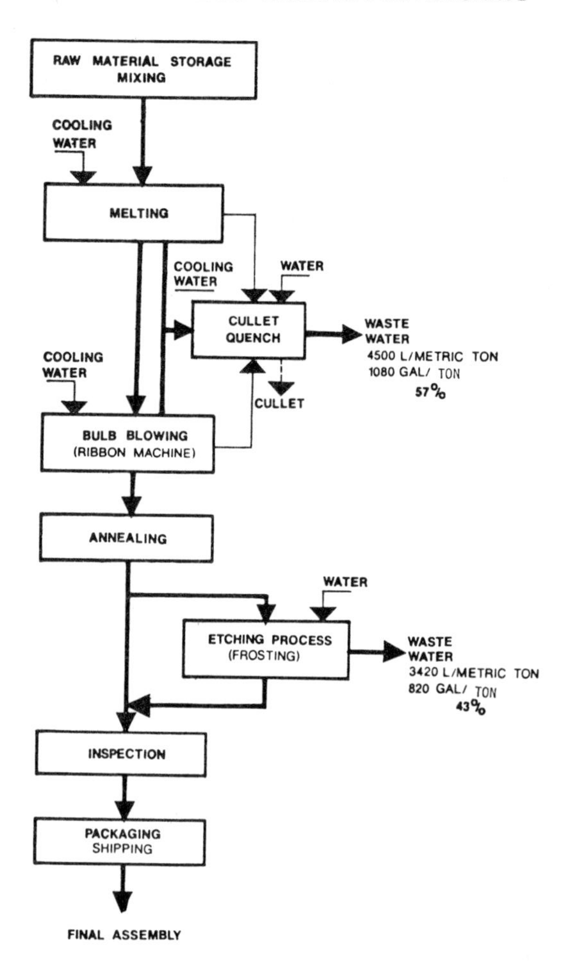

Source: EPA (47)

The flow is variable in accordance with water conservation practices and the quantity of once-through cooling water used. Reported combined quenching and frosting wastewater flows range from 5,420 l/metric ton pulled (1,300 gal/ton pulled) to 8,340 l/metric ton pulled (2,000 gal/ton pulled) or 570 to 1,510 cu m/day (0.15 to 0.4 mgd).

Suspended solids are generated by cullet quenching and by frosting of lamp envelopes. Fine glass particles are discharged with the cullet quench water and a significant concentration of suspended solids is contributed by the frosting rinse water. The typical suspended solids produced by cullet quenching is 0.11 kg/metric ton (0.23 lb/ton) and by frosting is 0.34 kg/metric ton frosted (0.68 lb/ton frosted).

Oil is contained in significant concentrations only in the cullet quench water and results from the residual emulsified oil used to spray the ribbon machine blowtips and from lubricating oil leaks. The typical loading is 0.11 kg/metric ton (0.23 lb/ton).

TABLE 84: RAW WASTEWATER[a]—INCANDESCENT LAMP GLASS MANUFACTURE

		Cullet Quenching			
Flow	4500	l/metric ton	1080	gal/ton	
Temperature	8°C		14°F		
pH	8.6				
COD	0.11	kg/metric ton	0.23	lb/ton	25 mg/l
Suspended Solids	0.11	kg/metric ton	0.23	lb/ton	25 mg/l
Oil	0.11	kg/metric ton	0.23	lb/ton	25 mg/l
		Frosting			
Flow	3420	l/metric ton	820	gal/ton	
Temperature	38°C		100°F		
pH	3.0				
COD	0.085	kg/metric ton	0.17	lb/ton	25 mg/l
Suspended Solids	0.34	kg/metric ton	0.68	lb/ton	100 mg/l
Fluoride	9.6	kg/metric ton	19.2	lb/ton	2800 mg/l
Ammonia	2.2	kg/metric ton	4.4	lb/ton	650 mg/l

(a) Representative of typical incandescent lamp glass manufacturing waste water. Absolute value given for pH and frosting temperature; increase over plant influent level given for other parameters.

Source: EPA (47)

Fluoride is contributed to the wastewater by the frosting solution carry-over and the discharge of fume scrubbing equipment. Spent frosting solution is usually regenerated and reused or disposed of separately and is not discharged to the wastewater stream. The typical fluoride content of the frosting wastewater is 9.6 kg/metric ton (19.2 lb/ton).

Ammonia is added to the plant wastewater as a result of frosting solution carry-over and the discharge from fume scrubbing equipment. Ammonia is one of the major constituents of the frosting solution and is apparently necessary in order to get the desired frosted effect. A considerable amount of ammonia vapors are picked up by the frosting area fume scrubber and then discharged to the wastewater flow. The typical discharge is 2.2 kg per metric ton (4.4 lb/ton).

Some information pertaining to COD, pH, and temperature is also included in Table 84. The typical COD concentration in both the cullet quench and frosting wastewater streams is only 25 mg/l and is not considered significant. The temperature increase during cullet quenching is similar to that obtained in the other subcategories. Frosting rinse water is heated and the 38°C (100°F) discharge temperature remains fairly constant.

Lamp glass envelope plants usually operate 24 hr/day and 5 days/week. Clear bulb production is continuous throughout the week. The frosting operation is intermittent and is related to consumer demand. No significant variations in the wastewater volume or characteristics of cullet quench or frosting wastewaters are experienced during plant start-up or shutdown.

Fluoride and ammonia nitrogen discharged at the concentrations typical of the raw wastewater are toxic and must be reduced. The furnaces are drained every 3 to 5 years for rebuilding and require excessive cullet quench water during the draining period. Wastewaters are produced during both forming and frosting in the manufacture of incandescent lamp envelopes. Cullet quench waters contain small quantities of oil and suspended solids, and frosting wastewaters contain moderate concentrations of suspended solids and high concentrations of fluoride and ammonia. Frosting wastes are presently treated for fluoride removal, but ammonia removal techniques are currently not employed. Treatment methods that may be employed to reduce the level of pollutants discharged by the incandescent lamp glass industry are illustrated in Figure 82.

FIGURE 82: WASTEWATER TREATMENT—INCANDESCENT LAMP GLASS MANUFACTURING

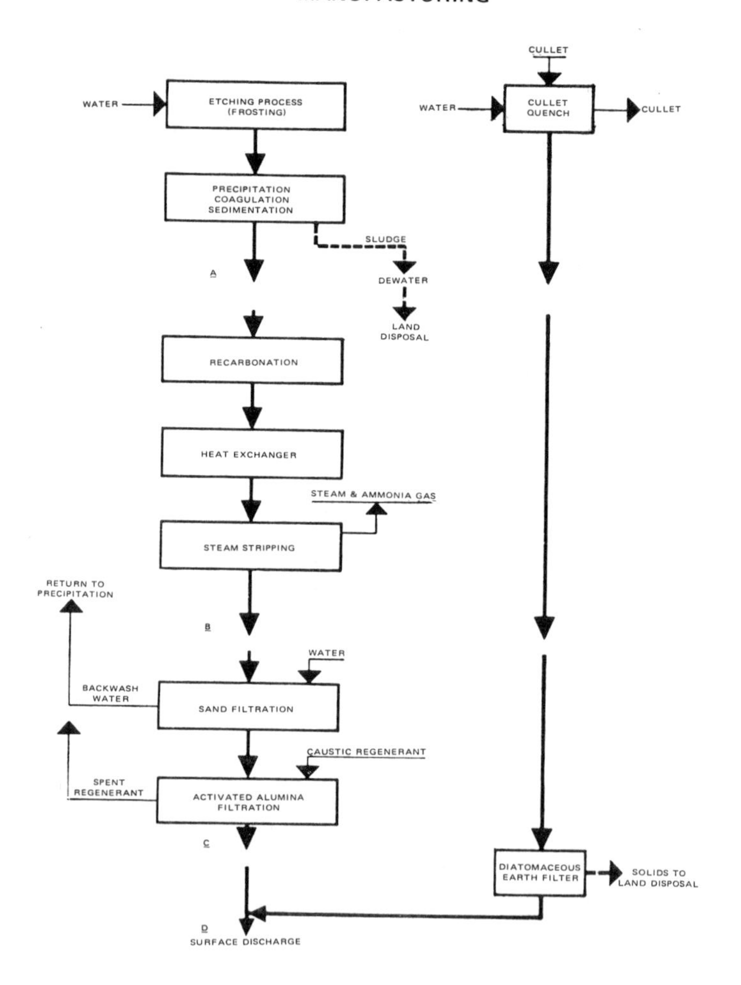

Source: EPA (47)

Existing Treatment and Control (Alternative A): Most treatment methods in the incandescent lamp glass industry are end-of-pipe methods. Cullet quench waters are discharged untreated or at some plants belt-type oil skimmers skim free oil from pump or discharged sumps. Frosting wastewaters are always treated using lime precipitation for fluoride and suspended solids removal, but this ineffective for ammonia removal. Some ammonia discharge is eliminated by separate disposal of the concentrated etching solution. At least one plant recovers the salts from this solution by evaporating most of the water and then allowing the sludge to air dry. Other plants truck the spent frosting solution to permanent storage. The percentage of lamps frosted varies from plant to plant, and therefore, the cullet quench and frosting wastewaters from this subcategory must be categorized separately.

Pollutants discharged in the cullet quench water as a result of forming will be expressed in terms of metric tons (tons) pulled from the furnace while the pollutant parameters contributed by frosting will be expressed in terms of the metric tons (tons) pulled for the frosting line. This value is calculated by multiplying the tons pulled by the percent of the plant output that is frosted. A plant frosting 85% of its production has been assumed for cost estimating purposes.

Little information is available on the quality of quench water, but it is apparent that all plants can achieve a level of 115 g/metric ton (0.23 lb/ton) suspended solids and oil. This is equivalent to 25 mg/l at the typical cullet quench water flow of 4,500 l/metric ton (1,080 gal/ton). Plants not presently achieving these levels can apply many of the methods for improved housekeeping described for the glass container manufacturing subcategory.

Much of the oil and suspended solids originates in the ribbon machine area. Careful attention to coolant spray and lubrication techniques should eliminate excessive oil discharges. It might be necessary, in some cases, to collect the highly contaminated wastewaters that occur during clean-up for separate disposal or treatment in the lime treatment system.

Frosting wastewaters are treated using lime to precipitate calcium fluoride, followed by flocculation and sedimentation. The effluent pH is lowered to at least 9.0 at most plants and neutralization is considered typical. It is possible, using existing equipment to treat frosting wastewaters to levels of 85 g/metric ton frosted (0.17 lb/ton frosted) for suspended solids and 68 g/metric ton frosted (0.14 lb/ton frosted) for fluoride.

These levels are equivalent to 25 mg/l, and 20 mg/l, respectively, at the typical flow of 3,420 l/metric ton frosted (820 gal/ton frosted). The typical concentration of 20 mg/l is higher than can be achieved with equivalent treatment technology in the television and handmade subcategories because of apparent interference by one or more constituents of the frosting solution. The data indicate consistently higher effluents from incandescent lamp envelope plants than are obtained in television picture tube envelope plants.

A typical plant that frosts 85% of its production would have a total effluent concentration of 25 mg/l for suspended solids, 15 mg/l for oil, 7.8 mg/l for fluoride, and 260 mg/l for ammonia. When the combined forming and frosting wastewaters are considered, one of the five plants for which data are available is presently achieving the recommended level for suspended solids and oil, two are achieving the recommended level for fluoride, and no plant significantly reduces ammonia.

Plants that are not achieving these effluent levels can upgrade their treatment systems using the methods discussed earlier in the treatment technology section. It is likely that, in many cases, excessive fluoride discharge is associated with poor suspended solids removal. Improvements to optimize suspended solids removal such as careful control of flocculation, addition of polyelectrolytes or other coagulant aids, sludge recycle, and reduced weir overflow rates may be employed in an existing wastewater treatment plant with a minimum of modification.

Ammonia Removal (Alternative B): The ammonia in the frosting wastewater can be reduced to a more acceptable level by steam stripping. This and other ammonia removal technologies are discussed in detail in the treatment technology section. One possible configuration is recarbonation, followed by a heat exchanger, and then the stripping column. Recarbonation will stabilize the excess calcium in the lime treatment discharge and control pH. Further experimentation will be required to determine the optimum location of the recarbonation step. Ammonia removal efficiency increases as the pH increases, but the calcium may precipitate in the stripping column and heat exchanger and form calcium carbonate scale. It is probable that a trade-off exists between ammonia removal efficiency and scaling. It is possible that recarbonation will be more advantageous subsequent to steam stripping. Purchased CO_2 is assumed in the cost estimate, but the melting furnace stack gas is rich in CO_2 and should be considered as a possible source. The heat exchanger will preheat the water entering the stipper while cooling the water being discharged, thus minimizing fuel requirements.

A packed or tray type column can be used. It is estimated that one pound of steam will be required for each gallon of water treated. Additional plant boiler capacity to meet this requirement is assumed to be a necessary expense. The waste heat discharged up the melting tank stack may be a potential source of heat, but this possibility can only be hypothesized pending further investigation by the industry.

The stripped ammonia vapor discharge may be above the threshold of odor, in which case it should be vented to the atmosphere through the melting tank stacks. Frosting wastewater ammonia levels can be reduced from 2.2 kg/metric ton frosted (4.4 lb/ton frosted) to 0.100 kg/metric ton frosted (0.20 lb/ton frosted) using this technology. This corresponds to an effluent concentration of 30 mg/l at the typical flow.

The alternative methods of ammonia removal discussed earlier in this section should also be carefully investigated before an ammonia removal system is chosen. Air stripping has been employed with some success in several domestic sewage treatment plants and may have potential in the glass industry.

Ion-exchange appears to have potential as a polishing step following air or steam stripping, but is still in the experimental stage and, therefore, has not been recommended. Steam stripping is a demonstrated technology and is presently being successfully used for ammonia removal in both the petroleum and fertilizer industries.

Sand and Activated Alumina Filtration (Alternative C): Fluoride in the frosting wastewater may be further reduced using sand filtration followed by activated alumina filtration. It may be possible for the activated alumina to serve the dual function of filtering suspended solids and adsorbing fluoride, but this is doubtful at the anticipated suspended solids loading.

Both the filter backwash and activated alumina regenerants can be returned to the head of the lime precipitation system for treatment and disposal. Suspended solids can be reduced to 17 g/metric ton frosted (0.034 lb/ton frosted) and fluoride to 7 g/metric ton frosted (0.014 lb/ton frosted) using this technology. These loadings are equivalent to 5 mg/l and 2 mg/l, respectively, at the typical frosting wastewater flow rate. This technology is not presently employed in the pressed and blown glass industry, but has been used for many years for potable water treatment.

Diatomaceous Earth Filtration (Alternative D): The oil and suspended solids in the cullet quench water can be reduced using diatomaceous earth filtration. The cullet quench water troughs can be intercepted and the water filtered through an oil-absorptive diatomaceous earth medium. A dry discharge-type filter will produce a sludge cake suitable for landfill. Approximately 0.54 cu m/day (0.7 cu yd/day) of 15% solids sludge will be produced. With this technology, the oil and suspended solids concentrations can be reduced to 5 mg per l or 23 g/metric ton (0.045 lb/ton). A similar technology is presently practiced in at least one glass container plant.

The typical incandescent lamp glass manufacturing plant may be located in any part of the country and is at least 50 years old. Daily production is 159 metric tons (175 tons). Frosted envelopes account for 85% of the plant production, and clear envelopes make up the remainder of the plant production. Cost and effluent quality for the four treatment alternatives are summarized in Table 85. Effluent characteristics are given for the combined cullet quench and frosting wastewater flows.

TABLE 85: WATER EFFLUENT TREATMENT COSTS–INCANDESCENT LAMP ENVELOPE MANUFACTURING

Alternative Treatment or Control Technologies:		($1000)			
		A	B	C	D
Investment		0	470	624	697
Annual Cost:					
Capital Costs		0	37.6	49.9	55.7
Depreciation		0	23.5	31.2	34.8
Operating & Maintenance Costs (excluding energy & power costs)		0	63	81.6	90.9
Energy & Power Costs		0	116	116.5	118.4
Total Annual Costs		0	240	279	300
Effluent Quality:					
Effluent Constituents	Raw Waste Load	Resulting Effluent Levels			
Flow(l/metric ton formed)	4500	4500	4500	4500	4500
(l/metric ton frosted)	3420	3420	3420	3420	3420
Oil (g/metric ton formed)	115	115	115	115	23
Suspended Solids					
(g/metric ton formed)	115	115	115	115	23
(g/metric ton frosted)	340	85	85	17	17
Fluoride(g/metric ton)	9600	68	68	7	7
Ammonia(g/metric ton)	2200	2200	100	100	100
Flow (l/sec)	13.7	13.7	13.7	13.7	13.7
Oil (mg/l)	15	15	15	15	3
Suspended Solids (mg/l)	54	25	25	17	5
Fluoride (mg/l)	1110	8	8	1	1
Ammonia (mg/l)	260	260	12	12	12

Source: EPA (47)

Alternative A – Existing Treatment and Control

Alternative A involves no additional treatment. Lime addition, coagulation, precipitation, and sedimentation are presently used throughout the industry for removal of fluoride and suspended solids from frosting wastes. Oil skimmers are employed for oil removal from cullet quench water. Some plants may have to improve housekeeping techniques to meet these effluent levels.

Costs – No additional costs
Reduction Benefits – Total reductions of suspended solids and fluoride are 56 and 99.3%, respectively.

Alternative B – Ammonia Removal

Alternative B involves ammonia removal by steam stripping of the effluent from Alternative A. This alternative also includes recarbonation and a heat exchanger. Recarbonation may be required for pH adjustment and also to prevent scaling in the stripping unit. A heat exchanger is used in conjunction with the steam stripping unit to maximize the efficiency of stripping and to reduce the discharge temperature of the treated wastewater.

Costs – Incremental investment costs are $470,000 and total annual costs are $240,000 over Alternative A.

Reduction Benefits – The incremental reduction of ammonia compared to Alternative A is 95%. The treated wastewater pH is adjusted to 9.0. Oil, suspended solids, and fluoride remain at the levels achieved in Alternative A.

Alternative C – Sand and Activated Alumina Filtration

This alternative includes sand filtration of the effluent from Alternative B. Following sand filtration, the wastewater is passed through a bed of activated alumina to reduce the remaining fluoride in the wastewater.

Costs – Incremental investment costs are $154,000 and total annual costs are $39,000 over Alternative B.

Reduction Benefits – Incremental reductions are 90% for fluoride and 34% for suspended solids. Total reductions of fluoride and suspended solids are 99.9 and 71%, respectively.

Alternative D – Diatomaceous Earth Filtration

Alternative D employs diatomaceous earth filtration of the cullet quench water. The frosting wastewaters are not treated above the level of Alternative C.

Costs – Incremental investment costs are $73,000 and total annual costs are $21,000 over Alternative C.

Reduction Benefits – Incremental reductions are 70% for suspended solids and 80% for oil. Total reductions of oil, suspended solids, fluoride, and ammonia are 80, 91, 99.9, and 95%, respectively.

Hand Pressed and Blown Glass Manufacturing

Hand pressed and blown glass manufacturing consists of melting raw materials and forming the molten glass with hand presses or by hand blowing to make high-quality stemware, tableware, and decorative glass products. The major process steps and points of water usage are listed in Figure 83.

Process water and wastewater are used almost entirely for finishing in the hand pressed and blown glass industry. Negligible quantities of water are used for forming; noncontact cooling water is not required. There are at least eight finishing steps that may be employed in the handmade industry. Some plants employ several finishing steps while others use only one or two.

Finishing steps that require water and produce wastewater include: crack-off and polishing, grinding and polishing, machine cutting, alkali washing, acid polishing and acid etching. Several handmade plants also have machine presses. Wastewaters resulting from machine forming are covered in the machine pressed and blown subcategory. Some of the machine pressed products are finished using the methods covered under this subcategory.

Data on wastewater volumes and characteristics from the hand pressed and blown glass industry are almost nonexistent. Almost all of the data presented in this report were collected during the sampling program. A negligible amount of water is required for quenching and for partial cooling of the glass during some types of forming. Small water-filled tanks are used at some plants to collect waste glass and rejects. Wheelbarrows with no water may be used at other plants. Some types of glassware are blown in a mold partially submerged in water. Small tanks, approximately 19 liters (5 gallons) in size, are used.

The quench tanks and forming tanks are drained periodically. Crack-off is required to remove excess glass left over from the hand blowing of stemware. Crack-off can be done either manually or by machine. The top portion of the stemware is scribed, the scribed surface heated, the excess glass removed, and the cut edge ground on a carborundum or other type abrasive surface.

The grinding surface is sprayed by a continuous stream of water for cooling and to remove grinding residue. Grinding may be followed by acid polishing to remove the scratches and is considered part of the crack-off operation in this presentation. Polishing is accomplished by two hydrofluoric acid rinses followed by two water rinses. The combined cut-off and acid polishing wastewater flow is 9,920 l/metric ton (2,380 gal/ton).

FIGURE 83: HAND PRESSED AND BLOWN GLASS MANUFACTURING

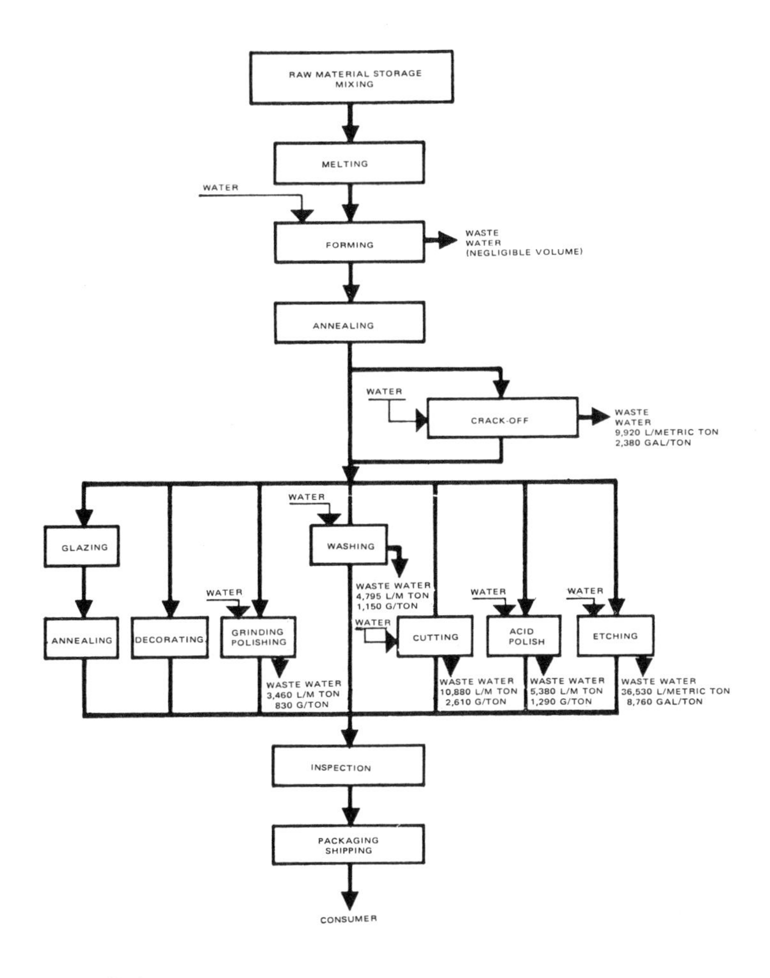

Source: EPA (47)

Abrasive grinding and polishing is a common finishing step and may be used to repair imperfect glassware. Water is required for cooling and lubrication when grinding wheel stones and belt polishers are used. Abrasive polishing may also be used and involves mechanical brushing with an abrasive slurry. The residual slurry is removed in a booth or wash sink. The grinding and polishing flow rates observed were 3,460 l/metric ton (830 gallons per ton). Designs may be cut into tableware or stemware. Water is required for lubrication and cooling of the cutting surface and to flush away glass particles. The observed flow from the machine cutting operation was 10,880 l/metric ton (2,610 gal/ton).

Acid polishing is another finishing operation that may be applied to handmade glassware. This improves the appearance of the glassware and removes rough edges. The glassware is dipped in hydrofluoric acid and then rinsed with water. The type of equipment used for acid polishing ranges from highly automated equipment to hand-dipped tubs and, consequently, the required volume of water varies. The observed acid polishing flow for a plant utilizing countercurrent rinsing was 5,380 l/metric ton (1,290 gal/ton).

Designs are etched onto some stemware. A pattern is stenciled on tissue paper using a proprietary mixture. The tissue is placed on the glass and then removed to leave the pattern. All parts of the ware, except for the pattern, are then coated with a wax mixture. At this point the glassware is ready for etching. Etching involves a number of steps including dipping in the etching solutions, rinsing, wax removal, additional rinsing, treatment with a cleaning solution, rinsing, nitric acid treatment to remove spots from the acid carryover, and final rinsing. The wastewaters result from the various rinsing steps. The acid and cleaning solution tanks are not drained. The observed flow from this type of system is 36,500 l/metric ton (8,760 gal/ton).

Final washing, prior to packing and shipment may be required for some products. An acid-alkali cleaning system is used for this purpose in at least one plant. The glassware first passes through an acid wash and then an alkali rinse followed by several hot water rinses. The flow from this unit is 4,795 l/metric ton (1,150 gal/ton). Finishing steps that do not involve water or wastewater are employed at many handmade glass plants and are generally referred to as glazing or decorating. Paint or some other coating is applied to the glassware and in many cases is baked onto the glass surface by reannealing.

Abrasive mold cleaning is employed at some plants. An abrasive slurry is sprayed on the molds at high pressure in a process similar to sandblasting. A small but undefined volume of high-suspended solids waste is produced. Following cleaning, the molds may be dipped in a rust preventative solution. This tank is not drained to the sewer. Fume scrubbers are required in the acid treatment areas and contribute significant fluoride to the acid polishing and etching wastewaters.

Observed wastewater characteristics for the finishing steps described above are listed in Table 86. In all cases, except for pH and some of the temperatures, the values listed are the quantities added to the water as a result of the manufacture of hand pressed and blown glassware. Concentrations in the influent water have been subtracted. Wastewater flows from hand pressed and blown glass manufacturing plants are highly variable and depend upon the quantity of glass finished and the finishing method employed. Reported values range from 0.15 cu m/day (40 gal/day) to 38 cu m/day (10,000 gal/day). Owing to the variation in finishing methods and the percentage of product finished, it is impossible to define a typical flow for the industry.

Grinding, polishing, and cutting are major sources of suspended solids. Lesser quantities are generated by the other finishing steps. Machine cutting and grinding and polishing contribute approximately 28 kg/metric ton (56 lb/ton) and 15 kg/metric ton (30 lb/ton) to the wastewaters respectively. Fluoride discharges result from crack-off and hydrofluoric acid polishing, hydrofluoric acid polishing, and hydrofluoric acid etching. Loadings expressed in terms of production vary significantly from 1.93 kg/metric ton (3.85 lb per ton) for crack-off and polishing to 10.6 kg/metric ton (21.3 lb/ton) for acid polishing, and 17 kg/metric ton (34 lb/ton) for acid etching.

TABLE 86: RAW WASTEWATER[a]–HAND PRESSED AND BLOWN GLASS MANUFACTURING

		Crack-Off and Polishing			
Flow	9920	l/metric ton	2380	gal/ton	
Temperature	2.8°C		5°F		
pH	3				
Suspended Solids	0.35	kg/metric ton	0.71	lb/ton	36 mg/l
Lead	0.010	kg/metric ton	0.019	lb/ton	0.96 mg/l
Fluoride	1.93	kg/metric ton	3.85	lb/ton	194 mg/l
		Grinding and Polishing			
Flow	3460	l/metric ton	830	gal/ton	
Temperature	2.8°C		5°F		
Suspended Solids	15	kg/metric ton	30	lb/ton	4350 mg/l
Lead	0.086	kg/metric ton	0.17	lb/ton	25 mg/l
		Machine Cutting			
Flow	10,880	l/metric ton	2610	gal/ton	
Temperature	1.6°C		3°F		
pH	10				
Suspended Solids	28	kg/metric ton	56	lb/ton	2580 mg/l
Lead	1.1	kg/metric ton	2.2	lb/ton	100 mg/l
		Alkali Washing			
Flow	4795	l/metric ton	1150	gal/ton	
Temperature (b)	57°C		135°F		
pH	11				
Suspended Solids	0.08	kg/metric ton	0.16	lb/ton	17 mg/l
		Acid Polishing			
Flow	5380	l/metric ton	1290	gal/ton	
Temperature (b)	46°C		114°F		
pH	2				
Suspended Solids	1.2	kg/metric ton	2.4	lb/ton	220 mg/l
Lead	0.17	kg/metric ton	0.33	lb/ton	31 mg/l
Fluoride	10.6	kg/metric ton	21.3	lb/ton	1980 mg/l
		Etching			
Flow	36,530	l/metric ton	8760	gal/ton	
Temperature (b)	33°C		91°F		
pH	4				
Suspended Solids	0.29	kg/metric ton	0.58	lb/ton	8 mg/l
Lead	0.29	kg/metric ton	0.58	lb/ton	8 mg/l
Fluoride	17	kg/metric ton	34	lb/ton	460 mg/l

(a) Representative of observed hand pressed and blown glass manufacturing waste water. Absolute value given for pH, increase over plant influent level for other parameters.

(b) Controlled temperature required for the process; therefore, absolute temperature given.

Source: EPA (47)

The differences are caused, at least in part, by variations in acid strength. The crack-off polishing solution is much less concentrated than the acid polishing or acid etching solutions. Lead is contained in all leaded glass finishing wastewaters. It is not clear if the lead in the abrasive waste streams is truly dissolved or is in the form of small glass particles, but standard analytical procedures show a significant concentration. Lead in the acid wastes is assumed to be in a soluble form. Other parameters that may be of significance include pH, temperature, dissolved solids, and nitrate. Raw wastewater pH values vary significantly depending on the source. No pH value is available for grinding and polishing but it is assumed the pH will be in the range of 8 to 10. Temperature increases are insignificant except where heated rinse waters are used. Dissolved solids are not reported, but significant concentrations may be anticipated in the acid polishing and etching wastewaters. Nitrates are discharged as a result of the rinsing steps following etching. Insufficient data are available to define the levels of discharge.

Hand pressed and blown glass manufacturing plants generally operate only one or two shifts per day, five days per week, and finishing is done only as necessary and varies with product demand. Rarely is all the finishing equipment available at a given plant in use at the same time. For these reasons, it is impossible to generalize the hand pressed and blown industry in terms of a typical plant.

Significant sources of wastewater in the hand pressed and blown glass manufacturing subcategory result from finishing operations. At least six wastewater-producing processes are presently used in the industry. These have been classified as crack-off and hydrofluoric acid polishing, grinding and polishing, machine cutting, alkali washing, hydrofluoric acid polishing, and hydrofluoric acid etching. Some plants employ all of the finishing steps, while others use only one or two, but grinding and polishing is probably the most frequently used. Owing to the variation in finishing steps, it is impossible to generalize the industry in terms of a typical plant.

The wastewater constituents requiring treatment are suspended solids, fluoride, and lead, but all of these are not contained in each type of wastewater. High and low pH values have also been observed, and neutralization may be required in some cases. Figure 84 illustrates the sequence of treatments that might be employed for a wastewater containing all of these constituents. This type of treatment system would apply to those plants which employ hydrofluoric acid finishing techniques to leaded or unleaded glass. Figure 85 illustrates the sequence of treatments that might be employed to a waste containing only suspended solids. This system would be applicable to those plants which produce leaded or unleaded glass and do not employ hydrofluoric acid finishing techniques.

Very limited data were available from the hand pressed and blown industry; therefore, the information presented in this subsection is almost entirely the result of plant visits and field sampling done as part of this study. Owing to the small size of the companies within the industry, the low wastewater volumes, the lack of significant quantities of cooling water that could be used for dilution, and the very limited data available, achievable effluent levels in the hand pressed and blown glass manufacturing subcategory are expressed in terms of milligrams per liter (mg/l).

Tables 87 and 88 present a summary of the current operating practices of the hand pressed and blown glass manufacturing subcategory. Forty-two plants were contacted with regard to treatment practices, type of glass produced, and finishing techniques employed. It should be noted that the majority (69%) of the plants either discharge to municipal systems or do not discharge process wastewater. Treatment practices for the remaining 31% of the subcategory vary from no treatment to sedimentation to batch lime precipitation.

Plants which employ hydrofluoric acid finishing techniques would have potential problems with regard to fluoride, suspended solids, and, in the case of leaded glass production, lead. Plants which do not employ hydrofluoric acid finishing techniques would have potential problems with regard to suspended solids. A treatment system for the removal of lead and fluoride from wastewater would include batch lime precipitation, sand filtration, and ion exchange, while for removal of suspended solids would include coagulation, sedimentation, and sand filtration. For this reason, two treatment schemes are discussed; the first is applicable to those plants which employ acid finishing techniques to leaded or unleaded glass, while the second is applicable to those plants which do not employ acid finishing techniques.

FIGURE 84: WASTEWATER TREATMENT–HAND PRESSED AND BLOWN GLASS MANUFACTURING

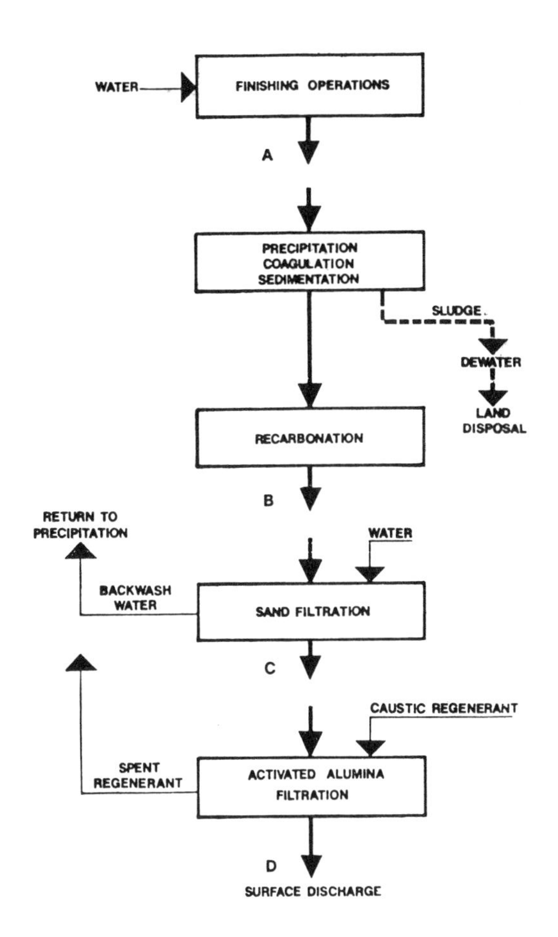

TABLE 87: CURRENT TREATMENT PRACTICES WITHIN THE HAND PRESSED AND BLOWN GLASS MANUFACTURING SUBCATEGORY

Treatment Practice	No. of Plants	Percentage of Subcategory
No Discharge	6	14.3
Treatment with Surface Discharge	7	16.7
No Treatment with Surface Discharge	6	14.3
Municipal Discharge	23	54.7
Total in Survey	42	100.0

Source: EPA (47)

FIGURE 85: WASTEWATER TREATMENT—HAND PRESSED AND BLOWN GLASS MANUFACTURING

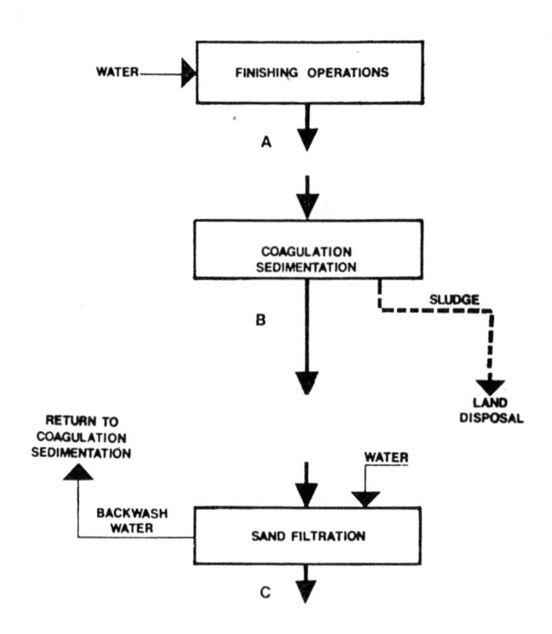

TABLE 88: CURRENT OPERATING PRACTICES WITHIN THE HAND PRESSED AND BLOWN GLASS MANUFACTURING SUBCATEGORY

Type of Glass Produced	No.	Percentage
Leaded Glass	4	9.5
Non-Leaded Glass	38	90.5
	42	100.0

Finishing Techniques	No.	Percentage
Employ HF	19	45.3
Do Not Employ HF	23	54.7
	42	100.0

Source: EPA (47)

First, treatment systems will be considered which are applicable to plants which employ hydrofluoric acid finishing techniques.

Existing Treatment and Control (Alternative A): Very few hand pressed and blown glass plants are presently treating wastewaters; however, a few plants have lime precipitation systems for fluoride and lead removal. In most cases, flows are low, less than 38 cu m per day (10,000 gpd). Significant quantities of pollutants may be discharged, however, and could have a detrimental effect on a small receiving stream.

Batch Precipitation and Recarbonation (Alternative B): Fluoride, lead, and suspended solids concentrations can be significantly reduced using batch lime precipitation followed by coagulation, sedimentation, and recarbonation for pH reduction and calcium stabilization. Using this system, the daily flow of wastewater might be collected in a tank equipped with a stirring mechanism. At the end of the day, lime and polyelectrolyte would be added to precipitate fluoride or lead where removal of these constituents was required. The tank would be slowly stirred for a sufficient time to allow optimum flocculation and then allowed to settle overnight. The following day the supernatant would be transferred to a second tank for recarbonation and additional sedimentation, and the sludge would be transferred to a holding tank where additional thickening would take place before the sludge was disposed of as landfill or to permanent storage.

Acid could be used in place of recarbonation for pH reduction, but dissolved solids levels would be increased rather than decreased. The achievable percent solids in the sludge would depend on the type of material treated and coagulant used. It is estimated that 10 to 15% solids can be achieved in a lime precipitation system. Effluent levels of 25 mg per l for suspended solids, 15 mg/l for fluoride, and 1.0 mg/l for lead are achievable using a batch system. At least one handmade plant is presently using batch lime precipitation, but is not neutralizing the effluent pH.

Sand Filtration (Alternative C): Precipitates and other particulates not removed by gravity separation can be further reduced by sand or graded media filtration. Additional suspended solids, fluoride, and lead can be removed using this technology. Effluent levels can be reduced to 10 mg/l for fluoride, 5 mg/l for suspended solids, and 0.1 mg/l for lead. Backwash waters can be returned to the batch treatment system for further treatment. No hand pressed and blown plants presently practice this technology, but filtration is widely used in the water treatment industry.

Activated Alumina Filtration (Alternative D): Activated alumina filtration is an available technology for further reducing fluoride concentrations. Following sand filtration, the wastewater may be passed through a bed of activated alumina to reduce the fluoride concentration to 2 mg/l. Sodium hydroxide may be used for regeneration and can be returned to the lime precipitation system for fluoride removal. This technology is not presently employed in the hand pressed and blown glass industry, but can be transferred from the water treatment industry.

Secondly, treatment systems will be considered which are applicable to plants which do not employ hydrofluoric acid finishing techniques.

Existing Treatment and Control (Alternative A): Many plants, where grinding and abrasive polishing are done, collect finishing wastewater in trenches with small traps which catch the gross solids. These are periodically cleaned and disposed of as a solid waste. In most cases flows are low, less than 11.4 cu m/day (3,000 gpd). The typical flow is 1.89 cu m/day (500 gpd).

Batch Coagulation and Sedimentation (Alternative B): Suspended solids concentrations can be significantly reduced using batch coagulation and sedimentation. Using this system, the daily flow of wastewater might be collected in a tank equipped with a stirring mechanism. Alum or some other coagulant would be added and the tank stirred slowly for a sufficient time to allow solids to settle. The following day the supernatant would be discharged and the sludge transferred to a holding tank where additional thickening would take place prior to sludge disposal. An effluent level of 25 mg/l for suspended solids is achieveable using a batch coagulation and sedimentation system. Many handmade glass plants employ sedimentation systems for solids control.

Sand Filtration (Alternative C): Particulates not removed by gravity separation can be further reduced by sand or graded media filtration. Additional suspended solids can be removed to an effluent level of 5 mg/l. Backwash waters can be returned to the batch treatment system for further treatment.

No typical plant can be developed for the hand pressed and blown glass manufacturing subcategory because of the wide variation in finishing steps applied to the handmade glass. The hypothetical plants assumed for cost estimating purposes may be located in any part of the country and are at least 50 years old. The first plant is one of the largest in the country and has a daily finished product output of 5.9 metric tons (6.5 tons). The plant employs all the finishing steps available at handmade glass plants and is representative of those plants which produce leaded or unleaded glass and employ hydrofluoric acid finishing techniques. The second hypothetical plant is representative of those handmade glass plants which produce leaded or unleaded glass and do not employ hydrofluoric acid finishing techniques. The cost and effluent quality for the treatment alternatives applicable to each hypothetical plant are listed in Tables 89 and 90.

TABLE 89: WATER EFFLUENT TREATMENT COSTS—HAND PRESSED AND BLOWN GLASS MANUFACTURING

Alternative Treatment or Control Technologies:		($1000)			
		A	B	C	D
Investment		0	284	325	371
Annual Costs:					
Capital Costs		0	22.7	26.0	29.7
Depreciation		0	14.2	16.2	18.5
Operating and Maintenance Costs (excluding energy and power costs)		0	15.6	18.6	21.4
Energy and Power Costs		0	2.6	2.7	2.7
Total Annual Cost		0	55.1	63.5	72.3
Effluent Quality:					
Effluent Constituents	Raw Waste Load	Resulting Effluent Levels			
Flow (l/sec)	0.61	0.61	0.61	0.61	0.61
pH (mg/l)	2	2	9	9	9
Suspended Solids (mg/l)	544	544	25	5	5
Fluoride (mg/l)	422	422	15	10	2
Lead (mg/l)	11.4	11.4	1	0.1	0.1

Source: EPA (47)

First, one can consider the economics of a treatment system applicable to plants which employ hydrofluoric acid finishing techniques.

Alternative A – Existing Treatment and Control

Alternative A is no wastewater treatment or control. Many plants do not need wastewater treatment or control because of the absence of waste-producing finishing steps or because of the low volume of discharge. Some plants have lime precipitation treatment facilities for the reduction of fluoride from hydrofluoric acid polishing and acid etching wastes. It is felt that for any plant discharging less than 0.19 cu m/day (50 gallons/day) of wastewater, treatment is impractical as other means of disposal are considerably less expensive (i.e., land retention or dust suppression).

Costs – None
Reduction Benefits – None

Alternative B – Batch Precipitation and Recarbonation

This alternative includes a batch lime precipitation system for reduction of suspended solids, fluoride, and lead from finishing wastewaters. The lime precipitation system effluent is recarbonated with carbon dioxide gas to adjust the treated wastewater to a neutral pH from the alkaline pH of the lime treatment process.

Costs – Incremental investment costs are $284,000 and total annual costs are $55,100 over Alternative A.

Reduction Benefits – Total reductions of suspended solids, fluoride, and lead are 95, 96, and 91%, respectively. The pH of the acidic waste is raised to an alkaline pH of 11 to 12 during lime treatment and then is lowered to a pH of 9 by recarbonation.

Alternative C – Sand Filtration

Alternative C involves the sand filtration of the effluent from Alternative B. The sand filtration system is similar to those employed at municipal water treatment works.

Costs – Incremental investment costs are $41,000 and total annual costs are $8,400 over Alternative B.

Reduction Benefits – Incremental reductions over Alternative B for suspended solids, fluoride, and lead are 80, 33, and 90%, respectively. Total reductions are 99.1% for suspended solids and lead, and 98% for fluoride. The wastewater pH is adjusted to 9.

Alternative D – Activated Alumina Filtration

This alternative includes activated alumina filtration of the effluent from Alternative C. Activated alumina filtration is employed for further reduction of the effluent fluoride concentration.

Costs – Incremental investment costs are $46,000 and total annual costs are $8,800 over Alternative C.

Reduction Benefits – The incremental reduction of fluoride is 80% over Alternative C. Total reductions of suspended solids, fluoride, and lead are 99.1, 99.5, and 99.1%, respectively.

TABLE 90: WATER EFFLUENT TREATMENT COSTS–HAND PRESSED AND BLOWN GLASS MANUFACTURING SUSPENDED SOLIDS REMOVAL

Alternative Treatment or Control		($1000)	
	A	B	C
Investment	0	48.7	54.3
Annual Costs:			
Capital Costs	0	3.9	4.3
Depreciation	0	2.4	2.7
Operating & Maintenance Costs (excluding energy & power costs)	0	5.3	8.0
Energy and Power Costs	0	0.3	0.3
TOTAL ANNUAL COST	0	11.9	15.3

Effluent Quality:

Effluent Constituents	Raw Waste Load	Resulting Effluent Levels		
Flow -- cu m/day	1.89	1.89	1.89	1.89
Suspended Solids(mg/l)	9600	9600	25	5

Source: EPA (47)

Then one can consider the economics of a treatment system applicable to plants which do not employ hydrofluoric acid finishing techniques.

Alternative A – Existing Treatment and Control

Alternative A does not involve any wastewater treatment or control. Many plants do not have any need for wastewater treatment or control because of the absence of waste-producing finishing steps or because of the low volume of discharge. Many plants employ some type of sedimentation system for solids control. It is believed that for any plant discharging less than 0.19 cubic meter/day (50 gallons/day) of wastewater, treatment is impractical as other means of disposal are considerably less expensive (i.e., land retention or dust suppression).

The raw wastewater suspended solids expressed in terms of grams (pounds) per production unit or concentration is impossible to typify, owing to the wide range of production methods employed in the subcategory. Approximately 9,600 mg/l were assumed for calculating sludge production, but the influent suspended solids concentration is not directly related to treatment costs. The typical flow is 1.89 cu m/day (500 gpd).

Costs – None.
Reduction Benefits – None.

Alternative B – Batch Coagulation and Sedimentation

The daily wastewater discharge is collected in one of two mixing tanks (one tank is treated and discharged while the other is filling). At the end of the day coagulants are added, and the mixture is flocculated. The treated wastewater is discharged following overnight sedimentation. Sludge is collected in a holding tank and eventually discharged as landfill.

Costs – Incremental investment costs are $48,700 and total annual costs are $11,900 over Alternative A.
Reduction Benefits – Suspended solids reduced to 25 mg/l.

Alternative C – Sand Filtration

Discharge from Alternative B is passed through sand filters for additional suspended solids reduction. Filter backwash is returned to the head of the system.

Costs – Incremental investment costs are $5,600 and total annual costs are $3,400 over Alternative B.
Reduction Benefits – Suspended solids reduced to 5 mg/l.

SOLID WASTE DISPOSAL

Landfilling of properly dewatered sludges from the flat glass industry is an appropriate means of disposal. The wastes are largely inorganic and incineration, composting, or pyrolysis would not be effective in reducing their volume. The dewatered solids are relatively dense and they are stable when used as fill material. If disposed of using proper sanitary landfill techniques, solids from flat glass manufacturing should cause no environmental problems.

With the exception of plate glass manufacturing, the volume of sludge associated with the various control and treatment technologies is relatively small. The lagoons used for plate glass suspended solids removal also serve as sludge disposal sites. The levees are generally raised to keep pace with the rising sediment level. At older plate plants large areas of low-lying land have been filled in. In some cases this is reclaimed as park land by spreading topsoil over the dry sludge solids.

Three types of waste solids are produced by the treatment systems indicated for the float, solid tempered automotive, and windshield manufacturing processes. These are (1) coagulation-sedimentation sludges associated with tempering wastewaters, and (2) spent diatomaceous earth, and (3) brine residue associated with at least one treatment alternative for each of the subcategories. The coagulation-sedimentation sludge is assumed to be dewatered by centrifuge to about 20% dry solids and the typical volume produced is estimated to be 0.38 cu m/day (13.5 cu ft/day).

Spent diatomaceous earth has an estimated moisture content of 85%, but is dry to the touch. This material is stable and should be suitable for landfill. Estimated production of diatomaceous earth waste is less than 0.23 cu m/day (8 cu ft/day) for each of the subcategories.

The salt residue that will be produced by a total recycle system will present the biggest disposal problem. To prevent ground water contamination, it must be permanently stored in lined basins. Only as much water as will evaporate can be allowed into the basin. The land used for salt storage will be permanently spoiled. The salt residue produced by the tempering and laminating processes is conservatively estimated to be 0.56 cu m/day (20 cu ft/day). Salt storage costs are directly related to the cost of land and the type of lining used.

The cost for hauling the coagulation sludge and diatomaceous earth to landfill, assuming a commercial disposal firm is used, is $60 to $100 a month. Disposal costs are variable depending on the equipment used and distance to the disposal site.

Three types of waste solids are produced by the treatment systems developed for the pressed and blown glass industry. These are: (1) gravity oil separator and dissolved air flotation skimmings, (2) spent diatomaceous earth, and (3) lime precipitation sludges associated with fluoride wastewater treatment.

The skimmings and spent diatomaceous earth result from the treatment of cullet quench wastewaters. The skimmings have a 3% solids content and the production of skimmings ranges from 21.4 to 49.1 kg/day (47 to 108 lb/day) or 720 to 1,630 l/day (190 to 430 gallons per day). The oily skimmings can be disposed of by an oil reclamation firm, used as road oil, or can be incinerated.

Spent diatomaceous earth has an estimated moisture content of 85%, but does not flow. This material is stable and should be suitable for landfill. Estimated production of diatomaceous earth waste ranges from 0.042 to 0.53 cu m/day (1.5 to 19 cu ft/day). The lower figure results from the treatment of the blowdown for the cullet quench system and the higher figure is the result of treating the entire cullet quench wastewater stream at an incandescent lamp envelope plant.

The lime precipitation process for fluoride removal produces the largest volume and most difficult sludge to handle. Vacuum filtration is used at almost all plants to reduce the sludge volume. The volume of sludge production ranges from 277 kg/day (610 lb/day) for a handmade glass plant to 20.9 metric tons/day (23 tons/day) at a television picture tube envelope manufacturing plant. The television picture tube envelope manufacturing plant is treating a combination of abrasive grinding wastes and fluoride containing rinse waters.

Most lime precipitation sludge is currently disposed of as landfill. Several attempts have been made to convert the sludge into a saleable material, but no markets have been found for these products. Currently, further research is being conducted to develop a saleable by-product from the sludge.

MINERAL WOOL INDUSTRY

The product mineral wool used to be divided into three categories: slag wool, rock wool, and glass wool. Today, however, straight slag wool and rock wool as such are no longer manufactured. A combination of slag and rock constitutes the charge material that now yields a product classified as a mineral wool, used mainly for thermal and acoustical insulation (15).

Mineral wool is made primarily in cupola furnaces charged with blast furnace slag, silica rock, and coke. The charge is heated to a molten state at about 3000°F (1650°C) and then fed to a blow chamber, where steam atomizes the molten rock into globules that develop long fibrous tails as they are drawn to the other end of the chamber. The wool blanket formed is next conveyed to an oven to cure the binding agent and then to a cooler.

Mineral wool is made today with a cupola by using blast furnace slag, silica rock, and coke (to serve as fuel). It has been produced in the past by using a reverberatory furnace charged with borax ore tailings, dolomite, and lime rock heated with natural gas. It is used mainly for domestic insulation in residential homes. Plants tend to be located near a source of slag such as a steel plant. Mineral wool plants operate 24 hours per day and 5 days a week. Typical plant capacity is about 50 tpd (18).

The cupola or furnace charge is heated to a molten state at about 3000°F, after which it is fed by gravity into a device at the receiving end of a large blow chamber. This device may be a trough-like arrangement with several drains, or a cup-like receiver on the end of a revolving arm. The molten material is atomized by steam and blasted horizontally towards the other end of the blow chamber. When the cup or spinner device is used, the action of the steam is assisted by centrifugal force. The steam atomizes the molten rock into small globules that develop and trail long, fibrous tails as they travel towards the other end of the blow chamber (14).

Phenolic resin or a mixture of linseed oil and asphalt are examples of binding agents that can be atomized at the center of the steam ring by a separate steam jet to act as a binder for the fibers. Annealing oil can also be steam atomized near the steam ring to incorporate a quality of resilience to the fibers that prevents breakage.

A temperature between 150° and 250°F is maintained in the blow chamber. Blowers, which take suction beneath the wire mesh conveyor belt in the blow chamber, aid the fibers in settling on the belt. The mineral wool blanket of fibers is conveyed to an oven for curing the binding agent. Normally gas fired, the oven has a temperature of 300° to 500°F.

The mineral wool is next programmed through a cooler, as shown in the flow diagram in Figure 86. Usually consisting of an enclosure housing a blower, the cooler reduces the temperature of the blanket to prevent the asphalt, which is applied later to the paper cover, from melting.

FIGURE 86: FLOW DIAGRAM OF MINERAL WOOL PROCESS

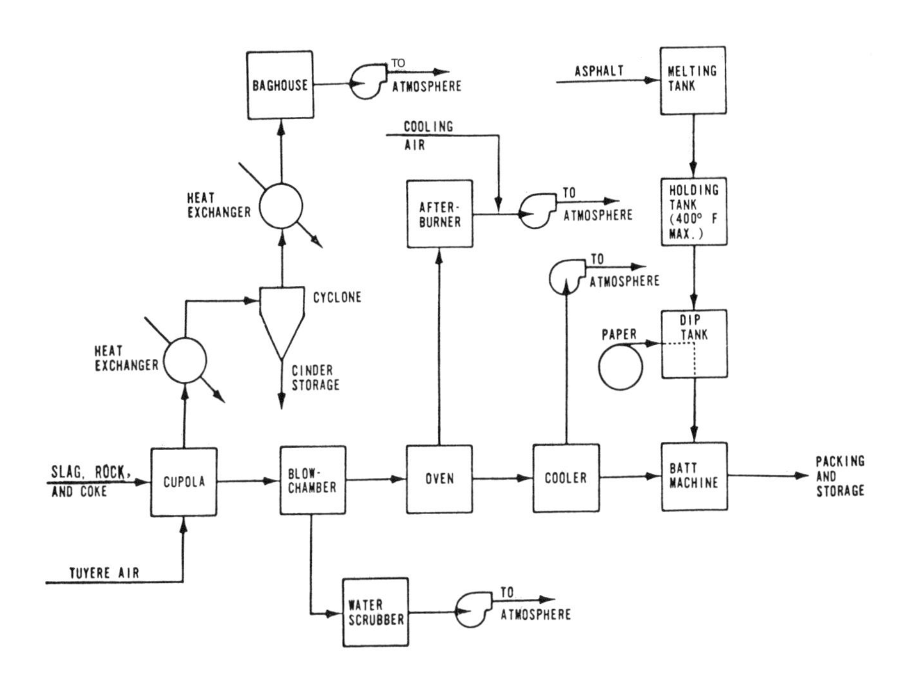

Source: J.A. Danielson (14)

To make batts, the blanket leaving the cooler is processed through a multibladed, longitudinal cutter to separate it into sections of desired widths. Brown paper and either asphalt-coated paper or aluminum foil are then applied to the sections of blanket. The asphalt-coated paper is passed through a bath of hot asphalt just before its application to the underside of each section. This asphalt film serves as a moisture barrier as well as a bonding agent against walls. The paper covered sections are cut to desired lengths by a transverse saw, after which the finished product is packed for storage and shipment. The two cutters, paper and asphalt applicators, and conveyor systems are sometimes referred to collectively as a batt machine.

A granulated wool production line differs from that just described in that the mineral wool blanket, after leaving the blow chamber, is fed to a shredder for granulation, then to a pelletizer.

The pelletizer serves two functions, namely, to form small 1-inch diameter wool pellets and to drop out small black particles called shot, which form as the molten slag cools in the blow chamber. A bagging operation completes the process. Since no binding agent is required, the curing oven is eliminated.

AIR POLLUTION

The major source of emissions is the cupola or furnace stack. Its discharge consists primarily of condensed fumes that have volatilized from the molten charge, and gases such as sulfur oxides and fluorides. Amounting to as much as 100 pounds per hour and submicron in size, condensed fumes create a considerable amount of visible emissions and can be a public nuisance.

Another source of air pollution is the blow chamber. Its emissions consist of fumes, oil vapors, binding agent aerosols, and wool fibers. In terms of weight, a blow chamber may also emit as much as 100 pounds of particulate matter per hour at a production rate of 2 tons per hour if the blow chamber vent is uncontrolled. Approximately 90% of these emissions consists of mineral wool fibers.

Types of air contaminants from the curing oven are identical to those from the blow chamber except that no metallurgical fumes are involved. These emissions amount to approximately 8 pounds per hour at a production rate of 2 tons per hour, since the amount of wool fibers discharged is much less than that for a blow chamber. From a visible standpoint, however, these pollutants may create opacities as high as 70%. Emissions from the cooler are only 4 or 5 pounds per hour at a production rate of 2 tons per hour. The asphalt applicator can also be a source of air pollution if the temperature of the melting or holding pot exceeds 400°F. Emission factors for various stages of mineral wool processing are shown in Table 91. The effect of control devices on emissions is shown in footnotes to the table.

TABLE 91: EMISSION FACTORS FOR MINERAL WOOL PROCESSING WITHOUT CONTROLS[a]

Type of process	Particulates		Sulfur oxides	
	lb/ton	kg/MT	lb/ton	kg/MT
Cupola	22	11	0.02	0.01
Reverberatory furnace	5	2.5	Neg[b]	Neg
Blow chamber[c]	17	8.5	Neg	Neg
Curing oven[d]	4	2	Neg	Neg
Cooler	2	1	Neg	Neg

[a]Emission factors expressed as units per unit weight of charge.
[b]Negligible.
[c]A centrifugal water scrubber can reduce particulate emissions by 60 percent.
[d]A direct-flame afterburner can reduce particulate emissions by 50 percent.

Source: EPA (15)

No special hooding arrangements as such are required in any of the exhaust systems employed in the control of pollution from mineral wool processes. The one possible exception is that canopy hoods may be used over the asphalt tanks if the emissions from these tanks are excessive and are vented to an air pollution control device. The ventilation requirements for the various individual processes in a mineral wool system are categorized as follows.

Cupolas – Based on test data, exhaust requirements can be estimated to be 5,000 to 7,000 standard cubic feet per minute for a cupola with a process weight of from 4,000 to 4,500 pounds per hour, on the assumption that no outisde cooling air is introduced. The charge door should be kept in the closed position to obtain maximum benefit from the capacity of the exhaust fan. A barometric damper in the line between the cupola and the blower can be used to control the amount of gases pulled from the cupola. The objective is to remove all tuyere air plus an additional amount of air to maintain a slight negative pressure above the burden.

Reverberatory Furnaces – Ventilation requirements are about 15,000 to 20,000 cubic feet per minute (at 600°F) for a furnace sized to produce 1,500 to 3,000 pounds of mineral wool per hour. The heat in these furnace gases can be used in making steam before filtration.

Blow Chambers – For a blow chamber with a size of about 4,500 cubic feet and with a capacity for processing 4,000 pounds of wool an hour, the minimum ventilation requirements are 20,000 to 25,000 standard cubic feet per minute. All duct takeoffs must be located at the bottom of the blow chamber beneath the conveyor to create downdraft, which packs the newly formed wool fibers onto the conveyor. From this viewpoint, 35,000 standard cubic feet per minute would be more desirable.

In addition, this increased ventilation holds the blow chamber temperature down to tolerable limits, which determine the type of air pollution control equipment to be selected. If the plant is processing granulated wool instead of batts, downdraft is less important and satisfactory operation can be achieved with a 25,000 standard cubic feet per minute exhaust system. If a lint cage is used to trap wool fibers in the discharge gases, frequent cleaning (four times an hour) of the cage is imperative for proper ventilation.

Curing Ovens – Exhaust requirements for a 2,500 cubic foot oven operating at 300° to 500°F and capable of processing 4,000 to 6,000 pounds of mineral wool an hour are about 5,000 standard cubic feet per minute. Sufficient oven gases must be removed to prevent a pressure buildup so that leakage does not occur. In sizing the fan, consideration must be given to temperature rises and possibly also to the introduction of outside cooling air for proper fan operation, particularly if the oven discharge gases are incinerated.

Coolers – Coolers normally do not require air pollution control devices. If outside ambient air is used as the cooling medium, the ventilation requirements are 10,000 to 20,000 cubic feet per minute for a cooler whose area is about 70 square feet.

Asphalt Tanks – If temperature regulators are successfully used to control emissions, the ventilation requirements for melting, holding, and dip tanks will be about 75 cubic feet per minute for each square foot of surface area. This value is for open tanks and for hoods having one open side. If the melting and holding tanks are closed, natural draft stacks may be used.

Baghouses have proved to be an effective and reliable means of controlling the discharge from mineral wool cupolas. Dacron or Orlon bags, which can withstand temperatures up to 275°F, should be used. Of these two synthetic fabrics, Dacron is now the more common, and features several advantages over Orlon. Glass fabric bags cannot be used, owing to the fluorides in the cupola effluent. (Results of a baghouse catch analysis disclose fluorides in a concentration of 9.85% by weight in the particulate matter discharged from a cupola. The life of glass bags under these conditions is about one week.)

Provisions for automatic bag shaking should be included in the baghouse design. Sufficient cloth area should be provided so that the filtering velocity does not exceed 2.5 feet per minute.

Since the discharge temperature of the gas is about 1000°F, heat removing equipment must be used to prevent damage to the cloth bags. This can be accomplished with heat exchangers, evaporative coolers, radiant cooling columns, or by dilution with ambient air. The cooling device should not permit the temperature in the baghouse to fall below the dewpoint. Safety devices should be included to divert the gas stream and thus protect the baghouse from serious damage in the event of failure of the cooling system. In some instances it may also be desirable to include a cyclone or knockout trap someplace upstream of the baghouse to remove large chunks of hot metal that can burn holes in the bags even after passing through the cooling system.

Baghouses should be equally effective in controlling emissions from reverberatory furnaces. The comments made about cupolas are generally applicable to these furnaces. Excelsior packed water scrubbers have been tried in Los Angeles County (14) but did not comply with air pollution statutes relating to opacity limitations.

The effluent from the curing oven is composed chiefly of oil and binder particles. These emissions, while not a great contributor to air pollution in terms of weight, are severe in terms of opacity. Since they are combustible, a possible method of control is incineration. This method, in fact, has proved practical for the mineral wool plant.

Generally, afterburners are divided into two categories, depending upon the method of oxidation. These are direct flame and catalytic. Important considerations for the direct flame type (see Table 92) are flame contact, residence times, and temperature. The afterburner should be designed so that a maximum of mixing is obtained with the flame. The design should also provide sufficiently low gas stream velocities to achieve a minimum retention time of 0.3 second. An operating temperature of 1200°F is the minimum requirement for efficient incineration.

TABLE 92: DATA FOR A MINERAL WOOL CURING OVEN CONTROLLED BY A DIRECT FLAME AFTERBURNER

Oven Data	
Type	gas fired, conveyorized
Operating temperature	350° to 450°F
Heat input	4 million Btu/hr
Afterburner Data	
Type	direct flame, gas fired, two-pass
Flame contact device	deflector plate
Heat input	5 million Btu/hr
Size	4 foot diameter by 9 foot length with 3 foot diameter by 10 foot length insulated retention tube
Gas temperature inlet	270°F
Operating temperature	1240°F
Gas velocity	37 feet per second
Retention time	0.3 second
Collection efficiency at 1230°F	
On particulate matter	50%
On aldehydes	59%
On combustibles	52%
On solvent soluble material	68%

Source: J.A. Danielson (14)

If a catalytic afterburner is used, the gas stream must be preheated to about 1000°F. Some type of precleaner must be used to remove the mineral wool fibers and thus prevent fouling of the catalytic elements. Because of this problem, catalytic afterburners have not proved very satisfactory for this service.

Electrical precipitators have been used as an alternative means of controlling emissions from mineral wool curing ovens. The precipitator is, however, preceded by a water scrubber and high velocity filter to remove the gummy material that would normally foul the ionizer and plate sections.

If the blow chamber temperature is maintained below 175°F to preclude the formation of oil mist, then the major air pollution problem is posed by wool fibers. The most practical means of collecting these fibers is an efficient water scrubber. If, however, the blow chamber temperature rises above 250°F, the feasibility of using a water scrubber is diminished.

A deflector plate at the blow chamber entrance can be used to deflect a large portion of the molten shot and thereby reduce the blow chamber temperature as well as reduce the chance for contact with oil mists. Water injection at the receiving end of the blow chamber combined with adequate ventilation air can further reduce this temperature to 150°F or less.

A simple wire mesh lint cage collects as much as 90 pounds of large pieces of fibrous material per hour. Constant cleaning of the lint cage is, however, required; otherwise lack of ventilation results in a temperature rise in the blow chamber.

Large water content in the blow chamber effluent precludes classifying the baghouse as a practical control device for the blow chamber. In addition, the resin binder would plug the pores of the bags, resulting in a severe maintenance problem.

Asphalt vapors emitted by the asphalt applicator can become a serious source of air pollution if the asphalt temperature is permitted to exceed 400°F. The simplest and most economical method of reducing these emissions to the atmosphere is to control the temperature. The temperature can sometimes be held to a maximum of 325°F by proper asphalt selection, thermostatic control, and use of a holding pot separate from the melting pot. (Asphalts made from different crude oils have different vaporizing points.)

If temperature control is used, best results can be obtained by using three separate tanks: melting tank, holding tank, and dip tank. All three should be provided with individual heating facilities, which thereby permit minimum temperature differentials between tanks. In this manner, the holding tank temperature can be held to a minimum (about 400°F) without regard to heat loss at the dip tank. Automatic temperature controls are necessary for the holding tank. An asphalt feed control bar installed on the asphalt roller in the dip tank permits the temperature to be reduced even further. This feed control bar, which is adjustable against the roller, controls the thickness of the asphalt film applied to the paper; otherwise this thickness would have to be controlled by controlling temperature and asphalt viscosity.

If control of asphalt temperature proves impractical, then a collection device should be used to prevent the fumes from escaping to the atmosphere. This can be done effectively with a two-stage, low voltage electrical precipitator, and sometimes with a high efficiency water scrubber. If a scrubber is used, recirculation of the water is not advised, since plugging of the water nozzles may occur unless the asphalt particles are somehow removed, say by flotation.

Proprietary Control Processes

A process developed by J.H. Tarazi (51) is one in which mineral wool fibers still hot from being formed by high speed centrifugation are coated with a composition comprising 1 to 30% of a nonionic surfactant and 70 to 99% of a polyalkylene glycol having a viscosity in the range of 35 to 180,000 cs at 100°F. As a result, smoke and dust are reduced, the fibers are annealed and a wetting action is imparted. The preferred surfactants are alkyl phenoxy polyethoxy ethanols.

SAND AND GRAVEL INDUSTRY

The production of sand and gravel represents the largest nonfuel mining operation in the United States. Production is reported from every state with the distribution pattern concentrated in areas of high population density. Sand and gravel production is expected to double by 1980, and forecast demand for this commodity ranges from 300 to 400% increase by the year 2000 (52).

Surface mining for sand and gravel has disturbed over 1,000,000 acres. A recent study by the Soil Conservation Service, as printed in the Congressional Record of January 1, 1974, estimated that 4,418,710 acres had been disturbed by all surface mining activities. Of the total 2,542,682 acres requiring reclamation, sand and gravel mining operations were responsible for 35%; coal, 37%; and all other mining, 28%. Unlike coal and iron surface mining that are confined to specific geographical areas, sand and gravel operations are widely distributed throughout the United States; hence, more people are directly affected by the adverse effects than any other mining operation.

Great Britain, Japan and the Netherlands are currently recovering appreciable amounts of ocean-dredged sand and gravel from depths of 20 to 100 feet. Ocean dredging in the United States is minimal at the present time. In 1972, 25% of the national production of sand and gravel was consumed by the urban areas of the 21 states bordering the ocean. Estimates by the Corps of Engineers indicate 75,000 square miles are suitable for sand and gravel recovery from the coastal zones of the United States. With improved technology, greater demand, urban encroachment, and the increasing cost of land transportation, it is reasonable to assume that ocean mining for sand and gravel will become a significant domestic source of supply within ten years.

Deposits of sand and gravel, the consolidated granular materials resulting from the natural disintegration of rock or stone, are found in banks and pits and in subterranean and subaqueous beds.

Depending upon the location of the deposit, the materials are excavated using power shovels, draglines, cableways, suction dredge pumps, or other apparatus; light-charge blasting may be necessary to loosen the deposit. The materials are transported to the processing plant by suction pump, earth mover, barge, truck, or other means. The processing of sand and gravel for a specific market involves the use of different combinations of washers; screens and classifiers, which segregate particle sizes; crushers, which reduce oversize material; and storage and loading facilities (15).

The raw materials for sand and gravel plants may be dredged from a river or quarried and then transferred by vehicle to the crushing and screening equipment. Material is frequently washed prior to processing to obtain a product which meets users' specifications. Preliminary screening, prior to crushing, is also practiced in some plants. Wet and dry screening may be used. Following processing and classification, the material is loaded for shipment or stockpiled in storage areas.

Three different methods of sand and gravel excavation are practiced:

(1) dry pit, sand and gravel removed is above the water table;

(2) wet pit, raw material extracted by means of a dragline or barge-mounted dredging equipment both above and below the water table; and

(3) dredging, sand and gravel is recovered from public waterways, including lakes, rivers, and estuaries. Figure 87 illustrates a typical sand and gravel processing system.

FIGURE 87: TYPICAL SAND AND GRAVEL PROCESSING SYSTEM

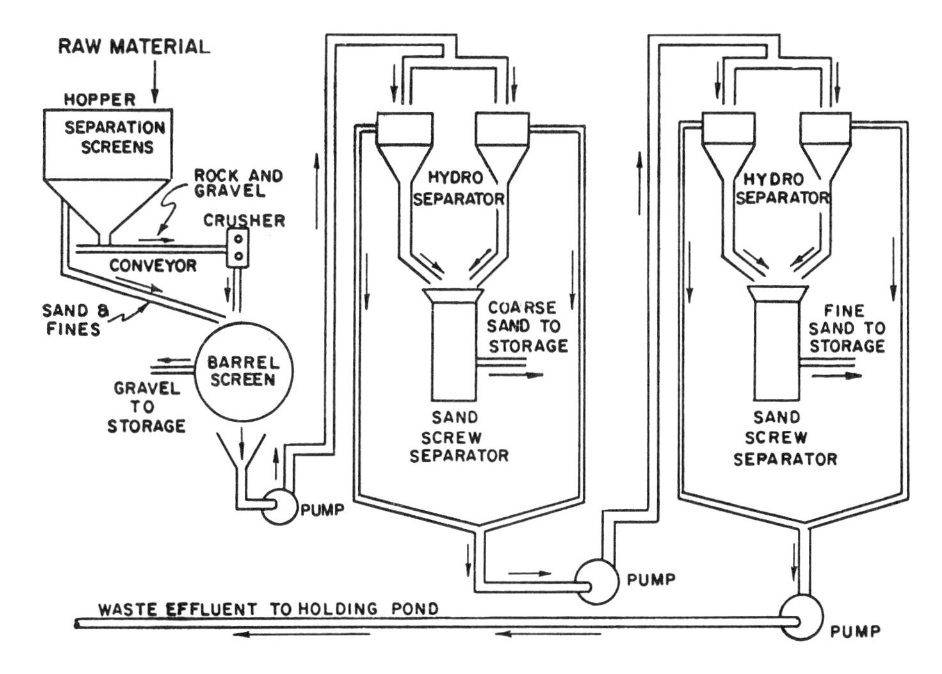

Source: B.F. Newport and J.E. Moyer (52)

Over 5,000 domestic plants in the United States fall into the above categories. A breakdown of their percent contribution of the total production is as follows: dry pit, 50%; wet pit, 30 to 40%; and dredging on public waterways, 10 to 20%. Considerable production variations exist within the industry, with the larger operations producing over 3.5 million tons per year. The smaller, part-time operations may produce less than 1,000 tons. Capital outlay required ranges from $20,000 for marginal producers to larger investments in excess of $10 million.

Although some of the larger operations are still releasing detrimental silt loads to public waterways, many of the major installations are now using totally closed systems or releasing only a small percentage of their process water after effective treatment. Many of the low volume or part-time producers, due to their relatively small or intermittent effluents, have continued to operate without treatment. Release of this sediment-laden plant process water has been overlooked by the environmentalists and many state agencies whose energies have been directed toward the more sensational forms of pollution.

Sand and gravel production in the United States in 1969 totaled 937 million short tons valued at $1,070,000,000. Crushed stone production in the United States in 1969 totaled 861 million short tons valued at $1,326,047,000.

Many sand and gravel and crushed stone plants need no washing in producing their finished product. Other plants must use up to 800 gallons of water per ton of production. Because many plants use no water, the average water demand per plant is only 70 gal/ton. About 126 billion gallons of water is used per year in processing sand and gravel and crushed stone (40).

AIR POLLUTION

Dust emissions occur during conveying, screening, crushing, and storing operations. Because these materials are generally moist when handled, emissions are much lower than in a similar crushed stone operation. Sizeable emissions may also occur as vehicles travel over unpaved roads and paved roads covered by dirt. Although little actual source testing has been done, an estimate has been made (15) for particulate emissions from a plant using crushers: Particulate emissions: 0.1 lb/ton (0.05 kg/MT) of product.

Particulate emission sources in sand and gravel processing parallel those in crushed stone preparation. The crushing, screening, and transfer operations can all generate significant quantities of dust. Emission rates are affected by moisture content of processed materials, degree of size reduction required, and type of equipment used for processing.

Observations from numerous plants indicate that a major source of dust, in addition to those associated with the plant equipment, is from vehicle traffic over unpaved roads or paved roads covered with dust. Stockpile losses would also contribute to the dust burden. Stockpiles of fine sand would be susceptible to wind loss (20).

No information has been found on emission factors from sand and gravel plants. One sampling report, furnished by a state agency, listed overall emissions as 0.06 lb of dust/ton of material through the plant. This report listed the discharge of the secondary and reducing crushers and the elevator boot on the dry side as the dust sources. Seventy-five percent of the dust was estimated to come from the crushers. Based on this limited information, an overall emission factor of 0.1 lb of dust/ton of product was assumed for sand and gravel plants.

Stockpile losses due to wind erosion have been estimated at 1% of finished product for sand and 0.5% for gravel. Most stockpiles are merely ground piles; however, silos and bins are also used. Potential fugitive dust emissions from stockpiles in sand and gravel plants are estimated to be about 6.5 million tons/year. No estimate has been made of dust emissions from quarrying or from vehicle traffic over unpaved or paved plant roads.

Limited data were found on the characteristics of effluents from sand and gravel plants. Stock sampling data for sand and gravel dryers, obtained from state control agencies, indicated that outlet grain loadings ranged from 5.8 to 38 grains/cubic foot. Mass median particle size of the particulates emitted from the dryers varied from 3.5 to 9.4μ.

Specific information on control practices and equipment utilized in sand and gravel plants is limited. Since many of the operations parallel those of the crushed stone industry, similar control practices and equipment are undoubtedly employed (20).

WATER POLLUTION

Storm runoff and plant process water are the two main sources of water pollution associated with sand and gravel production (52). Sediment loads, from several hundred to several thousand milligrams per liter, are detrimental to:

(1) aesthetic values,
(2) stream biota,
(3) downstream water quality for domestic and commercial uses, and
(4) the ability of a natural body of water to purify itself.

Contrary to most industrial effluents, sand and gravel wash waters contain essentially one component, sediment, that exerts a detrimental ecological impact on the environment. Due to the greatly increased concentration of suspended solids, sand and gravel processing waters may not be discharged under present regulations without prior treatment. Unfortunately, in many instances, wash waters from active installations, as well as storm runoff from both active and abandoned facilities, are being released directly to surrounding surface waters.

This situation has occasioned complaints from environmentalists and other concerned parties with regard to degradation of water quality for other users, deleterious effects on the biota of the receiving waters, and deterioration of the environment from an aesthetic standpoint.

Following siltation, an unsightly turbid body of water offers few recreational opportunities. Gravel shallows, once providing nesting areas for trout, bass, salmon, and other sport fish are covered and not available for these purposes. The benthic population is severely reduced, with some species disappearing. Rocky areas harboring organisms, while providing protective cover for fish larvae and nesting areas, no longer exist.

Turbidity, by affecting light penetration, reduces the thickness of the euphotic zone, thus seriously affecting the productivity of the planktonic and benthic community. Reduced numbers of organisms result in a significant reduction of fish production and carrying capacity of this water. The natural ability of a stream to purify itself is dependent upon the existence of viable communities of bacteria, benthic, and planktonic organisms. Solids that settle from suspension also carry organisms plus unstable organic matter. Consequently, the characteristic population increase response to organic waste discharges will not exist in silt-laden waters.

Three different methods of sand and gravel excavation are practiced:

(1) dry pit; sand and gravel is removed above the water table;
(2) wet pit; raw material is extracted by means of a dragline or barge-mounted dredging equipment both above and below the water table; and
(3) dredging; sand and gravel is recovered from public waterways, including lakes, rivers, estuaries, and oceans.

All methods require approximately 600 gallons of process water per ton to rid this product of mud, clay, and other debris. The total volume of process waters utilized represents over 500 billion gallons per year.

The most common treatment method practiced in the industry today is the retention of wash waters in settling ponds. Treatment by the ponding method requires construction of new ponds or utilization of an area previously excavated during the mining process. The size and number of treatment ponds vary considerably; usually, any configuration that enables the suspended matter to settle satisfactorily before wasting or reuse is considered adequate.

One of the major problems confronting the sand and gravel industry is the availability of sufficient land area to construct adequate holding ponds. Many of the operations are located near urban areas where additional land is either not available or prohibitively exorbitant in cost.

However, if the land requirements can be met, the settling characteristics of the waste are then determined to insure that adequate clarification will occur naturally within the allotted detention time of the treatment ponds. Determination of the physical characteristics of the effluent is vital before attempting full-scale pond treatment. Regardless of the detention time available, adequate suspended solid and turbidity reduction in some effluents cannot be attained without additional treatment.

Settling ponds are generally used where there is sufficient low ground or worked out areas available to form such ponds. The sediment will fill the pond areas and restore the land. Settling ponds are ponds that allow solids to settle and the clarified water overflows. Those ponds that will allow seepage are referred to as filter ponds and will be discussed in the next section. Most settling ponds are also filter ponds but are separated here for clarity (40).

Settling ponds have limited efficiency, especially in deposits with clay or light fines that do not settle quickly by the natural process. Yet, because this system takes very little effort on the part of the operator and because abandoned pits are frequently available, it is widely used. Proper planning and sizing of ponds result in an efficient and economical system. Water will rarely clear up enough to be discharged into waterways, but it normally will be clear enough for washing the aggregate in a recycling operation.

Correspondence from many plant operators (40) discussed effectiveness of settling ponds. A number of plants were visited during the course of the study. Often settlement rates are found to be much slower than Stokes law would suggest, probably due to varying negative charge of the suspended particles. The operator must decide what size particles can be permitted in the recycled water without damaging equipment and still achieve the necessary washing effect. The specific gravity must also be known and can be determined by using a hydrometer giving direct specific gravity reading.

The specific gravity of a body is defined as the ratio of its weight to an equal volume of water. The specific gravity of effluent water from an aggregate plant runs from 1.000 to 1.060. The weight of the dry sediment can be determined by evaporating the water from a sample of effluent and weighing the dry sample and measuring its volume.

It has been found that ponds will short circuit, that is, form a stream from inlet to outlet, causing great increases in horizontal velocities and a reduction in settlement. Turbulance, sediment and buildup variations in specific gravity can greatly reduce the settling pond efficiency.

A second pond is required to skim off the cleanest water with a weir. The second pond is normally the same size as the primary pond to allow for additional settling, thus it can be used as a future primary pond. The water must run from the pumping pond back to the plant. If clearer water is desired, coagulants can be added just ahead of the secondary ponds.

Several factors must be included in a well-designed settling pond.

(1) Water velocities must remain low, including inlet and outlet.

(2) Inlet water must be introduced into the pond across the full width to avoid short circuiting and to get full pond usage. This is accomplished by piping into the pond in several places across the width.

(3) Outlet water must flow out slowly to avoid scouring, that is, picking up the particles that are supposedly being removed. This can be done by maintaining a minimum outlet of one foot of width for every 25 gpm of water being used. This has not been the practice in most plants, but would be necessary to maintain pond efficiency. The depth of outlet water can be controlled by the use of gates for regulating the secondary pond level.

Before an operator makes a decision to adopt this system a sample of water should be tested to determine the length of time required to become sufficiently clear to permit recycling to the plant as wash water. If settling time is too lengthy, a small amount of coagulant could be used to neutralize the particles and aid in settling.

Filter ponds are those systems which allow the plant discharge water to flow into a pond that has no outlet or complete recirculation. The water seeps out the bottom and side of the pond and is clarified by this filtering process. Many factors affect the success of such a system, so they must be considered.

(1) The water table must be low enough that the water will filter out, not into, the pond at any time the plant is running.

(2) The walls and bottom of the pond must be porous through which to allow the water to flow.

(3) The pond must be large enough so it will not seal too rapidly.

(4) There must be enough water used so the suspended particles will not seal the bottom and sides too quickly.

Most of the plants which use this system have used old mined out pits or natural canyons for filter ponds. This means the pond was not designed for minimum size, but rather the land that was available was used. However, if the size and depth of these ponds can be determined, the bottom and side areas required to achieve equilibrium will be known. An example of a typical plant layout is cited below.

The plant in Figure 88, discharging 200 gpm into a 50 ft by 150 ft by 4 ft primary pond over a weir and into an 80 ft by 1,800 ft by 4 ft filter pond, has operated for 7 years without any signs of the pond sides or bottom sealing. The soil has heavy vegetation and on first appearance, would not seem to be porous enough to allow sufficient seepage. However, by inspecting the dredgings from the primary pond, which must be cleaned every 2 weeks, it was noted that the settled material is mostly granular.

Most of the fines in the natural deposit are granular, thus porous. The filter pond wall and bottom surface total about 150,000 ft^2 and provides an area of 750 ft^2 for each gallon per minute of inflow. The pond is at equilibrium at this point, giving an idea of the area required for filtering in this type of soil.

From this and other examples, areas to take into consideration in designing filter ponds can be established.

(1) Water table must be low enough all year to allow water to flow out during operation and rains. Pond berms must be high enough to guard against flood conditions carrying out turbid water.

(2) Pond walls and bottom areas must be inspected to determine the porosity of the soil. The wastewater solids should also be evaluated as light or heavy concentration of sand, silt or clay. The examples given show the wide variation of results caused by porosity.

(3) It is best to use two ponds. The primary pond is used for heavy particle settlement, so the filter bed will not fill with solids too fast. The weir from the primary pond to the secondary pond should be as wide as possible to reduce the velocity of water. This skims the cleanest water off the top and minimizes short circuiting.

(4) Filter pond walls are more efficient than the bottom, so deep steep walled filter ponds are desirable. Ponds will seal more slowly if they are kept full so all the area of walls and bottom can be working. Allowing the pond to rest every couple of months also improves its filter life.

(5) The use of flocculants will shorten the filter pond life, therefore their use should be restricted to the primary pond; or three ponds could be used and

flocculants added in the secondary pond. Flocculants should not be used in the filter pond.

(6) If the filter bed is above the water table, lowlands or wells close by could be endangered by seepage from the pond. This should be considered in the location and design of the filter pond.

FIGURE 88: PLANT LAYOUT—FILTER POND

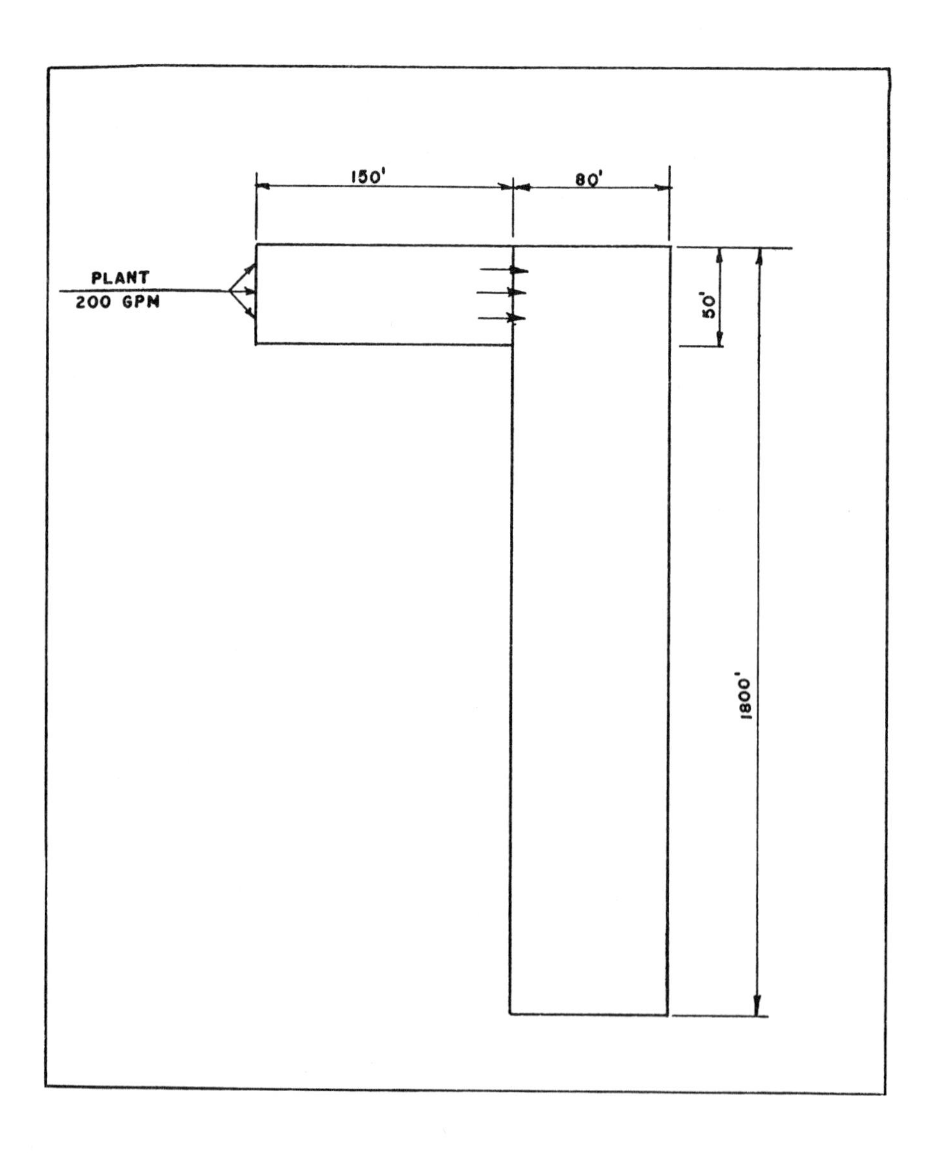

Source: R.G. Monroe (40)

To expedite settling and minimize the necessity for large settling ponds, some operators have installed systems that introduce flocculating agents such as alum to the effluent stream to assist in clarification. Capital expenditures for these systems have varied from elaborate systems in excess of $100,000 to extremely simplified systems consisting of a mixing barrel with an attached hose that only roughly meters chemicals into the effluent (52).

Prior to implementation of full-scale treatment, a plant survey to determine all points of suspended solids entry into the process waters is conducted. Generally, attempts are made to reduce or concentrate suspended solids normally discharged through the use of physical methods; e.g., the manipulation of plant processing procedures.

Following optimization of inplant procedures, technical expertise is usually solicited to obtain assistance in determining the most effective means of chemical treatment. Several manufacturers are now producing flocculating agents, consequently considerable marketing competition exists. Some companies, in an effort to merchandise their product, will send specialists on request to assist in developing an optimum treatment system. This service particularly benefits the small operators who may not have the available personnel or required technology to perform the necessary preliminary testing.

Maximum information concerning the physical and chemical characteristics of the waste to be treated should be obtained before progressing to the full scale process. Engineering parameters taken into consideration prior to chemical treatment are: total flow through the plant; flow variations due to production fluctuations; and most important, flow characteristics of settling ponds. Chemical metering, mixing and detention times must be optimized to attain efficient treatment. Proper construction and utilization of settling ponds can mean the difference between an efficient economical treatment system and an inadequate expensive one.

Quite often, despite extensive laboratory testing, types of chemicals or chemical concentrations may require altering before satisfactory treatment is attained. Once the proper conditions are determined, sediment removal by means of flocculent addition is relatively uncomplicated with many operators able to control the process in spite of varying effluent loadings.

The cost of chemical treatment for sand and gravel effluents ranges from 1 to 5 cents per ton of product produced. This variation is due to the initial outlay cost for equipment, the chemical selected, the amount of chemical used, and labor required to maintain the treatment system. In many cases, due to fixed labor cost, treatment expense will decrease as production increases. Considerable variations in the price of chemicals necessitate a critical assessment of chemicals employed.

In some instances clarification by the use of flocculent aids has been successful to the extent that the total wastewater from the final holding pond can be reused, thus, creating a closed system of treatment whereby no process waters leave the premises. In other applications a high percentage of the total effluent can be recycled. The determining factors on effluent reuse are the purity of the product desired and the amount of suspended solids in the water to be recycled. In many operations concentrations of suspended solids can be as high as 500 ppm and still have a recycling capability.

Fortunately for many sand and gravel companies, either through extensive planning or coincidence, effluents never leave company property. As has been pointed out, some effluents enter large holding ponds or go through extensive treatment before ultimate reuse. Natural containment of waste fines exists in areas where effluents enter low lands or marsh areas owned by the producing company. In instances where silt from effluents or storm runoff is contained on the premises, damage to the environment is eliminated.

Aggregate plant clarification costs vary even more than the ready-mix operation costs described earlier under Concrete Industry. Some reasons for additional variations are: Solids to be settled vary widely in specific gravity, negative charge and size. Land availability varies more widely—some plants are rehabilitating land with these solids and the clarification can be profitable while others have little land and must pay to haul and dump the waste. The natural deposits vary—some plants need no washing at all while others use several million gallons of water per day.

While an attempt was made in an EPA survey (40) to determine some average cost of clarification, no representative figures could be established. It appears that the price of the finished product will be increased substantially to include the cost of water clarification.

SOLID WASTE DISPOSAL

Currently, one of the most serious problems facing the sand and gravel industry is ultimate disposal of waste fines. From 1 to 20% of the total raw material processed will be classified as waste fines. Using a realistic figure of 5% waste fines and production figures quoted earlier, it can be estimated that larger operations will have an accumulation of 500 tons per day of solid waste.

Fortunately, many operations have sufficient land area available for disposal of this waste. Some use previously mined areas, obsolete sedimentation ponds, or open land areas to disperse the sludge for drying. Even with favorable space accommodations, however, waste fines handling can be financially quite burdensome. Periodically, in many operations, sediment basins fill to capacity, requiring the use of dragline and trucks to remove and dispose of sediment accumulations.

Problems concerning waste fines handling are compounded in operations that lack the necessary land for convenient disposal. Many sand and gravel companies continue to operate in areas where the marketing of waste fines for top soil and fill material is not economical. For these operators, it becomes of prime necessity to extract as much marketable material as possible down to the 100 mesh range. In this manner large amounts of fines are eliminated from accumulation in settling ponds.

Cyclone separators, widely used throughout the industry for the purpose of solid separation and material gradation, have proven highly successful in reducing the amount of waste fines released to settling ponds. One plant was originally cleaning a settling pond every six months at a cost of $10,000; with the addition of one cyclone separator (cost $2,000), cleaning intervals were extended to 18 months.

One additional problem of waste fines handling concerns the drying characteristics of the recovered sludge. Due to the varying nature of this material and the different disposal techniques utilized, drying times can range from a week to several years. Factors affecting drying times are: sludge thickness and permeability, disposal site drainage, and climatic conditions affecting rate of evaporation. Generally, assuming adequate space is available, sludge thicknesses of two feet will dry within 1 to 3 weeks.

If the sediment is of the quality of topsoil or fill dirt, and a readily accessible market exists in the immediate area, sediment recovery can prove to be a profitable operation. In many instances, it is advantageous to mix coarser material with the sludge to facilitate drying and enhance the quality of the finished product.

Some operators have added commercial fertilizer to waste fines to yield a profitable product from this once burdensome material. Since may sand and gravel operations are located near metropolitan areas, the economic feasibility of combining municipal sludge with waste fines to produce a marketable fertilizer or soil conditioner is a possibility.

Waste fines have also been utilized for the production of building bricks. With the increased demand for construction materials, activity in this area is expected to increase. Some operators are currently stockpiling suitable material for this eventuality.

It is misleading to imply that all waste fines can eventually be channeled into useful or profitable products, since the sand and gravel industry generates roughly 90 million tons per year. If only a portion of this material can be converted into useful products, however, the effort would be of benefit from an environmental standpoint.

STONE QUARRYING AND PROCESSING

Rock and crushed stone products are loosened by drilling and blasting them from their deposit beds and are removed with the use of heavy earth-moving equipment. This mining of rock is done primarily in open pits. The use of pneumatic drilling and cutting, as well as blasting and transferring, causes considerable dust formation. Further processing includes crushing, regrinding, and removal of fines. Dust emissions can occur from all of these operations, as well as from quarrying, transferring, loading, and storage operations. Drying operations, when used, can also be a source of dust emissions (15).

A simplified flow diagram for a typical plant is shown in Figure 89. Incoming material is routed through a jaw crusher, which is set to act upon rocks larger than about 6 inches and to pass smaller sizes. The product from this crusher is screened into sizes smaller and larger than 1½ to 2 inches, the undersize going to a screening plant, and the oversize to the crushing plant. These next crushers are of the cone or gyratory type. In a large plant, two or three primary crushers are used in parallel followed by two to five secondary crushers in parallel.

FIGURE 89: SIMPLIFIED FLOW DIAGRAM OF A TYPICAL ROCK GRAVEL PLANT

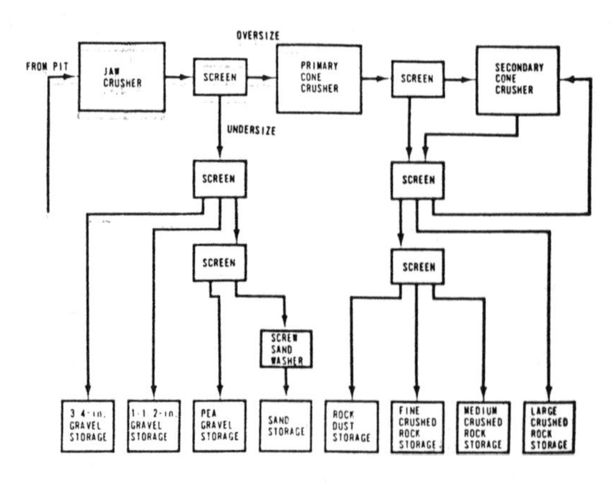

Source: J.A. Danielson (14)

The initial step in the processing of crushed stone occurs at the quarry site. Rock and stone products are loosened by drilling and blasting from their deposit sites. Primary drilling, primary blasting, and secondary blasting or breakage comprise the principal steps in the quarry operation. The secondary blasting operation in many quarries is now either eliminated by better fragmentation during primary blasting, or by the use of drop ball cranes. Tractor-mounted air or hydraulic operated rock-splitters have proven satisfactory for some operations.

The broken rock or stone is transported from the quarry to the processing plant. Transport is usually by truck or heavy earth moving equipment. The processing of stone includes such operations as drying, crushing, pulverizing, screening, and conveying. Primary crushers will normally reduce stone to 1 to 3 inches in size. Secondary crushers are used to reduce stone to sizes below 1 inch. Following the processing operations, the stone or rock is loaded for shipment to the customer or sent to storage (20).

AIR POLLUTION

Particulate emission sources in the crushed stone industry (20) can be divided into two categories; first, sources associated with the actual quarrying, crushing, screening, and processing operations; and second, fugitive sources involving reentrainment of previously settled dust. These fugitive sources include vehicle traffic on temporary roads, transfer operations, and stockpiles.

The use of pneumatic drilling and cutting as well as blasting and transferring can cause considerable dust formation at quarry sites. The transport of stone by vehicle creates dust from unimproved roads. No quantitative data is available on emission rates from these sources.

Rock processing operations (i.e., crushing, pulverizing, classifying) are potential dust sources. Dust is discharged from crushers and grinder inlet and outlet ports. Factors affecting emissions include moisture content of the rock, type of rock processed, amount of rock processed, and type of crusher employed.

Some minerals require drying prior to processing. Dryers are usually direct-fired, either parallel or counter flow, rotary units. Particulate emissions from dryers can be significant, and the amount emitted varies with type of mineral processed, degree of drying, and dryer type. A major fugitive dust source is stockpiles (i.e., windblown dust). Losses vary with size of stored material, density of material, moisture content, and wind speed.

Actual test data on emissions from the above sources are limited. The limited emission data are presented in Table 93. Table 93 does not include dust blown from stockpiles. Stockpile losses due to wind erosion have been estimated at 0.5% of finished product for crushed stone. The potential amount of dust that could arise from stockpiles in crushed stone storage areas is about 3.1 million tons per year.

Factors which would reduce this estimate are: (1) the amount of product which is loaded for shipment directly from processing without being sent to stockpiles; (2) the amount stored in bins and silos; and (3) amount of product which has sufficient moisture content at the time of discharge to stockpiles to inhibit the formation of dust by wind erosion. No estimate has been made of dust emissions from quarrying or from vehicle traffic over unpaved or paved plant roads.

The sand and rock, as it comes from the pit, is usually moist enough to remain nondusting throughout the sand- and uncrushed-rock-screening stages. When the pit material is not sufficiently moist, it must be wetted before it leaves the pit. As the larger rocks are crushed, dry surfaces are exposed and airborne dust can be created. An inventory of sources of dust emissions usually begins with the first crusher and continues with the conveyor transfer points to and including the succeeding crushers. Here the rock is more finely ground, and

TABLE 93: PARTICULATE EMISSIONS–QUARRYING AND CRUSHED STONE

Source	Quantity of Material	Emission Factor	Efficiency of Control C_c	Application of Control C_t	Net Control $C_c \cdot C_t$	Emissions (tons/year)
Crushed Stone						
A. Primary crusher	681,000,000 tons (cement, lime & dolomite not included)	0.5 lb/ton of rock[a]				
B. Secondary crushing & screening		1.5[a]				
C. Tertiary crushing & screening		6.0[a]				
D. Fines milling		6.0[a]				
E. Re-crushing & screening		1.0[b]				
		15.0[a]	0.80	0.25	0.20	4,100,000
F. Conveying, general screening, etc.		1.7 lb/ton of product	0.80	0.25	0.20	454,000
G. Dryers		--	--	--	--	--
Quarrying						
A. Drilling		--	--	--	--	--
B. Blasting		--	--	--	--	--
C. Loading - unloading		--	--	--	--	--
In-Plant Vehicle Traffic		--	--	--	--	--

a/ Pounds/ton of rock through the primary crusher.

b/ Listed emission factor is 5 lb/ton of rock re-crushed. Twenty percent of product is assumed to be re-crushed.

Source: A.E. Vandegrift, L.J. Shannon, E.W. Lawless, P.G. Gorman, E.E. Sallee and M Reichel (20)

dust emissions become greater. As the process continues, dust emissions are again prevalent from sources at conveyor transfer points and at the final screens (14). The points that require hooding and ventilation are the crusher discharge points, all elevator and belt conveyor transfer points, and all screens.

All these dust sources should be enclosed as nearly completely as possible and a minimum indraft velocity of 200 fpm should be maintained through all open areas. The following rules are also a guide to the amount of ventilation air required (Committee on Industrial Ventilation, 1960):

(1) Conveyor transfer points – 350 cfm per foot of belt width for speeds of less than 200 fpm; 500 cfm per foot of belt width for belt speeds over 200 fpm;

(2) Bucket elevators – tight casing required with a ventilation rate of 100 cfm per square foot of casing cross-section;

(3) Vibrating screens – 50 cfm per square foot of screen area, no increase for multiple decks.

One method of suppressing the dust emissions consists of using water to keep materials moist at all stages of processing; the other, of using a local exhaust system and a dust collector to collect the dust from all sources.

If the use of water can be tolerated, then water can be added with spray nozzles, usually at the crusher locations and the shaker screens. Figure 90 shows nozzle arrangements for control of emissions from the outlet of the crushers. Figure 91 shows nozzle arrangements at the inlet to the shaker screens. The amount of water to be used can best be determined by trial under normal operating conditions. Water quantities vary with crusher size, crusher setting, feed rate, type of feed, and initial moisture content of the feed.

Adding water in the described manner tends to cause blinding of the finest size screens used in the screening plants, which thereby reduces their capacity. It also greatly reduces the amount of rock dust that can be recovered, since most of the finest particles adhere to larger particles. Since rock dust is in considerable demand, some operators prefer to keep the crushed material dry and collect the airborne dust with a local exhaust system.

The preferred dust collector device is a baghouse. Standard cotton sateen bags can be used at a filtering velocity of 3 fpm. For large plants that maintain continuous operation, compartmented collectors are required to allow for bag shaking. Most plants, however, have shutdown periods of sufficient frequency to allow the use of a noncompartmented collector. Virtually 100 percent collection can be achieved, and as mentioned previously, the dust is a salable product.

A combination of a dry centrifugal collector and a wet scrubber is sometimes used. In this case, only the centrifugal device collects material in a salable form. A centrifugal collector alone would allow a considerable amount of very fine dust to be emitted to the atmosphere. A scrubber of good design is required, therefore, to prevent such emissions.

FIGURE 90: NOZZLE ARRANGEMENT FOR CONTROL OF DUST EMISSIONS UPON DISCHARGE OF CRUSHER

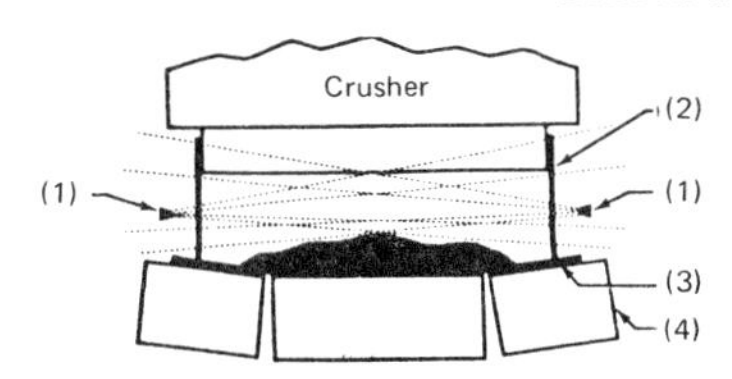

(1) Two flat atomizing type spray nozzles One each end of rubber shield
(2) Hard rubber shield
(3) Conveyor belt
(4) Belt conveyor rollers

Source: J.A. Danielson (14)

FIGURE 91: NOZZLE ARRANGEMENT FOR CONTROL OF DUST EMISSIONS FROM THE INLET TO THE SHAKER SCREENS

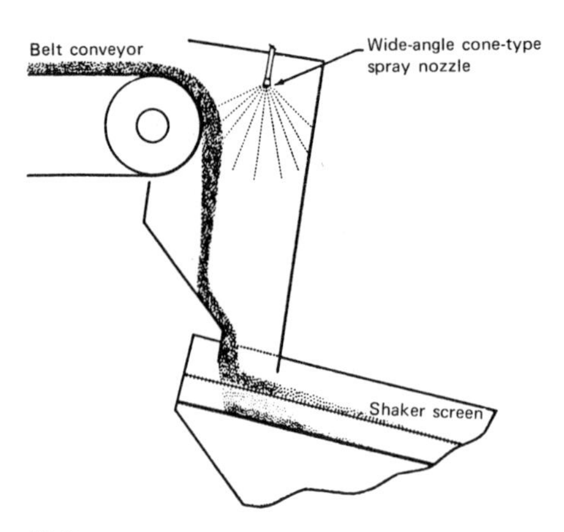

Source: J.A. Danielson (14)

As enumerated above, dust emissions occur from many operations in stone quarrying and processing. Although a big portion of these emissions is heavy particles that settle out within the plant, an attempt has been made to estimate the suspended particulates. These emission factors are shown in Table 94. Factors affecting emissions include the amount of rock processed; the method of transfer of the rock; the moisture content of the raw material; the degree of enclosure of the transferring, processing, and storage areas; and the degree to which control equipment is used on the processes.

TABLE 94: PARTICULATE EMISSION FACTORS FOR ROCK-HANDLING PROCESSES

Type of process	Uncontrolled total[a]		Settled out in plant,	Suspended emission	
	lb/ton	kg/MT	%	lb/ton	kg/MT
Dry crushing operations[b,c]					
Primary crushing	0.5	0.25	80	0.1	0.05
Secondary crushing and screening	1.5	0.75	60	0.6	0.3
Tertiary crushing and screening (if used)	6	3	40	3.6	1.8
Recrushing and screening	5	2.5	50	2.5	1.25
Fines mill	6	3	25	4.5	2.25
Miscellaneous operations[d]					
Screening, conveying, and handling	2	1			
Storage pile losses	10	5			

[a]Typical collection efficiencies: cyclone, 70 to 85 percent; fabric filter, 99 percent.
[b]All values are based on raw material entering primary crusher, except those for recrushing and screening, which are based on throughput for that operation.
[c]Based on units of stored product.
[d]The factor assigned here is the author's estimate for uncontrolled total emissions. Use of this factor should be tempered with knowledge about the size of materials stored, the local meteorological factors, the frequency with which the piles are disturbed, etc.

Source: EPA (15)

Capital and operating costs have been estimated by R.J. Evans (53) using a wet suppression, fabric filter, and a combination system of the two for hypothetical hard rock stone crushing plants located in central Pennsylvania, operating at 300, 600, and 1,000 tons per hour.

Fixed capital expenditure for the dust abatement systems ranged from $49,129 to $249,162. Annual operating costs for the dust abatement systems ranged from $8,394 to $46,606 (0.9 to 3.2 cents per ton). If these dust abatement systems are properly sized, installed, and maintained, efficiencies approaching 95 percent can be expected for reduction in weight of dust particles emitted before control.

FUTURE TRENDS

ASBESTOS INDUSTRY

Research efforts directed toward the establishment of a dose-response relationship for human exposure to airborne asbestos are in progress (5). The only measure available at this time to protect the public health from airborne asbestos is to control asbestos emissions to the greatest degree practicable for the following reasons:

A safe exposure level to asbestos has not been established;

Exposure to asbestos in community air may produce disease; and

The consequences of asbestos caused disease can be extremely serious.

Research is in progress to find substitute materials for the asbestos contained in spray fireproofing for steel and reinforced concrete structures and for the asbestos in pipe insulating materials (6). This research has already produced some asbestos-free spray fireproofing materials, which have been marketed. In another application, amosite has been replaced by fiber glass in boiler blankets for naval vessels. On the other hand, interest in expanding the already vast number of applications for asbestos fibers has increased.

There are available gas-cleaning devices of the fabric filter type, which can reduce fiber counts to below levels presently required by industrial hygiene standards. Should fibers of submicron diameter be discovered to contribute significantly to adverse health effects, however, it would be necessary to determine fractional collection efficiencies of fabric filters for this range of fiber sizes.

Fractional collection efficiency data, required for a more complete evaluation of the effectiveness of fabric filters as control devices, are apparently not available at present. As an initial measure, standardized laboratory tests for the total mass collection efficiency of filter fabrics should be developed.

The treatment of surfaces of mill tailings dumps to promote the growth of vegetation, thereby securing the material from atmospheric entrainment and dispersion, is presently under investigation. Only limited success has been achieved. Also, methods to revegetate and to reforest exposed mining lands are undergoing study and development.

In the manufacture and field fabrication of products containing asbestos, emphasis in abatement activities centers on either the containment or the airborne entrainment and subsequent collection of potential emissions at the sources.

New applications of dust-capturing hoods of both the low velocity, high volume and high velocity, low volume types are expected. Also anticipated are further development and more extensive use of dust elimination methods, such as pulpable bags for utilization in the manufacture of asbestos paper and wet mix shipping bags for asbestos fiber products.

Finally, research is in progress and clear technical definition must be made as to the effects of waterborne asbestos fibers on human health. Such contamination will undoubtedly have to be controlled in the future.

CEMENT INDUSTRY

Based on the EPA estimates, The Boston Consulting Group (21) projects the incremental investment and operating costs will have the following major impacts on the industry:

Cement prices will average about 2% higher than would have prevailed in the absence of the additional investments.

Total future United States cement consumption will not be reduced significantly, since little or no substitution will take place due to higher costs resulting from pollution control expenditures.

Imports, which currently represent about 3% of United States consumption, will increase; precise estimates of import growth would require a more complete analysis of production, transportation, and depollution costs in key foreign countries.

The additional impact on cement industry employment, which has already trended steadily downward in this industry, will be minimal and not concentrated geographically; the Lehigh Valley of Pennsylvania, however, will suffer somewhat more than other areas. Most such displaced employees will have transferable skills (electricians, maintenance, etc.) which will lessen the impact.

Closings of already marginal plants will accelerate substantially, as both the capital and the operating costs of depollution will fall more heavily (on a per barrel basis) on the older, less efficient plants.

The relative advantages of larger, newer plants will increase, and virtually all plants of greater than three million barrels will install the required controls. Most plants of greater than 1.8 million barrels located in remote and isolated markets will depollute rather than expand or shut down.

The combined effects of plant size advantage, required pollution control expenditures on existing plants, and labor cost pressure will motivate a rate of plant construction in excess of the rate of demand growth, which itself will exceed that of the past decade. The consequence will be a substantial increase in plant construction, virtually all of which will be in plants with kiln sizes of 2.5 million barrels and above.

To finance the added investment, the industry will tend to reduce dividends to about one-half of earnings; increase debt/equity ratios, nearly doubling present industry debt in the next five years; and increase prices wherever possible, with overall prices perhaps 5% above their otherwise probable levels. Increased prices will be used to finance growth in preference to increased debt, where allowed, but both foreign and domestic competition will hold back the rates of price increase.

The differentiating effects of alternative strategies will become rapidly more apparent, as those companies with withdrawal patterns accelerate their retirements and the more aggressive investors (and probably new companies as well) continue to expand both market coverage and market share.

If industry estimates of total and anticipated pollution control costs are used, instead of EPA figures, the foregoing conclusions would not be changed, but would be intensified. In particular, the threshold limit for unprofitably small plant size would be increased, and a larger number of plants would be closed. The rate of plant closings would be only moderately affected, however, due to the lead time to bring new capacity into being.

The economics of the cement industry appear to be changing with some rapidity. In part, pollution control costs give this trend more impetus. But the economics of cement manufacture are being changed as well by the rising costs of labor versus the declining costs of automation; the improving technology in operation of giant kilns; the changing costs of distribution resulting from terminals and deepwater transport; and the investment economics of large plants, especially those which replace small plants at an existing site.

The cement industry has characteristically been conservative in its financial policies. Dividend payouts have been high, debt ratios low, and prices kept sufficiently high to provide profits even with substantial unused capacity. Yet there is evidence of a new school of thought among some executives in traditional cement companies and companies not historically in this business. New and much more aggressive financial policies are being adopted which place severe pressure on those companies which continue to operate with the traditional patterns. Foreign competition has become a significant factor, particularly in coastal and border markets. The rate of import growth is accelerating even as prices soften in response to the increased supply.

Some technological change has taken place in recent years, and a more rapid rate of introducing new technologies seems to be occurring. Up to now, most new technological development has focused on the automated operation of large plants, and in the determination of efficient distribution patterns. Now the focus of technology seems to be on the cement manufacturing process itself, in finding more efficient ways to use fuel and raw materials. It is unlikely that the rotary kiln will be displaced in the foreseeable future, but new technologies may significantly lower process costs.

The combination of these and other factors means that investment opportunities of great magnitude are being created in the cement industry. New forms of competition will continue to force traditional producers out of the business, while those who capitalize on the opportunities in this pattern of change will be rewarded with substantial profits.

There is no question that the price of competition in the cement industry is rising. Not only are the initial capital costs higher, but the time horizons for profitable operations are lengthening. Only those firms with substantial capital resources and long staying power can afford to participate in most major cement markets.

Yet in some ways the critical resources—well positioned quarries, customer relationships, and experienced labor forces, among others—are possessed by smaller firms with inadequate plants. They do not have the financial resources required to grow and compete. That situation seems likely to lead to a good deal of acquisition and merger activity. This, associated with some withdrawals, will lead to a much more concentrated cement industry, at least in terms of plant ownership.

The potential for high return investments is not lost on many industry leaders and many prospective investors outside the industry. That potential will attract a good deal of investment into this industry during the next decade. The critical controls over success can be identified in advance, but the specifics of each investment opportunity require very sophisticated analysis.

They differ a great deal from market to market, because of the wide range of market and competitive characteristics throughout the United States. Some companies will be unwilling or unable to provide that level of planning insight, which will lead to some spectacular losses while other competitors are reaping large rewards. The upshot of this is that the cement industry may be one of the more exciting industries during the decade of the 1970s, not so much as a result of high growth, but due to high rates of change within the industry. The investment opportunity exists here, as always, in terms of the ability to change competitive positions by the exercise of new and energetic corporate strategies.

The aggregate and ready mixed concrete industries are primarily smaller businesses producing a low cost product. Probably for this reason, these industries have not had the development in water clarification technology as in many other industries. Thorough research needs to be made in handling, clarifying and reuse or recycling of the wastewater (40).

The one area that became apparent first and which appears to be one of the greatest importance is that of evaluating the effects of using wash water as mix water in concrete. The small amount of testing which has been done thus far, points to encouraging results. It is expected that all the wash water could be reused in this type of system and no wastewater would be discharged. Research and experimentation should be made to design and develop efficient methods of water clarification for the various types of plants.

The aggregate industry uses much more water and has a somewhat more difficult problem than does the ready mix concrete industry. Additional study is needed to develop inexpensive closed loop systems for the aggregate industry. The closed loop system is desirable because the water does not need to be completely clarified to be acceptable as wash water.

The natural or man-made filter ponds show a great deal of promise, but current information is not complete enough to be of measurable value to operators. Study in this area would require construction of test models as well as design research.

Plant operators are also in need of more knowledge regarding the handling of wastewater. The information needed includes proper pump selection, required slope on pipelines to avoid settlement, proper methods for handling the very wet sediment, as well as possible commercial uses for the wet sediment.

Water clarification equipment is being developed very rapidly at this time. The industry could benefit by having each product explained, evaluated and tested so the plant operators could choose the best equipment for their use. Chemicals are being used to settle suspended particles out of the water. The effects of these chemicals on the effluent water and on the product quality should be studied.

FIBER GLASS INDUSTRY

Presently, six of the nineteen operating plants are achieving no discharge of process wastewaters to navigable waters. It is concluded that the remainder of the industry can achieve the requirement as set forth herein by July 1, 1977. The aggregate capital needed for achieving such limitations and standards by all plants within the industry is estimated to be about $10 million, assuming that there are presently no treatment facilities.

These costs could increase the capital investment in the industry 1.2 to 3.8%. As a result, the increased costs of insulation fiber glass to compensate for pollution control requirements could range from 0.6 to 3.8% under present conditions. The application of achieving such limitations and standards will result in complete elimination of all toxic substances in the wastewaters (41).

GLASS INDUSTRY

The objective of an EPA study (46) was to provide an analysis of the economic impact of the water pollution control requirements anticipated under the Federal Water Pollution Control Amendments of 1972. The impact was analyzed for three levels of treatment, as shown on the following page.

Best Practicable Technology available (BPT), to be met by industrial dischargers by 1977

Best Available Technology economically achievable (BAT), to be met by 1983

New Source Performance Standards (NSPS), to be applied to all new facilities that discharge directly to navigable waters and constructed after the promulgation of guidelines.

Specifically the economic impacts analyzed are:

Price
Profitability, growth and capital availability
Employment
Community
Balance of payments
Related industries.

The information and data base for carrying out this analysis were developed primarily from three sources:

Information in the effluent guideline development document on the flat glass industry (46);

Background, knowledge and experience with the flat glass industry; and

Assistance of certain participating companies in the industry and other external sources.

Based on information generated in this study the economic impact of the cost of meeting the anticipated water pollution guidelines for the three levels of treatment are summarized below. The cost and effect on present prices for each segment are summarized in Table 95.

TABLE 95: SUMMARY OF INCREMENTAL UNIT COSTS TO MEET PROPOSED 1977 AND 1983 GUIDELINES

Segment	Present Average Price	Incremental Cost To Meet B.P.T. (1977)	% of Present Price	Incremental Cost To Meet B.A.T. (1983)	% of Present Price
Float	—	Neg.	—	Neg.	--
Plate	$0.13/lb	0.01¢/lb	0.1	0.04¢/lb	0.3
Tempered	$0.50/SF	0.06¢/SF	0.1	0.11¢/SF	∿ 0.2
Laminated	$1.50-1.80/SF	0.18¢/SF	< 0.1	0.41¢/SF	∿ 0.3

Source: EPA (54)

Sheet Glass: In the manufacture of sheet glass products, no process wastewaters are involved, therefore, no wastewater or wasteload need be treated and none should be discharged. In the light of these findings no additional costs are anticipated for sheet glass manufacturing operations and there will be no corresponding impact on price, profitability, production curtailment, employment or related community effects. These conclusions apply to all three levels of treatment.

Float Glass: To meet the BPT level of treatment no additional costs are required and therefore no economic impact in price, production, profitability, employment, etc., will result. The cost and investment of control technology for BAT level of treatment in a float glass operation (no discharge) are minimal. Considering the healthy state of this segment of the industry no economic impacts are anticipated.

The proposed limit for new sources of flat glass manufacturing is the same as for BAT, i.e., no discharge. Considering the substantial investment required for new float glass facilities, the incremental investment will not impose restrictions on future plant construction.

Plate Glass: The anticipated economic effect of water pollution control treatment on the plate glass manufacturing segment of the industry will be minimal. The conclusions are based on two factors: the incremental costs to meet either BPT (1977) or BAT (1983) are small; and continuation of the manufacture of glass by the plate process will be limited by other factors, i.e., the replacement by float glass operations.

Due to the noncompetitive nature of plate glass manufacturing, future construction of plate glass plants will not take place and new sources treatment will not be a consideration.

Solid Tempered Automotive Glass: The incremental cost of meeting the proposed BPT (1977) for tempered automotive glass represents approximately 0.1% of present average prices. It was concluded that this incremental cost will be passed on by the glass fabricator and therefore would have no negative effects on current rates of profitability. The required investment will be available on a plant-by-plant and company-by-company basis and in no way restricts production or expansion.

It is further concluded that incremental costs to meet BAT treatment (1983) will be small (less than 0.2% of present prices) and this small increase will be passed on by the fabricator. Therefore no impacts are anticipated. The proposed control technology for new sources of tempered automotive glass are identical with those for BAT (1983) and therefore will impose no negative economic factors on the industry.

Laminated Windshield Fabrication: The unit costs for meeting the proposed BPT and BAT guidelines are 0.18¢/SF and 0.41¢/SF, respectively. This incremental cost represents less than 0.1% of the present price of laminated windshields in the former case and less than 0.3% for the latter. Thus it is concluded that the cost resulting from the recommended control and treatment technologies will have little or no economic effect on the laminated windshield fabrication industry.

Any incremental cost will be passed on and thus no negative effect on profitability or plant operations are anticipated. The required capital will be available to meet these costs. The proposed requirements for new sources are identical with those proposed for BAT (1983) and therefore the conclusions will not be altered for new plants.

Table 96 summarizes the results of an initial analysis of the pressed and blown glass industry (47). In that table, the five industry subcategories are analyzed to determine the effect of the proposed effluent guidelines based on the best practicable technology (BPT), and the best available technology (BAT). In this analysis, four important factors were considered, including two financial ratios (the amount of additional capital to meet these guidelines as a percent of the replacement value of fixed assets and the operating costs as a percent of annual sales) and two qualitative evaluations of the characteristics of the subcategories (the level of competition from other materials or from imports, and the ability of the industry to raise capital).

Table 96 forms the basis for selecting subcategories for more detailed study as to what may be the impact resulting from the implementing of the proposed water effluent guidelines. In reviewing this table, it is apparent that only the hand pressed and blown glassware could be adversely affected based on the two financial ratios described above. The amount of investment needed for this industry to meet guidelines based on both BPT and BAT represents a substantial portion of investment in equivalent new plant; that is 19 and 24.7% respectively. In addition, this is a competitive industry with limited possibilities of raising capital principally because the firms are small and privately owned.

TABLE 96: ESTIMATED WATER POLLUTION CONTROL IMPACT ON SEGMENTS OF THE PRESSED AND BLOWN GLASS INDUSTRY

	TYPICAL PLANT DESCRIPTION			INDUSTRY CHARACTER	
INDUSTRY	Day Production (tons)	Sales $MM/YR	Investment $MM	Market Competition	Capital Availability
GLASS CONTAINERS	500	18	18	HIGH	HIGH
TELEVISION PICTURE TUBES	250	35	30	LOW	HIGH
INCANDESCENT LAMPS	175	30	35	LOW	HIGH
MACHINE PRESSED AND BLOWN GLASSWARES	100	15	9	HIGH	MEDIUM
HAND PRESSED AND BLOWN GLASSWARES	5	3	1.5	HIGH	LOW

		POLLUTION CONTROL COSTS		POLLUTION CONTROL COSTS	
INDUSTRY	Level	Capital (THOUSAND $)	Annual Operating (THOUSAND $)	Capital as % of Plant Investment	Operating as % of Annual Sales
GLASS CONTAINERS	BAT	312	67	1.7	0.4
	BPT	0	0	0	0
TELEVISION PICTURE TUBES	BAT	231	68	0.8	0.2
	BPT	0	0	0	0
INCANDESCENT LAMPS	BAT	697	300	2.0	1.0
	BPT	470	240	1.3	0.8
MACHINE PRESSED AND BLOWN GLASSWARES	BAT	214	53	2.4	0.4
	BPT	0	0	0	0
HAND PRESSED AND BLOWN GLASSWARES	BAT	371	72	24.7	2.4
	BPT	284	55	19.0	1.8

Source: Arthur D. Little, Inc., estimates. Pollution Control Costs determined by Sverdrup & Parcel Associates as represented in EPA (55)

The selection of a second subcategory for inclusion in the final part of this study presented more difficulty and required judgments based more on qualitative factors. Considering the two financial ratios used in the table, there is little to choose between the other four subcategories. Capital requirements as a percent of plant investment ranges from 2.4% for machine pressed and blown glassware down to as low as 0.8% for TV tube blanks. Operating costs as a percent of annual sales is even less significant ranging from a high of 1% for incandescent lamps to as low as 0.2% for TV tube blanks.

Therefore, the choice of the second segment, machine made household glassware, was not based on the financial ratios but on the characteristics of the industry as summarized in the Market Competition and Capital Availability columns of Table 96. It is believed that incandescent light bulb blanks will not be impacted as this is a relatively noncompetitive industry and the possibility of substitution by other materials is very low. Capital is readily available as these products are manufactured by large corporations with ample capital.

Glass containers caused some concern because it is an extremely competitive market even though the segment has the capital to make the necessary alterations in plant. However, after reviewing the situation for other competitive materials particularly plastic, aluminum, and steel, it was concluded (55) that those industries will be impacted as much or perhaps considerably more than glass containers by new effluent guidelines and by air pollution requirements. Therefore, it was not felt that glass containers would suffer any disadvantage by the proposed effluent guideline regulations.

The television picture tube blank industry, although the analysis was not fully developed is also shown in Table 96 for sake of comparison. For this industry segment, capital and operating costs to meet the proposed BAT levels, are small in comparison to plant investment and sales, and the relatively low level of market competition and good capital availability also support the decision not to analyze this segment further at this time (55).

SAND AND GRAVEL INDUSTRY

The following conclusions have been arrived at as a result of an EPA study (52) of the water pollution problems of the sand and gravel industry:

> Research should be conducted on methods to remove finely dispersed colloidal fines (smaller than 200 mesh) that remain in suspension in sand and gravel effluents despite the utilization of settling aids. Methodologies to be considered for removal purposes should include gas flotation, tube or lamellae settlers, and microfiltration devices.
>
> A survey of sand and gravel producers currently utilizing advanced treatment procedures for the removal of suspended fines from their waste discharges should be conducted. The survey should be made the subject of a report delineating successful treatment technologies with cost considerations, and receive wide distribution among members of the industry.
>
> Studies should be undertaken to determine effective and economical means of dewatering refractory clay slimes from sand and gravel processing procedures, especially industrial glass-sand production.
>
> Research should be initiated to determine methods for the removal or containment of suspended solids generated from dredging operations for sand and gravel on public waterways. While past efforts to contain these sediments utilizing diking techniques, silt curtains, and bubble barriers have not been totally successful, these procedures should be reinvestigated.
>
> Since ocean mining of sand and gravel in the United States is certain to become an important domestic source of this product within a few years, research should be undertaken to determine the impact of ocean dredging operations on the marine environment. An investigation such as the NOMES project (New England Off-Shore Mining Environmental Study) should be coordinated among the interested agencies and approved for funding.
>
> Governmental control bodies, Federal, State and local, should develop a uniform set of rules, regulations, and guidelines for sand and gravel operations to assist producers in planning their mining operations.

REFERENCES

(1) Tripler, A.B., Jr. and Smithson, G.R., Jr., "A Review of Air Pollution Problems and Control in the Ceramic Industries," paper before American Ceramic Society 72nd Annual Meeting, Philadelphia, Pa. (May 5, 1970).

(2) Sullivan, R.J. and Athanassiadis, Y.C., "Air Pollution Aspects of Asbestos," *Report PB-188,080,* Springfield, Va., Nat. Tech. Information Service (Sept. 1969).

(3) U.S. Dept. of Health, Education and Welfare, "Criteria for a Recommended Standard: Occupational Exposure to Asbestos," Wash., D.C., Nat. Inst. for Occupational Safety and Health (1972).

(4) Paddock, R.E., Ayer, F.A., Cole, A.B. and Le Sourd, D.A., "Comprehensive Study of Specified Air Pollution Sources to Assess the Economic Impact of Air Quality Standards: Asbestos, Beryllium and Mercury," *Report PB-222,858,* Springfield, Va., Nat. Tech. Information Service (Aug. 1972).

(5) U.S. Environmental Protection Agency, "Background Information—Proposed National Emission Standards for Hazardous Air Pollutants: Asbestos, Beryllium, Mercury," *Publication No. APTD-0753,* Office of Air Programs (Dec. 1971).

(6) U.S. Environmental Protection Agency, "Control Techniques for Asbestos Air Pollutants," *Publication No. AP-117,* Research Triangle Park, N.C., Office of Air Quality Planning and Standards (Feb. 1973).

(7) U.S. Environmental Protection Agency, "Background Information on Development of National Emission Standards for Hazardous Air Pollutants: Asbestos, Beryllium and Mercury," *Report PB-222,802,* Springfield, Va., Nat. Tech. Information Service (March 1973).

(8) Davis, W.E. and Assoc., "National Inventory of Sources and Emissions—Cadmium, Nickel and Asbestos," Durham, N.C., National Air Pollution Control Admin. (Feb. 1970).

(9) Sittig, M., *Pollution Detection and Monitoring Handbook,* Park Ridge, New Jersey, Noyes Data Corp. (1974).

(10) Heffelfinger, R.E., Melton, C.W., Kiefer, D.L. and Henry, W.M., "Development of a Rapid Survey Method of Sampling and Analysis for Asbestos in Ambient Air," *Report PB 209,477,* Springfield, Va., Nat. Tech. Information Service (Feb. 29, 1972).

(11) Hutcheson, J.R.M., "Environmental Control in the Asbestos Industry of Quebec," Paper before the 73rd Annual General Meeting of the Canadian Institute of Mining and Metallurgy, Quebec City (1971) as quoted in Reference (6) above.

(12) U.S. Environmental Protection Agency, "Development Document for Proposed Ef-

fluent Limitations Guidelines and New Source Performance Standards for the Textile, Friction Materials and Sealing Devices Segment of the Asbestos Manufacturing Point Source Category," *Report EPA 440/1-74/035,* Wash., D.C. (Aug. 1974).

(13) U.S. Environmental Protection Agency, "Development Document for Proposed Effluent Limitations Guidelines and New Source Performance Standards for the Building, Construction and Paper Segment of the Asbestos Manufacturing Point Source Category," *Report EPA 440/1-73/017,* Wash., D.C. (Oct. 1973).

(14) Danielson, J.A., Ed., "Air Pollution Engineering Manual," 2nd Ed., *Publication No. AP-40,* Research Triangle Park, N.C., Office of Air and Water Programs, U.S. Environmental Protection Agency (May 1973).

(15) U.S. Environmental Protection Agency, "Compilation of Air Pollutant Emission Factors," 2nd Ed., *Publ. No. AP-42,* Research Triangle Park, N.C., Office of Air and Water Programs (April 1973).

(16) Johnson, H.B., U.S. Patent 3,662,695; May 16, 1972; assigned to GAF Corp.

(17) Hardison, L.C. and Greathouse, C.A., "Air Pollution Control Technology and Costs in Nine Selected Areas," *Report PB 222,746,* Springfield, Va., Nat. Tech. Information Service (Sept. 1972).

(18) Engineering Science, Inc., "Exhaust Gases from Combustion and Industrial Processes," *Report PB-204,861,* Springfield, Va., Nat. Tech. Information Service (Oct. 2, 1971).

(19) Robinson, J.M., Gruber, G.I., Lusk, W.D. and Santy, M.J., "Engineering and Cost Effectiveness Study of Fluoride Emissions Control," *Report PB-207,506,* Springfield, Va., Nat. Tech. Information Service (Jan. 1972).

(20) Vandegrift, A.E., Shannon, L.J., Lawless, E.W., Gorman, P.G., Sallee, E.E. and Reichel, M., "Particulate Pollutant System Study, Vol. III, Handbook of Emission Properties," *Report PB-203,522,* Springfield, Va., Nat. Tech. Information Service (May 1, 1971).

(21) Boston Consulting Group, "The Cement Industry: Economic Impact of Pollution Control Costs, Vol. I, Executive Summary," *Report PB 207,150,* Springfield, Va., Nat. Tech. Information Service (Nov. 1971).

(22) U.S. Environmental Protection Agency, "Development Document for Proposed Effluent Limitations Guidelines and New Source Performance Standards for the Cement Manufacturing Point Source Category," *Report EPA 440/1-73/005* (Aug. 1973).

(23) Boston Consulting Group, "The Cement Industry: Economic Impact of Pollution Control Costs, Vol. II," *Report PB-207,151,* Springfield, Va., Nat. Tech. Information Service (Nov. 1971).

(24) Kreichelt, T.E., Kemnitz, D.A. and Cuffe, S.T., "Atmospheric Emissions from the Manufacture of Portland Cement," *Report PB-190,236,* Springfield, Va., Nat. Tech. Information Service (1967).

(25) U.S. Environmental Protection Agency, "Air Pollution Aspects of Emission Sources: Cement Manufacturing, a Bibliography with Abstracts," *Publ. No. AP-94,* Research Triangle Park, N.C., Office of Air Programs (May 1971).

(26) Oglesby, S. and Nichols, G.B., "A Manual of Electrostatic Precipitator Technology, Part II, Application Areas," *Report PB-196,381,* Springfield, Va., Nat. Tech. Information Service (Aug. 25, 1970).

(27) U.S. Dept. of Health, Education and Welfare, "Control Techniques for Particulate Air Pollutants," *Publ. No. AP-51,* Wash., D.C., National Air Pollution Control Admin. (Jan. 1969).

(28) Le Sourd, T.A. and Bunyard, F.L., Eds., "Comprehensive Study of Specified Air Pollution Sources to Assess the Economic Impact of Air Quality Standards, Vol. I," *Report PB-222,857,* Springfield, Va., Nat. Tech. Information Service (Aug. 1972).

(29) U.S. Environmental Protection Agency, "EPA Response to Remand Ordered by U.S. Court of Appeals for the District of Columbia in Portland Cement Association v. Ruckelshaus (486 F. 2d 375, June 29, 1973)," *Report PB-237,952,* Springfield,

Va., Nat. Tech. Information Service (Nov. 1974).

(30) Burton, R.E., U.S. Patent 3,192,154; June 29, 1965.

(31) Barr, G.W., U.S. Patent 3,266,225; Aug. 16, 1966; assigned to Southwestern Portland Cement Co.

(32) Matsuda, S., U.S. Patent 3,444,668; May 20, 1969; assigned to Onoda Cement Co., Ltd.

(33) Deussner, H., U.S. Patent 3,485,012; Dec. 23, 1969; assigned to Klockner-Humboldt-Deutz AG

(34) Deynat, G., U.S. Patent 3,503,187; Mar. 31, 1970; assigned to Societe de Forges et Ateliers du Creusot.

(35) Kraszewski, L. and Zulauf, G.A., U.S. Patent 3,507,482; Apr. 21, 1970.

(36) Hoad, J.G., U.S. Patent 3,577,709; May 4, 1971; assigned to John G. Hoad and Associates, Inc.

(37) Patzias, G., "Extraction of Potassium Oxide from Cement Kiln Flue Dust," Doctoral Dissertation, Wayne State University, Detroit, Mich. (1959).

(38) Strehlow, R.W., U.S. Patent 3,812,889; May 28, 1974; assigned to Rexnord, Inc.

(39) Mills, A.A., Jr. and Koerner, N.H., Sr., U.S. Patent 3,868,238; Feb. 25, 1975; assigned to the Columbus Bin Co., Inc.

(40) Monroe, R.G., "Wastewater Treatment Studies in Aggregate and Concrete Production," *Report EPA-R2-73-003,* Wash., D.C., U.S. Environmental Protection Agency (Feb. 1973).

(41) U.S. Environmental Protection Agency, "Development Document for Proposed Effluent Limitations Guidelines and New Source Performance Standards for the Insulation Fiberglass Manufacturing Segment of the Glass Manufacturing Point Source Category," Wash., D.C. (July 13, 1973).

(42) Warner, F.E. and Rice, A.P., U.S. Patent 3,528,220; Sept. 15, 1970; assigned to Fibreglass Ltd.

(43) Borst, J.A., U.S. Patent 3,762,896; Oct. 2, 1973; assigned to Owens-Corning Fiberglas Corp.

(44) Loeffler, R.E., U.S. Patent 3,865,540; Feb. 11, 1975; assigned to Johns-Manville Corp.

(45) Etzel, J.E., Helbing, C.H. and Justus, C.A., U.S. Patent 3,791,807; Feb. 12, 1974; assigned to Certain-Teed Products Corp.

(46) U.S. Environmental Protection Agency, "Development Document for Proposed Effluent Limitations Guidelines and New Source Performance Standards for the Flat Glass Segment of the Glass Manufacturing Point Source Category," *Report EPA 440/1-73/001-a,* Wash., D.C. (Oct. 1973).

(47) U.S. Environmental Protection Agency, "Development Document for Proposed Effluent Limitations Guidelines and New Source Performance Standards for the Pressed and Blown Glass Segment of the Glass Manufacturing Point Source Category," Wash., D.C. (Aug. 1974).

(48) Swift, H.R. and O'Connell, T.B., U.S. Patent 3,123,458; Mar. 3, 1964; assigned to Libbey-Owens-Ford Glass Co.

(49) Bowman, G.A., U.S. Patent 3,728,094; Apr. 17, 1973; assigned to Bowman and Associates, Inc.

(50) Mahoney, W.P., U.S. Patent 3,789,628; Feb. 5, 1974; assigned to Ball Corp.

(51) Tarazi, J.H., U.S. Patent 3,861,895; Jan. 21, 1975; assigned to Arthur C. Withrow Co.

(52) Newport, B.D. and Moyer, J.E., "State of the Art: Sand and Gravel Industry," *Report EPA-660/2-74-066,* Corvallis, Oregon, National Environmental Research Center (June 1974).

(53) Evans, R.J., "Methods and Costs of Dust Control in Stone Crushing Operations," *Information Circular 8669,* Wash., D.C., U.S. Bureau of Mines (1975).

(54) U.S. Environmental Protection Agency, "Economic Analysis of Proposed Effluent Guidelines: Flat Glass Industry," *Report EPA-230/1-73-013,* Wash., D.C., Office of Planning and Evaluation (Aug. 1973).

(55) U.S. Environmental Protection Agency, "Economic Analysis of Proposed Effluent Guidelines: The Pressed and Blown Glass Industry," *Report EPA-230/1-74-037* (Aug. 1974).

ENVIRONMENTAL SOURCES AND EMISSIONS HANDBOOK 1975

by Marshall Sittig

Environmental Technology Handbook No. 2

Environmental pollution with its far-reaching effects is only now beginning to be understood. In this practical handbook the various modifications a given pollutant can assume, are traced back to their origins, and the intermedia transfers are discussed fully. Intermedia transfers include direct transfer (removal of a pollutant from one medium and its disposal in another) or indirect (pollution created in another medium and usually in another form by a basic change in a process or industry).

Also, an example of insidious conversion is mercury. Metallic mercury is relatively innocuous, but when organic molecules or dead organisms are present in rivers and lakes, mercury reacts with these molecules to form toxic methylmercury compounds which are excreted very slowly by fish and man.

The implications are formidable: Chemical, biochemical, technological, statistical, mineralogical, zoological, pharmacological, medical, legal, legislative and public health involvements are all too obvious.

This volume therefore surveys the origins of both air and water pollution. Significant emphasis is placed on altered pollution which can result when an air pollutant is intentionally or accidentally transferred to an aqueous stream or vice versa.

Sometimes pollutants react with each other or initiate deleterious chain reactions. Yet this same reactivity may be the key to efficient removal: by adding certain chemicals or flocculants or microorganisms, toxic substances may be carried off physically or converted not only to nontoxic substances, but also into useful products.

The why and where is in this book, commensurate with present-day technology, without becoming too sophisticated or losing sight of the all important economic considerations.

Operators or potential operators of processes which produce pollutants will find this volume quite useful. Besides discussing all sorts of pollution sources, it should help to define industry-wide emission practices and magnitudes.

A partial and condensed table of contents follows here. The book contains 200 subject entries and 382 tables with figures and graphs. Descriptions are mostly based on studies conducted by industrial and engineering firms or university research teams under the auspices of various government agencies. As is the case with other handbooks in this series the entries are arranged in an alphabetical and encyclopedic fashion:

SOURCES OF SPECIFIC TYPES OF POLLUTANTS

Acids & Alkalis
Aldehydes
Ammonia
Arsenic
Asbestos
Barium
Beryllium
Biological Materials
Boron
Cadmium
Carbon Monoxide
Chlorides
Chlorine
Chromium
Copper
Cyanides
Ethylenes
Fluorides
Heavy Metals
Hydrocarbons
Hydrogen Chloride
Hydrogen Cyanide
Hydrogen Sulfide
Iron
Lead
Magnesium
Manganese
Mercury
Molybdenum
Nickel
Nitrogen Compounds
Nitrogen Oxides
Odorous Compounds
Oily Wastes
Organics
Particulates
Pesticides
Phenols
Phosphorus (Elemental)
Phosphorus Compounds
Pollens
Polynuclear Aromatics
Radioactive Materials
Selenium
Silver
Solid Waste Incineration Products
Sulfur Oxides
Surfactants
Suspended Solids
Tin
Titanium Dioxide
Total Dissolved Solids
Vanadium
Zinc

EMISSIONS FROM SPECIFIC PROCESSES

Adipic Acid Manufacture
Aircraft Operation
Alfalfa Dehydrating
Aluminum Chloride Manufacture

Aluminum Production, Primary
Aluminum Industry, Secondary
Aluminum Sulfate Manufacture
Ammonia Manufacture
Ammonium Nitrate Manufacture
Ammonium Sulfate Production
Asbestos Products Industry
Asphalt Roofing Manufacture
Asphaltic Concrete Plants (Asphalt Batching)
Automobile Body Incineration
Automotive Vehicle Operation
Bauxite Refining
Beet Sugar Manufacture
Boat & Ship Operation
Brass & Bronze Ingot Prod. (Secondary Copper Industry)
Brick & Related Clay Products Manufacture
Builders Paper Manufacture
Calcium Carbide Manufacture
Calcium Chloride Manufacture
Carbon Black Manufacture
Castable Refractory Manufacture
Cement Manufacture
Ceramic Clay Manufacture
Charcoal Manufacture
Chlor-Alkali Manufacture
Clay & Fly Ash Sintering
Coal Cleaning
Coal Combustion, Anthracite
Coal Combustion, Bituminous
Coffee Roasting
Coke (Metallurgical) Manufacture
Combustion Sources
Concrete Batching
Conical Burners
Copper (Primary) Smelting
Cotton Ginning
Dairy Industry
Dry Cleaning
Electroplating
Explosives Manufacture
Feedlots
Fermentation Industry
Ferroalloy Production
Fertilizer Industry
Fiberglass Manufacture
Fish Processing
Frit Manufacture
Fruit & Vegetable Industry
Fuel Oil Combustion
Glass Manufacture
Gold & Silver Mining & Production
Grain & Feed Mills & Elevators
Gypsum Manufacture
Hydrochloric Acid Manufacture
Hydrofluoric Acid Manufacture
Hydrogen Peroxide Manufacture
Incineration, Municipal Refuse
Inorganic Chemical Industry
Iron Foundries
Iron & Steel Mills
Lead Smelting, Primary
Lead Smelting, Secondary
Leather Industry
Lime Manufacture
Liquefied Petroleum Gas Combustion
Magnesium Smelting, Secondary
Meat Industry
Mercury Mining & Production
Mineral Wool Manufacturing
Natural Gas Combustion
Nitric Acid Manufacture
Nitrogen Fertilizer Manufacture
Nonferrous Metals Industry
Open Burning
Orchard Heating
Organic Chemicals Manufacture
Paint & Varnish Manufacture
Paperboard Manufacture
Perlite Manufacture
Pesticide Manufacture
Petroleum Marketing & Transportation
Petroleum Refining
Petroleum Storage
Phosphate Chemicals Manufacture
Phosphate Fertilizer Manufacture
Phosphate Rock Processing
Phosphoric Acid Manufacture
Phosphorus Manufacture
Phosphorus Oxychloride Manufacture
Phosphorus Pentasulfide Manufacture
Phosphorus Pentoxide Manufacture
Phosphorus Trichloride Manufacture
Phthalic Anhydride Manufacture
Plastics Industry
Potassium Chloride Production
Potassium Dichromate Manufacture
Potassium Sulfate Manufacture
Poultry Processing
Power Plants
Printing Ink Manufacture
Pulp & Paper Industry
Railway Operation
Refractory Metal (Mo, W) Production
Rubber Industry, Synthetic
Sand & Gravel Processing
Seafood Processing Industry
Sewage Sludge Incineration
Soap & Detergent Manufacture
Sodium Bicarbonate Manufacture
Sodium Carbonate Manufacture
Sodium Chloride Manufacture
Sodium Dichromate Manufacture
Sodium Metal Manufacture
Sodium Silicate Manufacture
Sodium Sulfite Manufacture
Starch Manufacture
Stationary Engine Operation
Steel Foundries
Stone Quarrying & Processing
Sugar Cane Processing
Sulfur Production
Sulfuric Acid Manufacture
Surface Coating Application
Synthetic Fiber Manufacture
Terephthalic Acid Manufacture
Textile Industry
Timber Industry
Tire & Inner Tube Manufacture
Titanium Dioxide Manufacture
Urea Manufacture
Wood Veneer & Plywood Products Industry
Wood Waste Combustion
Zinc Processing, Secondary
Zinc Smelting, Primary

ISBN 0-8155-0568-X **521 pages**

POLLUTION DETECTION AND MONITORING HANDBOOK 1974

by Marshall Sittig

Environmental Technology Handbook No. 1

This handbook contains methods for the detection and monitoring of pollutants in industrial effluents, notably air and water.

Such methods are prerequisites for any kind of management of environmental quality. Of equal importance are the data handling systems that must be used to collect and interpret the measurements, regardless of whether manual or automated identifying and monitoring techniques are chosen.

Generally the most important single factor in determining whether manual or automated methods of analysis and level monitoring should be used, is the required frequency interval. In the manual approach costs vary almost directly with the measurement frequency. At monitoring intervals oftener than once a day, or when a continuous watch must be maintained, installation of automatic sensing is more economical, provided reliable automatic sensors are available.

For meaningful control and prevention of pollution the accepted standards and parameters for air and water quality must be known, whether they have been legislated or not. Acceptable levels for each type of impurity from methylmercury to dissolved chlorides or carbon particles must be known by the industrialist or public health official, and this book, with its 1633 references makes a massive attempt to communicate such levels.

As a guideline for the industrial user, data on the toxicity of the various pollutants are included. Such data on the deleterious effects were taken from various government publications. This book is not a treatise of pharmacology or industrial toxicology, but a guide to the accurate evaluation of pollutant levels in a given effluent or ambient substrate.

Whenever applicable, the name of the pollutant is followed by chemical and other information. After this come directions for sampling in air or water or both, as the case may be. Thereafter are given qualitative and quantitative analytical methods suitable for identification and measurement of the degree or level of pollution. This comprehensive information is followed by measurement techniques suitable for repeated or continuous monitoring. Preference is given to methods giving reproducible results in environmental quality control, as recommended by the EPA and other U.S. government agencies. This is followed in each case by many up-to-date references to government publications, patents, and journal articles.

Arrangement is encyclopedic. The book is thus a worthy companion to the author's POLLUTANT REMOVAL HANDBOOK available from the same publisher.

Pollution control is a serious business, and in the design of a monitoring program the specific objectives to be served must be clearly in mind from the very beginning. Otherwise the effort most probably will not result in an efficient and meaningful program. The length of the initial survey, selection of monitoring points, parameter coverage, and sampling frequency all depend on the specific objectives and the timeframe in which the objectives must be achieved. These variables also have a very significant impact on monitoring costs.

Monitoring for the purpose of long-term trend identification, evaluation of standards compliance, and total management of environmental quality must, of course, be a continuing program without end.

It is hoped sincerely that this book, together with its companion volume POLLUTANT REMOVAL HANDBOOK, will do much to establish and sustain such antipollution programs.

A partial and condensed table of contents follows here in which subheadings are indicated below for only the first few entries. The book contains a total of 88 subject entries arranged in an alphabetical and encyclopedic fashion. The subject name refers to the polluting substance, and the text underneath each entry tells how to measure and monitor pollution by said substance:

INTRODUCTION

DEFINITION OF PROBLEMS
Terms Used

POLLUTANT EFFECTS
On Health
On the Environment

AMBIENT QUALITY STANDARDS

EMISSION STANDARDS
For Stationary Sources
For Mobile Sources

MONITORING

ANALYTICAL SYSTEMS
Manual
Automated (General)
Automated (Air)
Automated (Water)
Remote
Public

POLLUTION INSTRUMENTATION
Size of the Market
References

SAMPLING
Air and Gas Sampling
Water Sampling
References

SUBSTANCES AND MATERIALS

ACIDS AND ALKALIS

ALDEHYDES

ALUMINUM

AMMONIA

ANTIMONY

ARSENIC

ASBESTOS

AUTOMOTIVE EXHAUST

BACTERIA

BARIUM

BERYLLIUM

BISMUTH

BORON

BROMIDES

CADMIUM

CALCIUM

CARBOHYDRATES

CARBON

CARBON DIOXIDE

CARBON MONOXIDE

CHLORIDES

CHLORINATED HYDROCARBONS

CHLORINE

CHROMIUM

COBALT

COPPER

CYANIDES

ETHYLENE

FLUORIDES

HALOGENS

HALOGEN COMPOUNDS

HYDROCARBONS

HYDROGEN CHLORIDE

HYDROGEN SULFIDE

IODIDES

IRON

LEAD

LEAD ALKYLS

MAGNESIUM

MANGANESE

MERCAPTANS

MERCURY

MOLYBDENUM

NICKEL

NICKEL CARBONYL

NITRATES

NITRIC ACID

NITRITES

NITROGEN COMPOUNDS

NITROGEN OXIDES

ODOROUS COMPOUNDS
Measurement in Air
Sampling Methods
Qualitative Analyses
Quantitative Methods (Organoleptic)
Quantitative Methods (Instrumental)
Measurement in Water
References

OIL AND GREASE

ORGANICS

OXYGEN DEMAND

OZONE AND OXIDANTS

PARTICULATES

PEROXYACETYL NITRATE (PAN)

PESTICIDES

PHENOLS

PHOSGENE

PHOSPHATES

PHOSPHITES

PHOSPHORIC ACID

PHOSPHORUS

PHOSPHORUS COMPOUNDS (ORG.)

POLLEN

POLYNUCLEAR AROMATICS

POTASSIUM

RADIOACTIVE MATERIALS

SEDIMENTS

SELENIUM

SILVER

SODIUM

STRONTIUM

SULFATES

SULFIDES

SULFITES

SULFUR COMPOUNDS (ORG.)

SULFUR DIOXIDE

SULFUR TRIOXIDE

SULFURIC ACID

SURFACTANTS

TIN

TITANIUM

URANIUM

VANADIUM

ZINC

ZIRCONIUM

FUTURE TRENDS

This book should prove to be very useful as an overall reference volume. The organization is clear, and the text is especially well arranged. The tables and figures are extensive and will be of value to students, engineers, and pollution control authorities.

ISBN 0-8155-0529-9

401 pages

POLLUTION CONTROL IN THE PLASTICS AND RUBBER INDUSTRY 1975

by Marshall Sittig

Pollution Technology Review No. 18

Air pollution of the plastics industry are assuming greater importance as more becomes known about the toxicity of such monomers as vinyl chloride, vinylidene chloride and styrene. Stricter standards are being set by OSHA (Occupational Safety and Health Administration, U.S. Dept. of Labor). Design and control techniques are being improved to permit conformance to these standards.

Water pollution problems of the plastics industry are being given increased attention by EPA (Environmental Protection Agency). Waste effluents are being more strictly characterized and treatment methods adapted to meet zero discharge standards.

Solid waste problems of the plastics industry consist of the twin problems of reuse and ultimate disposal. Reuse of plastics waste is as yet of little commercial value, but deserves continuing research and economic evaluation. Incineration requires most careful manipulation and furnace design to avoid the generation of noxious and corrosive gases. Research thus far has not produced plastic packaging materials which are truly biodegradable. High pressure compaction of solid waste for sanitary landfills, however, represents a temporary working solution.

The solid waste problem of the rubber industry is truly monstrous. The problem of handling discarded tires has become worse with the introduction of the steel-belted tire, adding weight and impossibility of comminution. The most practical disposal of waste rubber appears to be pyrolysis or energy recovery, but such technology is unknown and unproven in the market place. In addition, a critical cost and logistics problem is the collection of tires to central processing locations. Fiscal subsidies are clearly indicated here.

In the meantime much can be done by industry during manufacture of raw materials, intermediates and end products. Therefore the present volume is not just a book on pollution control in the plastics and rubber industries, but also a comprehensive treatise on plastics and rubber manufacturing processes and product finishing. The point of view is an integrated one: New and timely reviews of industrial practice are given together with constant attention to air pollution, water pollution, and solid waste problems.

INTRODUCTION

PLASTICS MANUFACTURE

Air Pollution
Polyvinyl Chloride Manufacture
Polyurethane Resin Manufacture

Water Pollution from Specific Processes
Acrylic Resins Manufacture
Amino Resin Manufacture (Urea and Melamine)
Cellulose Acetate Resin Manufacture
Cellulose Ether Manufacture
Cellulose Nitrate Manufacture
Epoxy Resin Manufacture
Ethylene-Vinyl Acetate Copolymer Manufacture
Fluorocarbon Polymer Manufacture
Nitrile Barrier Resin Manufacture
Phenolic Resin Manufacture
Polycarbonate Manufacture
Polyester Resin (Unsat.) & Alkyd Resin Manufacture
Polyolefin Manufacture
Polyphenylene Sulfide Manufacture
Polysulfone Resin Manufacture
Polyvinyl Acetate, Polystyrene & Related Polymers
Polyvinyl Butyral Manufacture
Polyvinyl Chloride Manufacture
Polyvinyl Ether Manufacture
Polyvinylidene Chloride Manufacture
Silicone Manufacture

Aqueous Waste Characterization

Water Pollution Control & Treatment Methods
Alternative Treatment Technologies

Water Pollution Control Economics

SYNTHETIC RUBBER MANUFACTURE

Air Pollution

Water Pollution from Specific Processes
Emulsion Crumb Production
Solution Crumb Production
Latex Production

Water Pollution Control & Treatment Methods
In-Plant Control
End-of-Pipe Treatment
Polymer Corp. Process
Petro-Tex Corp. Process

Water Pollution Control Economics

PLASTIC PRODUCTS INDUSTRY

Air Pollution

Water Pollution

Solid Waste Generation From Specific Operations
Compounding
Injection Molding
Film & Sheet Extrusion
Extrusion Coating & Coating of Wire & Cable
Pipe, Rod, Tubing & Special Profile Extrusion
Coextrusion
Blow Molding
Rotational Molding
Compression & Transfer Molding
Dip & Slush Molding
Casting
Calendering
Cellular Plastics
Thermoforming
Laminating
Coating
Reprocessing

RUBBER PRODUCTS INDUSTRY

Air Pollution

Water Pollution
Tire & Inner Tube Manufacture
Molded Products
Extruded Products
Fabricated Products
Reclaimed Rubber Production
Latex-Based Products

Aqueous Waste Characterization
Tire & Inner Tube Area
Molded, Extruded & Fabricated Area
Reclaimed Rubber Area
Latex-Based Products Area

Water Pollution Control & Treatment Methods
Tire & Inner Tube Area
Fabricated & Reclaimed Products Area

Water Pollution Control Economics
Tire & Inner Tube Area
Molded, Extruded & Fabricated Area
Reclaimed Rubber Area
Latex-Based Products Area

Solid Waste Generation from Manufacture of Specific Products
Tire Manufacturing
Footwear Manufacturing
Belting Manufacturing
Hose Manufacturing
Sponge & Foam Rubber Product Manufacture
Mechanical Rubber Goods Manufacture
Wire & Cable Manufacture

RECOVERY OF PLASTICS PRODUCTS FROM SOLID WASTE

Separation of Plastics from Mixed Waste
Air Classification
Liquid Media Separation
Screening & Electrostatic Separation
Solvent Extraction
Proposed Combination Processes

Refabrication

Chemical Recovery by Pyrolysis
Polyvinyl Chloride
Polyvinylidene Chloride
Polytetrafluoroethylene
Polyurethanes
Polyolefins
Polystyrene
Mixed Plastics

Recovery by Solvent Extraction

Manufacture of Composite Materials from Waste Plastics

RECOVERY OF RUBBER PRODUCTS FROM SOLID WASTES

DISPOSAL OF PLASTIC PRODUCTS

Intermediate Disposal

Collection

Treatment Processes
Baling
Pulverizing, Shredding & Milling
Melting

Environmentally Disposable Plastic Products
Biodegradable Plastic Products
Water-Soluble Plastic Products
Photodegradable Plastic Products

Incineration

Composting

Landfills

DISPOSAL OF RUBBER PRODUCTS

Incineration

Chemical Recovery by Pyrolysis

Other Waste Rubber Utilization Techniques
Road Building
Reef Building

FUTURE TRENDS

LITERATURE REFERENCES

ISBN 0-8155-0572-8 **306 pages**

INDUSTRIAL WATER PURIFICATION 1974

by Louis F. Martin

Pollution Technology Review No. 16

The Federal Water Pollution Control Act (now made into law) will have far-reaching effects on all industry. The pretreatment standards, about to be published as a result of this act, will exert increasing constraints on industrial process design and expansion plans in the years ahead.

The *Guidelines for the Pretreatment of Discharges of Publicly-Owned Works* prohibit the introduction of industrial pollutants which would pass through a municipal facility "inadequately treated," therefore, all existing industrial facilities and new plant designs must include specific downstream water purification processes, before the effluent is sewered or discharged into public systems.

This book describes over 160 recently devised processes for the treatment of contaminated industrial waters.

A partial and condensed table of contents follows here. Numbers in parentheses indicate the number of processes per topic, chapter headings are given, followed by examples of important subtitles.

While conventional separation processes for solid-liquid and water-oil effluents are applicable to many industrial streams, the process literature clearly highlights the areas of greatest immediate concern as shown by this table of contents:

1. SOLID-LIQUID SEPARATION PROCESSES (29)

Flocculation
Methylamine-Epichlorohydrins
Polyelectrolytes
Coagulation and Clarification
Continuous Filtration
Other Separation Techniques
Ion Exchange
Electrostatic and Electromagnetic Methods
Suspended Solids Techniques

2. OIL-WATER SEPARATIONS (18)

Special Equipment Needs
Flotation Cells
Gravity Separation
Vortex Separator
Multiphase Liquid/Liquid Separations
Use of Absorbents
Polypropylene Oil Mops
Permeable Foams for Coalescence

3. REMOVAL OF METALS (30)

Separation of Heavy Metals
Mercapto-s-Triazines
N-Acylamino Acids
Ion-Selective Membranes
Electrolytic Methods
Removal of Metal Ions
Specific Mercury Removal Methods
Other Specific Processes
Nuclear Fuel Wastes
Metal-Containing Lubricants
Organic Lead Removal
Vanadium and Cobalt Removal

4. METAL FINISHING EFFLUENTS (18)

Cyanide Removal
Conversion to Non-Toxic Biodegradable Compounds
Chromium Removal
Control of Aluminum
Other Removal Processes
Use of Chelates

5. PULP AND PAPER PROBLEMS (18)

Closed Cycle Process for Treating Waste Liquors
Hydropyrolysis and Use of Activated Carbon
Wet Air Oxidation
Flocculation
Oxyaminated Polyacrylamides
Sulfur Recovery
Ultrafiltration

6. COAL, ORE AND TAR SAND PROCESSING (14)

Reverse Osmosis + Neutralization
Use of Waste Steam
Salt-Free Condensates
Use of Nucleoproteins
Cyclone Floats

7. REFINING and CHEMICAL PROCESSES (34)

Refinery Operations
Fertilizers
Paint
Phenol
Photography
Meat Processing
Other Specific Processes

ISBN 0-8155-0554-X

300 pages

CEMENT AND MORTAR ADDITIVES 1972

by L. F. Martin

Perhaps no other product that is outside of FDA jurisdiction has been subjected to such extensive and rigorous testing of additives than Portland cement. Additives range from simple inorganic compounds and minerals to highly sophisticated organic polymers.

This book describes 195 additive compositions which have been tested in mortar and concrete formulations, with major emphasis on the more critical areas of set modifiers, air entrainment, work-ability agents and general strength improvement.

The description of these additives has, for clarity and continuity, been divided into the main categories of cement, mortar, gypsum and well-cementing compositions. Other processes described the use of additives to provide water repellent, foamed and other cement mixtures and the aggregates or pebble conglomerates with which they can be used.

195 U.S. patents, issued since 1960, are reviewed. The numbers in ( ) indicate the number of processes for each topic. Chapter headings are given, followed by examples of important subtitles.

1. SET MODIFIERS (28)
- Retarders (12)
  - Zinc Borate-Gluconic Acid Reaction Products
  - Silicofluorides and Phosphoric Acid
- Accelerators (7)
  - Spodumene
  - Amine Formates
  - Sodium Aluminate Sinters
- Grinding Aids and Pack Set Inhibitors (9)
  - Urea
  - Calcium Acetate and Lignin Sulfonates

2. CEMENT ADDITIVES (52)
- Water Control (11)
  - Tricalcium Silicate
  - Polyvinylpyrrolidone and Sodium Naphthalene Sulfonate
- Air Entrainment (5)
  - Styrene Oxide-Sulfonated Lignin Products
  - Sulfite Liquors
- High Strength and Similar Characteristics (16)
  - Fatty Acid Esters of Polyglycols
  - Saccharide Polymers
  - Lime-Sugar Blends
- Water Repellency (7)
  - Oleic Acid
  - Methyl Polysiloxanes
- Efflorescence (2)
  - Polyvinyl Alcohol + Barium Hydroxide
  - Tall Oil, Rosin, Casein
- Fillers (11)
  - Blast Furnace Slag
  - Calcium Aluminum Sulfate in Prestressed Cement

3. GYPSUM AND MORTAR (42)
- Set Retarders (8)
  - Fatty Acid Ammonium Salts
  - Boric Acid and Ammonia
- Set Accelerators (3)
  - Sodium Aluminate and Tartaric Acid
- Water Control and Workability (8)
  - Methylcellulose
  - Exfoliated Vermiculite
- Air Entrainment (2)
  - Cellulosics and Na-Pentachlorophenate
- Water Absorption (3)
  - Maleic Anhydride + Rosin Reaction Products
- Other Additives (18)
  - Cyanamide
  - Fluid Coke

4. WELL-CEMENTING COMPOSITIONS (36)
- Set Modifiers (13)
  - Water-Soluble Organic Polymers
- Fluid Loss Control (8)
  - Xanthated Starch
  - Gilsonite
- Turbulence Inducers (6)
  - Polyalkylenepolyamine Reaction Products
- General Compositions (9)

5. FOAMED AND REINFORCED PRODUCTS (13)

6. COATINGS AND SPECIALTY PRODUCTS (24)
- Inorganic Waterproofing Compositions
- Protective Treatment for Alkali-Aggregate Reaction
- Lime-Cement Mixtures
- Elastomeric Sealants
- Floating Concrete
- Radiation Shield
- Coloring Agents
- Composition for Underwater Use
- Iron Plated Concrete Floors

274 pages